I0028925

Marat V. Markin
Elementary Operator Theory

Also of Interest

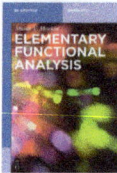

Elementary Functional Analysis
Marat V. Markin, 2018
ISBN 978-3-11-061391-9, e-ISBN (PDF) 978-3-11-061403-9,
e-ISBN (EPUB) 978-3-11-061409-1

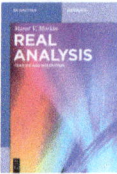

Real Analysis. Measure and Integration
Marat V. Markin, 2019
ISBN 978-3-11-060097-1, e-ISBN (PDF) 978-3-11-060099-5,
e-ISBN (EPUB) 978-3-11-059882-7

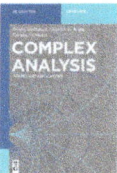

Complex Analysis. Theory and Applications
Teodor Bulboacă, Santosh B. Joshi, Pranay Goswami, 2019
ISBN 978-3-11-065782-1, e-ISBN (PDF) 978-3-11-065786-9,
e-ISBN (EPUB) 978-3-11-065803-3

Functional Analysis with Applications
Svetlin G. Georgiev, Khaled Zennir, 2019
ISBN 978-3-11-065769-2, e-ISBN (PDF) 978-3-11-065772-2,
e-ISBN (EPUB) 978-3-11-065804-0

Applied Nonlinear Functional Analysis. An Introduction
Nikolaos S. Papageorgiou, Patrick Winkert, 2018
ISBN 978-3-11-051622-7, e-ISBN (PDF) 978-3-11-053298-2,
e-ISBN (EPUB) 978-3-11-053183-1

Marat V. Markin

Elementary Operator Theory

—

DE GRUYTER

Mathematics Subject Classification 2010
47-01, 47A10, 47A30, 47A35, 47A56, 47A60, 47B07, 47B25, 46-01, 46A30, 46A35, 46A45, 46E15

Author
Dr. Marat V. Markin
California State University, Fresno
Department of Mathematics
5245 N. Backer Avenue, M/S PB 108
Fresno, California 93740-8001
USA
mmarkin@csufresno.edu

ISBN 978-3-11-060096-4
e-ISBN (PDF) 978-3-11-060098-8
e-ISBN (EPUB) 978-3-11-059888-9
DOI https://doi.org/10.1515/9783110600988

Library of Congress Control Number: 2019951807

Bibliographic information published by the Deutsche Nationalbibliothek
The Deutsche Nationalbibliothek lists this publication in the Deutsche Nationalbibliografie;
detailed bibliographic data are available on the Internet at http://dnb.dnb.de.

© 2020 Walter de Gruyter GmbH, Berlin/Boston
Cover image: Mordolff/Getty Images
Typesetting: VTeX UAB, Lithuania
Printing and binding: CPI books Lecks, GmbH

www.degruyter.com

To the beauty and power of mathematics.

Preface

Mathematics is the most beautiful and most powerful creation of the human spirit.
Stefan Banach

A Few Words on the Subject of Operator Theory

As was shrewdly observed by S. Krein and Yu. Petunin, "To study a problem, one must choose a space and study the corresponding functionals, operators, etc. in it. ... The choice of the space in which the problem is studied is partly connected with the subjective aims set by the investigator. Apparently, the objective data are only the operators that appear in the equations of the problem. On this account, it seems to us that the original and basic concept of functional analysis is that of an operator" (see, e. g., [44]). Thus, in the context of the ancient chicken-egg argument, the concept of space emerges as consequential to that of an operator.

Operator theory is a modern, vast, and rapidly developing branch of functional analysis, which addresses operators, most notably linear. Combining profoundly abstract nature with extensive applicability, which encompasses ordinary and partial differential equations, integral equations, calculus of variations, quantum mechanics, and much more, operator theory is a powerful apparatus for solving diverse problems. A course in it, most certainly, deserves to be a vital part of a contemporary graduate mathematics curriculum, increasing its value not only for graduate students majoring in mathematics but also for those majoring in physics, science, and engineering.

Book's Purpose and Targeted Audience

The book is intended as a text for a *one-semester* Master's level graduate course in operator theory to be taught within the existing constraints of the standard for the United States graduate curriculum (fifteen weeks with two seventy-five-minute lectures per week). Considering the above, this is an introductory text on the fundamentals of operator theory with prerequisites intentionally set not high, the students not being assumed to have taken graduate courses either in analysis (real or complex) or general topology, to make the course accessible and attractive to a wider audience of STEM (science, technology, engineering, and mathematics) graduate students or advanced undergraduates with a solid background in calculus and linear algebra. Designed to teach a one-semester operator theory course "from scratch", not as a sequel to a functional analysis course, this book cannot but have a certain nontrivial material overlap with the author's recent textbook [45] remaining, however, very distinct from the latter by the scope and the learning outcomes, with the basics of the spectral theory of linear operators taking the center stage.

https://doi.org/10.1515/9783110600988-202

Book's Scope and Specifics

The book consists of *six chapters* and an *appendix*, taking the reader from the fundamentals of abstract spaces (metric, vector, normed vector, and inner product), the *Banach Fixed-Point Theorem* and its applications, such as *Picard's Existence and Uniqueness Theorem*, through the basics of linear operators, two of the *three fundamental principles* (the *Uniform Boundedness Principle* and the *Open Mapping Theorem* and its equivalents: the *Inverse Mapping* and *Closed Graph Theorems*), to the elements of the spectral theory, including *Gelfand's Spectral Radius Theorem* and the *Spectral Theorem for Compact Self-Adjoint Operators*, and its applications, such as the celebrated *Lyapunov Stability Theorem* and the *Mean Ergodicity Theorem* [48, Theorem 4.1], the latter being a result of the author's own research.

The course is designed to be taught starting with Chapter 2, Chapter 1 outlining certain necessary preliminaries and being referred to whenever the need arises.

The Appendix gives a concise treatise of the *Axiom of Choice*, its equivalents (the *Hausdorff Maximal Principle, Zorn's Lemma*, and *Zermelo's Well-Ordering Principle*), and *ordered sets*, which underlie several fundamental statements of the course, such as the *Basis Theorem* (Theorem 3.2).

Conceived as a text to be used in a classroom, the book constantly calls for the student's actively mastering the knowledge of the subject matter. There are problems at the end of each chapter, starting with Chapter 2 and totaling 150. These problems are indispensable for understanding the material and moving forward. Many important statements, such as the *Fundamental Sequence with Convergent Subsequence Proposition* (Proposition 2.22, Section 2.18, Problem 25), are given as problems; a lot of them are frequently referred to and used in the book. There are also 432 Exercises throughout the text, including Chapter 1 and the Appendix, which require of the student to prove or verify a statement or an example, fill in certain details in a proof, or provide an intermediate step or a counterexample. They are also an inherent part of the material. More difficult problems, such as Section 4.7, Problem 25, are marked with an asterisk; many problems and exercises are supplied with "existential" hints.

The book is generous on Examples and contains numerous Remarks accompanying every definition and virtually each statement to discuss certain subtleties, raise questions on whether the converse assertions are true, whenever appropriate, or whether the conditions are essential.

As amply demonstrated by experience, students tend to better remember statements by their names rather than by numbers. Thus, a distinctive feature of the book is that every theorem, proposition, corollary, and lemma, unless already possessing a name, is endowed with a descriptive one, making it easier to remember, which, in this author's humble opinion, is quite a bargain when the price for better understanding and retention of the material is only a little clumsiness while making a longer reference. Each statement is referred to by its name and not just the number, e. g., the *Norm Equivalence Theorem* (Theorem 3.9), as opposed to merely Theorem 3.9.

With no pretense on furnishing the history of the subject, the text provides certain dates and lists every related name as a footnote.

Acknowledgments

First and foremost, I wish to express my heartfelt gratitude to my mother, Svetlana A. Markina, for her unfailing love and support, without which this and many other endeavors of mine would have been impossible.

My utmost appreciation goes to Mr. Edward Sichel, my pupil and graduate advisee, for his invaluable assistance with proofreading and improving the manuscript.

I am very thankful to Dr. Przemyslaw Kajetanowicz (Department of Mathematics, CSU, Fresno) for his kind aid with graphics.

My sincere acknowledgments are also due to Dr. Apostolos Damialis, formerly *Walter de Gruyter GmbH* Acquisitions Editor in Mathematics, for seeing value in my manuscript and making authors his highest priority, Ms. Nadja Schedensack, *Walter de Gruyter GmbH* Project Editor in Mathematics and Physics, for superb efficiency in managing all project related matters, as well as Ms. Ina Talandienė and Ms. Ieva Spudulytė, VTeX Book Production, for their expert editorial and LaTeX typesetting contributions.

Clovis, California, USA   Marat V. Markin
June–July 2019

Contents

1 Preliminaries

In this chapter, we outline certain terminology, notations, and preliminary facts essential for our subsequent discourse.

1.1 Set-Theoretic Basics

1.1.1 Some Terminology and Notations

- The logic *quantifiers* $\forall$, $\exists$, and $\exists!$ stand for *"for all"*, *"there exist(s)"*, and *"there exists a unique"*, respectively.
- $\mathbb{N} := \{1, 2, 3, \dots\}$ is the set of *natural numbers*.
- $\mathbb{Z} := \{0, \pm 1, \pm 2, \dots\}$ is the set of *integers*.
- $\mathbb{Q}$ is the set of *rational numbers*.
- $\mathbb{R}$ is the set of *real numbers*.
- $\mathbb{C}$ is the set of *complex numbers*.
- $\mathbb{Z}_+$, $\mathbb{Q}_+$, and $\mathbb{R}_+$ are the sets of *nonnegative* integers, rationals, and reals, respectively.
- $\overline{\mathbb{R}} := [-\infty, \infty]$ is the set of *extended real numbers* (*extended real line*).
- For $n \in \mathbb{N}$, $\mathbb{R}^n$ and $\mathbb{C}^n$ are the *n*-spaces of all *ordered n-tuples* of real and complex numbers, respectively.

Let X be a set. Henceforth, all sets are supposed to be subsets of X.
- $\mathscr{P}(X)$ is the *power set* of X, i. e., the collection of all subsets of X.
- 2^X is the set of all binary functions $f : X \to \{0, 1\}$, provided that $X \neq \emptyset$.
- Sets $A, B \subseteq X$ with $A \cap B = \emptyset$ are called *disjoint*.
- Let I be a nonempty indexing set. The sets of a collection $\{A_i\}_{i \in I}$ of subsets of X are said to be *pairwise disjoint* if

$$A_i \cap A_j = \emptyset, \ i, j \in I, i \neq j.$$

- For $A, B \subseteq X$, $A \setminus B := \{x \in X \mid x \in A, \text{ but } x \notin B\}$ is the *difference* of A and B, in particular, $A^c := X \setminus A = \{x \in X \mid x \notin A\}$ is the *complement* of A and $A \setminus B = A \cap B^c$;
- Let I be a nonempty indexing set and $\{A_i\}_{i \in I}$ be a collection of subsets of X. *De Morgan's laws* state that

$$\left(\bigcup_{i \in I} A_i\right)^c = \bigcap_{i \in I} A_i^c \quad \text{and} \quad \left(\bigcap_{i \in I} A_i\right)^c = \bigcup_{i \in I} A_i^c.$$

More generally,

$$B \setminus \bigcup_{i \in I} A_i = \bigcap_{i \in I} B \setminus A_i \text{ and } B \setminus \bigcap_{i \in I} A_i = \bigcup_{i \in I} B \setminus A_i.$$

https://doi.org/10.1515/9783110600988-001

– The *Cartesian product* of sets $A_i \subseteq X$, $i = 1, \ldots, n$ ($n \in \mathbb{N}$),

$$A_1 \times \cdots \times A_n := \{(x_1, \ldots, x_n) \mid x_i \in A_i, \ i = 1, \ldots, n\}.$$

1.1.2 Cardinality and Countability

Definition 1.1 (Similarity of Sets). Sets A and B are said to be *similar* if there exists a one-to-one correspondence (bijection) between them.

Notation. $A \sim B$.

Remark 1.1. Similarity is an *equivalence relation* (*reflexive*, *symmetric*, and *transitive*) on the power set $\mathscr{P}(X)$ of a nonempty set X.

Exercise 1.1. Verify.

Thus, in the context, we can use the term *"equivalence"* synonymously to *"similarity"*.

Definition 1.2 (Cardinality). Equivalent sets are said to have the same number of elements or *cardinality*. Cardinality is a characteristic of an equivalence class of similar sets.

Notation. $\mathscr{P}(X) \ni A \mapsto |A|$.

Remark 1.2. Thus, $A \sim B$ *iff* $|A| = |B|$. That is, two sets are equivalent iff they share the same cardinality.

Examples 1.1.
1. For a nonempty set X, $\mathscr{P}(X) \sim 2^X$.
2. $\forall n \in \mathbb{N} : |\{1, \ldots, n\}| = n$, $|\{0, \pm 1, \ldots, \pm n\}| = 2n + 1$.
3. $|\mathbb{N}| = |\mathbb{Z}| = |\mathbb{Q}| := \aleph_0$.
4. $|[0, 1]| = |\mathbb{R}| = |\mathbb{C}| := \mathfrak{c}$.

See, e. g., [29, 33].

Definition 1.3 (Domination). If sets A and B are such that A is equivalent to a subset of B, we write

$$A \preceq B$$

and say that B *dominates* A. If, in addition, $A \not\sim B$, we write

$$A \prec B$$

and say that B *strictly dominates* A.

Remark 1.3. The relation $\preceq$ is a *partial order* (*reflexive, antisymmetric,* and *transitive*) on the power set $\mathscr{P}(X)$ of a nonempty set X (see Appendix A).

Exercise 1.2. Verify *reflexivity* and *transitivity*.

The *antisymmetry* of $\preceq$ is the subject of the following celebrated theorem:

Theorem 1.1 (Schröder–Bernstein Theorem)**.** *If, for sets A and B, $A \preceq B$ and $B \preceq A$, then $A \sim B$.*[1]

For a proof, see, e. g., [29].

Remark 1.4. The set partial order $\preceq$ defines a partial order $\le$ on the set of cardinals:

$$|A| \le |B| \;\Leftrightarrow\; A \preceq B.$$

Thus, the *Schröder–Bernstein theorem* can be equivalently reformulated in terms of cardinalities as follows:

If, for sets A and B, $|A| \le |B|$ and $|B| \le |A|$, then $|A| = |B|$.

Theorem 1.2 (Cantor's Theorem)**.** *Every set X is strictly dominated by its power set $\mathscr{P}(X)$*[2]*:*

$$X \prec \mathscr{P}(X).$$

Equivalently,

$$|X| < \big|\mathscr{P}(X)\big|.$$

For a proof, see, e. g., [29].

In view of Examples 1.1, we obtain the following corollary:

Corollary 1.1. *For a nonempty set X, $X \prec 2^X$, i. e., $|X| < |2^X|$.*

Definition 1.4 (Countable/Uncountable Set)**.** A *countable set* is a set with the same *cardinality* as a subset of the set $\mathbb{N}$ of natural numbers, i. e., *equivalent* to a subset of $\mathbb{N}$.

A set that is not *countable* is called *uncountable*.

Remarks 1.5.
- A countable set A is either *finite*, i. e., equivalent to a set of the form $\{1, \dots, n\} \subset \mathbb{N}$ with some $n \in \mathbb{N}$, in which case, we say that A has n elements, or *countably infinite*, i. e., equivalent to the entire $\mathbb{N}$.

---

[1] Ernst Schröder (1841–1902), Felix Bernstein (1878–1956).
[2] Georg Cantor (1845–1918).

– For a finite set A of n elements ($n \in \mathbb{N}$),

$$|A| = |\{1, \ldots, n\}| = n.$$

For a countably infinite set A,

$$|A| = |\mathbb{N}| = \aleph_0$$

(see Examples 1.1).
– In some sources, the term *"countable"* is used in the sense of *"countably infinite"*. To avoid ambiguity, the term *"at most countable"* can be used when finite sets are included in consideration.

The subsequent statement immediately follows from *Cantor's theorem* (Theorem 1.2).

Proposition 1.1 (Uncountable Sets). *The sets $\mathscr{P}(\mathbb{N})$ and $2^{\mathbb{N}}$ (the set of all binary sequences) are uncountable.*

Theorem 1.3 (Properties of Countable Sets).
(1) *Every infinite set contains a countably infinite subset (based on the Axiom of Choice (see Appendix A)).*
(2) *Any subset of a countable set is countable.*
(3) *The union of countably many countable sets is countable.*
(4) *The Cartesian product of finitely many countable sets is countable.*

Exercise 1.3. Prove that
(a) the set $\mathbb{Z}$ of all *integers* and the set of all *rational numbers* are countable;
(b) for any $n \in \mathbb{N}$, $\mathbb{Z}^n$ and $\mathbb{Q}^n$ are countable;
(c) the set of all *algebraic numbers* (the roots of polynomials with integer coefficients) is countable.

Subsequently, we also need the following useful result:

Proposition 1.2 (Cardinality of the Collection of Finite Subsets). *The cardinality of the collection of all finite subsets of an infinite set coincides with the cardinality of the set.*

For a proof, see, e. g., [33, 41, 52].

1.2 Terminology Related to Functions

Let X and Y be nonempty sets, $\emptyset \neq D \subseteq X$, and

$$f : D \to Y.$$

– The set D is called the *domain (of definition)* of f.
– The *value* of f corresponding to an $x \in D$ is designated by $f(x)$.
– The set

$$\{f(x) \mid x \in D\}$$

of all values of f is called the *range* of f (also *codomain* or *target set*).
– For a set $A \subseteq D$, the set

$$f(A) := \{f(x) \mid x \in A\}$$

of values of f corresponding to all elements of A is called the *image* of A under the function f.
Thus, the *range* of f is the image $f(D)$ of the whole domain D.
– For a set $B \subseteq Y$, the set

$$f^{-1}(B) := \{x \in D \mid f(x) \in B\}$$

of all elements of the domain that map to the elements of B is called the *inverse image* (or *preimage*) of B.

Example 1.2. For $X = Y := \mathbb{R}$ and $f(x) := x^2$ with $D := [-1, 2]$,
– $f([-1, 2]) = [0, 4]$ and $f([1, 2]) = [1, 4]$,
– $f^{-1}([-2, -1]) = \emptyset$, $f^{-1}([0, 1]) = [-1, 1]$, and $f^{-1}([1, 4]) = \{-1\} \cup [1, 2]$.

Theorem 1.4 (Properties of Inverse Image). *Let X and Y be nonempty sets, $\emptyset \neq D \subseteq X$, and*

$$f : D \to Y.$$

Then, for an arbitrary nonempty collection $\{B_i\}_{i \in I}$ of subsets of Y,
(1) $f^{-1}(\bigcup_{i \in I} B_i) = \bigcup_{i \in I} f^{-1}(B_i)$,
(2) $f^{-1}(\bigcap_{i \in I} B_i) = \bigcap_{i \in I} f^{-1}(B_i)$, and
(3) *for any $B_1, B_2 \subseteq Y$, $f^{-1}(B_1 \setminus B_2) = f^{-1}(B_1) \setminus f^{-1}(B_2)$.*

That is, the preimage preserves all set operations.

Exercise 1.4.
(a) Prove.
(b) Show that image preserves unions. That is, for an arbitrary nonempty collection $\{A_i\}_{i \in I}$ of subsets of D:

$$f\left(\bigcup_{i \in I} A_i\right) = \bigcup_{i \in I} f(A_i),$$

and unions *only*. Give corresponding counterexamples for intersections and differences.

1.3 Upper and Lower Limits

Definition 1.5 (Upper and Lower Limits). Let $(x_n)_{n\in\mathbb{N}}$ (another notation is $\{x_n\}_{n=1}^{\infty}$) be a sequence of real numbers.

The *upper limit* or *limit superior* of $(x_n)_{n\in\mathbb{N}}$ is defined as follows:

$$\overline{\lim_{n\to\infty}} \, x_n := \lim_{n\to\infty} \sup_{k\geq n} x_k = \inf_{n\in\mathbb{N}} \sup_{k\geq n} x_k \in \overline{\mathbb{R}}.$$

The *lower limit* or *limit inferior* of $(x_n)_{n\in\mathbb{N}}$ is defined as follows:

$$\underline{\lim_{n\to\infty}} \, x_n := \lim_{n\to\infty} \inf_{k\geq n} x_k = \sup_{n\in\mathbb{N}} \inf_{k\geq n} x_k \in \overline{\mathbb{R}}.$$

Alternative notations are $\limsup_{n\to\infty} x_n$ and $\liminf_{n\to\infty} x_n$, respectively.

Example 1.3. For

$$x_n := \begin{cases} n, & n \in \mathbb{N} \text{ is odd,} \\ -1/n, & n \in \mathbb{N} \text{ is even,} \end{cases}$$

$$\overline{\lim_{n\to\infty}} \, x_n = \infty \quad \text{and} \quad \underline{\lim_{n\to\infty}} \, x_n = 0.$$

Exercise 1.5.

(a) Verify.

(b) Explain why the upper and lower limits, unlike the regular limit, are guaranteed to exist for an arbitrary sequence of real numbers.

(c) Show that

$$\underline{\lim_{n\to\infty}} \, x_n \leq \overline{\lim_{n\to\infty}} \, x_n.$$

Proposition 1.3 (Characterization of Limit Existence). *For a sequence of real numbers* $(x_n)_{n\in\mathbb{N}}$,

$$\lim_{n\to\infty} x_n \in \overline{\mathbb{R}}$$

exists iff

$$\underline{\lim_{n\to\infty}} \, x_n = \overline{\lim_{n\to\infty}} \, x_n,$$

in which case

$$\underline{\lim_{n\to\infty}} \, x_n = \lim_{n\to\infty} x_n = \overline{\lim_{n\to\infty}} \, x_n.$$

1.4 Certain Facts from Linear Algebra

Throughout this section, $\mathbb{F}$ stands for the scalar field of real or complex numbers (i. e., $\mathbb{F} = \mathbb{R}$ or $\mathbb{F} = \mathbb{C}$).

1.4.1 Coordinate Vector Mapping

Let X be an *n-dimensional* vector space $(n \in \mathbb{N})$ over $\mathbb{F}$ with an (ordered) *basis B* $:= \{x_1, \ldots, x_n\}$.

By the *Basis Representation Theorem* (Theorem 3.3), each vector $x \in X$ has a unique representation relative to basis B:

$$x = \sum_{k=1}^{n} c_k x_k,$$

where $c_j \in \mathbb{F}, j = 1, \ldots, n$, are the *coordinates of x relative to basis B* (see Section 3.1.5.2). The *coordinate vector mapping*

$$X \ni x = \sum_{k=1}^{n} c_k x_k \mapsto Tx := [x]_B := (c_1, \ldots, c_n) \in \mathbb{F}^n,$$

where

$$[x]_B := (c_1, \ldots, c_n) \in \mathbb{F}^n,$$

is the *coordinate vector of x relative to basis B* (see, e. g., [34, 49, 54]), being an *isomorphism* (see Section 3.1.2) between the vector spaces X and $\mathbb{F}^n$.

Exercise 1.6. Verify that the coordinate mapping is an isomorphism between X and $\mathbb{F}^n$.

In this sense, every n-dimensional vector space is identical to $\mathbb{F}^n$.

1.4.2 Matrix Representations of Linear Operators

Theorem 1.5 (Matrix Representation). *Let X be an n-dimensional vector space $(n \in \mathbb{N})$ over $\mathbb{F}$ and let $B := \{x_1, \ldots, x_n\}$ and $B' := \{x'_1, \ldots, x'_n\}$ be (ordered) bases for X. Then, for an arbitrary linear operator $A : X \to X$ (see Section 4.1),*

$$[Ax]_{B'} = {}_{B'}[A]_B [x]_B, \ x \in X,$$

where $[x]_B \in \mathbb{F}^n$ is the coordinate vector of x relative to basis B, $[Ax]_{B'} \in \mathbb{F}^n$ is the coordinate vector of Ax relative to basis B', and

$$_{B'}[A]_B := [[Ax_1]_{B'} \quad [Ax_2]_{B'} \quad \cdots \quad [Ax_n]_{B'}]$$

is the $n \times n$ matrix with entries from $\mathbb{F}$ whose columns are the coordinate vectors $[Ax_j]_{B'}$, $j = 1, \ldots, n$, of $Ax_j, j = 1, \ldots, n$, relative to basis B'.

For a proof, see, e. g., [34, 49, 54].

Remarks 1.6.

- The matrix $_{B'}[A]_B$ is called the *matrix representation of A relative to bases B and B'*.
- If $B' = B$, the matrix representation $_B[A]_B$ is denoted by $[A]_B$ and called the *matrix representation of A relative to basis B*, in which case, we have

$$[Ax]_B = [A]_B[x]_B, \; x \in X.$$

In particular, if $X = \mathbb{F}^n$ and if

$$B := \{e_1 := (1, 0, \ldots, 0), e_2 := (0, 1, \ldots, 0), \ldots, e_n := (0, 0, \ldots, 1)\}$$

is the *standard basis*:

$$\forall x := (x_1, \ldots, x_n) \in \mathbb{F}^n : \; [x]_B = (x_1, \ldots, x_n) = x,$$

and hence, by the *Matrix Representation Theorem*, we have

$$\forall x \in \mathbb{F}^n : \; Ax = [Ax]_B = [A]_B[x]_B = [A]_B x.$$

In this case, the matrix representation of A relative to the standard basis

$$[A]_B := \begin{bmatrix} Ae_1 & Ae_2 & \cdots & Ae_n \end{bmatrix}$$

is the $n \times n$ matrix whose columns are the vectors Ae_j, $j = 1, \ldots, n$.
Hence, any linear operator $A : \mathbb{F}^n \to \mathbb{F}^n$ is an operator of multiplication by an $n \times n$ matrix. The converse is true as well (see Examples 4.1).

Examples 1.4.

1. For the *zero* and *identity operators* on a finite-dimensional vector space X, relative to an arbitrary basis B:

$$[0]_B = 0 \text{ and } [I]_B = I,$$

where 0 and I stand for both the zero and identity operators and matrices, respectively.

Exercise 1.7. Verify.

2. On $\mathbb{F}^2$, let A, C, and D be the linear operators of multiplication by the matrices

$$\begin{bmatrix} 1 & 0 \\ 0 & 2 \end{bmatrix}, \begin{bmatrix} 0 & 1 \\ 0 & 0 \end{bmatrix}, \text{ and } \begin{bmatrix} 0 & 1 \\ -1 & 0 \end{bmatrix},$$

respectively (see Examples 4.1).
Then, relative to the standard basis $B := \{[\begin{smallmatrix} 1 \\ 0 \end{smallmatrix}], [\begin{smallmatrix} 0 \\ 1 \end{smallmatrix}]\}$,

$$[A]_B = \begin{bmatrix} 1 & 0 \\ 0 & 2 \end{bmatrix}, [C]_B = \begin{bmatrix} 0 & 1 \\ 0 & 0 \end{bmatrix}, \text{ and } [D]_B = \begin{bmatrix} 0 & 1 \\ -1 & 0 \end{bmatrix}$$

and, relative to bases B and $B' := \{[\begin{smallmatrix}1\\1\end{smallmatrix}], [\begin{smallmatrix}0\\1\end{smallmatrix}]\}$,

$$_{B'}[A]_B = \begin{bmatrix} 1 & 0 \\ -1 & 2 \end{bmatrix}, \; _{B'}[C]_B = \begin{bmatrix} 0 & 1 \\ 0 & -1 \end{bmatrix}, \text{ and } _{B'}[D]_B = \begin{bmatrix} 0 & 1 \\ -1 & -1 \end{bmatrix}.$$

Exercise 1.8. Verify.

3. For the *differentiation operator*

$$P_3 \ni p \mapsto Dp := p' \in P_3,$$

where P_3 is the four-dimensional space of polynomials with coefficients from $\mathbb{F}$ of degree at most 3 (see Examples 3.1), the matrix representation of D relative to the *standard basis*

$$B := \{1, x, x^2, x^3\},$$

is

$$[D]_B = \begin{bmatrix} 0 & 1 & 0 & 0 \\ 0 & 0 & 2 & 0 \\ 0 & 0 & 0 & 3 \\ 0 & 0 & 0 & 0 \end{bmatrix}.$$

Exercise 1.9. Verify.

For any polynomial

$$p(x) = c_0 + c_1 x + c_2 x^2 + c_3 x^3 \in P_3 \quad \text{with} \quad [p]_B = \begin{bmatrix} c_0 \\ c_1 \\ c_2 \\ c_3 \end{bmatrix},$$

we have

$$(Dp)(x) = c_1 + 2c_2 x + 3c_3 x^2 \quad \text{with} \quad [Dp]_B = \begin{bmatrix} c_1 \\ 2c_2 \\ 3c_3 \\ 0 \end{bmatrix},$$

which is consistent with what we get by the *Matrix Representation Theorem*:

$$[Dp]_B = [D]_B[p]_B = \begin{bmatrix} 0 & 1 & 0 & 0 \\ 0 & 0 & 2 & 0 \\ 0 & 0 & 0 & 3 \\ 0 & 0 & 0 & 0 \end{bmatrix} \begin{bmatrix} c_0 \\ c_1 \\ c_2 \\ c_3 \end{bmatrix} = \begin{bmatrix} c_1 \\ 2c_2 \\ 3c_3 \\ 0 \end{bmatrix}.$$

1.4.3 Change of Basis, Transition Matrices

Theorem 1.6 (Change of Basis). *Let X be an n-dimensional vector space (n ∈ ℕ) over*
$\mathbb{F}$ *and let* $B := \{x_1,\ldots,x_n\}$ *and* $B' := \{x'_1,\ldots,x'_n\}$ *be (ordered) bases for X. The change of*
coordinates from basis B to basis B' is carried out by the following change-of-coordinates
formula:

$$[x]_{B'} = {}_{B'}[I]_B [x]_B,$$

where

$${}_{B'}[I]_B := [[x_1]_{B'} \quad [x_2]_{B'} \quad \cdots \quad [x_n]_{B'}]$$

is the n × n matrix whose columns are the coordinate vectors $[x_j]_{B'}$, $j = 1,\ldots,n$, *of the*
vectors x_j, $j = 1,\ldots,n$, *of basis B relative to basis B'.*

For a proof, see, e. g., [34, 49, 54].

Remark 1.7. The matrix ${}_{B'}[I]_B$, called the *transition matrix from basis B to basis B'*, is,
in fact, the matrix representation of the identity operator $I : X \to X$ relative to bases B
and B'.

Thus, the *Change of Basis Theorem* is a particular case of the *Matrix Representation*
Theorem (Theorem 1.5) for $A = I$, which explains the use of the notation ${}_{B'}[I]_B$.

Examples 1.5.
1. Consider the *standard basis*

$$B := \left\{ e_2 := \begin{bmatrix} 1 \\ 0 \end{bmatrix}, e_1 := \begin{bmatrix} 0 \\ 1 \end{bmatrix} \right\}$$

in the space $\mathbb{F}^2$.
(a) The *transition matrix* from B to the basis

$$B' := \left\{ \begin{bmatrix} 0 \\ 1 \end{bmatrix}, \begin{bmatrix} 1 \\ 0 \end{bmatrix} \right\}$$

is

$${}_{B'}[I]_B = \begin{bmatrix} 0 & 1 \\ 1 & 0 \end{bmatrix}.$$

Indeed, since

$$e_1 = \begin{bmatrix} 1 \\ 0 \end{bmatrix} = 0 \begin{bmatrix} 0 \\ 1 \end{bmatrix} + 1 \begin{bmatrix} 1 \\ 0 \end{bmatrix}$$

and

$$e_2 = \begin{bmatrix} 0 \\ 1 \end{bmatrix} = 1 \begin{bmatrix} 0 \\ 1 \end{bmatrix} + 0 \begin{bmatrix} 1 \\ 0 \end{bmatrix},$$

we have

$$[e_1]_{B'} = \begin{bmatrix} 0 \\ 1 \end{bmatrix} \quad \text{and} \quad [e_2]_{B'} = \begin{bmatrix} 1 \\ 0 \end{bmatrix}.$$

Therefore, the *change-of-coordinates formula* from basis B to basis B' is

$$\begin{bmatrix} y \\ x \end{bmatrix} = \begin{bmatrix} 0 & 1 \\ 1 & 0 \end{bmatrix} \begin{bmatrix} x \\ y \end{bmatrix}.$$

(b) The *transition matrix* from B to the basis

$$B' := \left\{ \begin{bmatrix} 1 \\ 1 \end{bmatrix}, \begin{bmatrix} 0 \\ 1 \end{bmatrix} \right\}$$

is

$$_{B'}[I]_B = \begin{bmatrix} 1 & 0 \\ -1 & 1 \end{bmatrix}.$$

Indeed, since

$$e_1 = \begin{bmatrix} 1 \\ 0 \end{bmatrix} = 1 \begin{bmatrix} 1 \\ 1 \end{bmatrix} + (-1) \begin{bmatrix} 0 \\ 1 \end{bmatrix}$$

and

$$e_2 = \begin{bmatrix} 0 \\ 1 \end{bmatrix} = 0 \begin{bmatrix} 1 \\ 1 \end{bmatrix} + 1 \begin{bmatrix} 0 \\ 1 \end{bmatrix},$$

we have

$$[e_1]_{B'} = \begin{bmatrix} 1 \\ -1 \end{bmatrix} \quad \text{and} \quad [e_2]_{B'} = \begin{bmatrix} 0 \\ 1 \end{bmatrix}.$$

Therefore, the *change-of-coordinates formula* from basis B to basis B' is

$$\begin{bmatrix} x \\ y - x \end{bmatrix} = \begin{bmatrix} 1 & 0 \\ -1 & 1 \end{bmatrix} \begin{bmatrix} x \\ y \end{bmatrix}.$$

2. In the space P_2 of polynomials with coefficients from $\mathbb{F}$ of degree at most 2 (see Examples 3.1), the *transition matrix* from the standard basis $B := \{1, x, x^2\}$ to the basis $B' := \{1 - x, 1 - x^2, 1 + 2x\}$ is

$$_{B'}[I]_B = \begin{bmatrix} 2/3 & -1/3 & 2/3 \\ 0 & 0 & -1 \\ 1/3 & 1/3 & 1/3 \end{bmatrix}.$$

Indeed, since

$$1 = (2/3)(1 - x) + 0(1 - x^2) + (1/3)(1 + 2x),$$
$$x = (-1/3)(1 - x) + 0(1 - x^2) + (1/3)(1 + 2x),$$
$$x^2 = (2/3)(1 - x) + (-1)(1 - x^2) + (1/3)(1 + 2x),$$

we have

$$[1]_{B'} = \begin{bmatrix} 2/3 \\ 0 \\ 1/3 \end{bmatrix}, \quad [x]_{B'} = \begin{bmatrix} -1/3 \\ 0 \\ 1/3 \end{bmatrix}, \quad \text{and} \quad [x^2]_{B'} = \begin{bmatrix} 2/3 \\ -1 \\ 1/3 \end{bmatrix}.$$

Therefore, for the polynomial $p(x) := 3 - x + x^2 \in P_2$ with $[p]_B = \begin{bmatrix} 3 \\ -1 \\ 1 \end{bmatrix}$, the *change-of-coordinates formula* from basis B to basis B' yields

$$[p]_{B'} = \begin{bmatrix} 2/3 & -1/3 & 2/3 \\ 0 & 0 & -1 \\ 1/3 & 1/3 & 1/3 \end{bmatrix} \begin{bmatrix} 3 \\ -1 \\ 1 \end{bmatrix} = \begin{bmatrix} 3 \\ -1 \\ 1 \end{bmatrix}.$$

Remark 1.8. Interestingly enough, for the given polynomial, the coordinates relative to both bases are identical.

Theorem 1.7 (Inverse of Transition Matrix). *Let X be an n-dimensional vector space ($n \in \mathbb{N}$) over $\mathbb{F}$ and let $B := \{x_1, \ldots, x_n\}$ and $B' := \{x'_1, \ldots, x'_n\}$ be ordered bases for X. The $n \times n$ transition matrix from basis B to basis B' is invertible, and*

$$_{B'}[I]_B^{-1} = {_B}[I]_{B'} := \begin{bmatrix} [x'_1]_B & [x'_2]_B & \cdots & [x'_n]_B \end{bmatrix}.$$

That is, the inverse of the transition matrix $_{B'}[I]_B$ from basis B to basis B' is the transition matrix $_B[I]_{B'}$ from basis B' to basis B whose columns are the coordinate vectors $[x'_j]_B$, $j = 1 \ldots, n$, of the vectors $x'_j, j = 1 \ldots, n$, of the basis B' relative to the basis B.

For a proof, see, e. g., [34, 49, 54].

Example 1.6. As follows from Examples 1.5, in the space P_2, the *transition matrix* from the standard basis $B := \{1, x, x^2\}$ to the basis $B' := \{1 - x, 1 - x^2, 1 + 2x\}$ is

$$_{B'}[I]_B = \begin{bmatrix} 2/3 & -1/3 & 2/3 \\ 0 & 0 & -1 \\ 1/3 & 1/3 & 1/3 \end{bmatrix}.$$

By the *Inverse of Transition Matrix Theorem*,

$$_B[I]_{B'} = {_{B'}}[I]_B^{-1} = \begin{bmatrix} 1 & 1 & 1 \\ -1 & 0 & 2 \\ 0 & -1 & 0 \end{bmatrix}.$$

Exercise 1.10. Verify the last equality.

Theorem 1.8 (Change of Basis for Linear Operators). *Let X be an n-dimensional vector space (n ∈ ℕ) over 𝔽 and let B := {$x_1, \ldots, x_n$} and B' := {$x'_1, \ldots, x'_n$} be ordered bases for X. Then, for an arbitrary linear operator A : X → X, the matrix representations $[A]_B$ and $[A]_{B'}$ are similar and relate as follows:*

$$[A]_{B'} = {}_{B'}[I]_B [A]_{BB} [I]_{B'} = {}_B[I]_{B'}{}^{-1}[A]_{BB}[I]_{B'},$$

where

$$_{B'}[I]_B := [[x_1]_{B'} \quad [x_2]_{B'} \quad \cdots \quad [x_n]_{B'}]$$

and

$$_B[I]_{B'} := [[x'_1]_B \quad [x'_2]_B \quad \cdots \quad [x'_n]_B]$$

are the transition matrices from basis B to basis B' and from basis B' to basis B, respectively.

For a proof, see, e. g., [34, 49, 54].

Remarks 1.9.
- The *similarity* of matrices is an *equivalence relation* on the vector space $M_{n \times n}$ (n ∈ ℕ) of all n × n with entries from 𝔽 (see Examples 3.1).
- Similar matrices with complex entries have the same *determinant, characteristic polynomial, characteristic equation,* and *eigenvalues* with identical *algebraic* and *geometric multiplicities,* i. e., the multiplicity as a zero of the characteristic polynomial and the dimension of the corresponding eigenspace.

(See, e. g., [34, 49, 54].)

1.4.4 Cayley–Hamilton Theorem

Theorem 1.9 (Cayley–Hamilton Theorem). *For an arbitrary n × n (n ∈ ℕ) matrix A,*

$$p(A) = 0,$$

where $p(\lambda) := \det(A - \lambda I)$ is the characteristic polynomial of A.[3]

For a proof, see, e. g., [34, 49, 54].

---

3 Arthur Cayley (1821–1895), William Rowan Hamilton (1805–1865).

2 Metric Spaces

In this chapter, we study abstract sets endowed with a notion of *distance*, whose properties mimic those of the regular distance in three-dimensional space. Distance brings to life various topological notions such as *limit*, *continuity*, *openness*, *closedness*, *compactness*, and *denseness*, the geometric notion of *boundedness*, and the notions of *fundamentality* of sequences and *completeness*. Here, we consider all these and beyond in depth.

2.1 Definition and Examples

Definition 2.1 (Metric Space). A *metric space* is a nonempty set X with a *metric* (or *distance function*), i. e., a mapping

$$\rho(\cdot,\cdot) : X \times X \to \mathbb{R}$$

subject to the following *metric axioms*:

1.	$\rho(x,y) \geq 0,\ x,y \in X.$	*Nonnegativity*
2.	$\rho(x,y) = 0$ iff $x = y.$	*Separation*
3.	$\rho(x,y) = \rho(y,x),\ x,y \in X.$	*Symmetry*
4.	$\rho(x,z) \leq \rho(x,y) + \rho(y,z),\ x,y,z \in X.$	*Triangle Inequality*

For any fixed $x,y \in X$, the number $\rho(x,y)$ is called the *distance* of x from y, or from y to x, or between x and y.

Notation. (X,ρ).

Remark 2.1. A function $\rho(\cdot,\cdot) : X \times X \to \mathbb{R}$ satisfying the metric axioms of *symmetry*, *triangle inequality*, and the following weaker form of the *separation axiom*:
2w. $\rho(x,y) = 0$ if $x = y$,

also necessarily satisfies the axiom of *nonnegativity* and is called a *pseudometric* on X (see the examples to follow).

Exercise 2.1. Verify that 2w, 3, and 4 imply 1.

Examples 2.1.
1. Any nonempty set X is a *metric space* relative to the *discrete metric*

$$X \ni x,y \mapsto \rho_d(x,y) := \begin{cases} 0 & \text{if } x = y, \\ 1 & \text{if } x \neq y. \end{cases}$$

2. The *real line* $\mathbb{R}$ and the *complex plane* $\mathbb{C}$ are metric spaces relative to the regular distance function $\rho(x,y) := |x - y|$.

https://doi.org/10.1515/9783110600988-002

3. Let $n \in \mathbb{N}$ and $1 \le p \le \infty$. The real/complex n-space, $\mathbb{R}^n$ or $\mathbb{C}^n$, is a *metric space* relative to *p-metric*

$$\rho_p(x,y) = \begin{cases} [\sum_{k=1}^{n} |x_k - y_k|^p]^{1/p} & \text{if } 1 \le p < \infty, \\ \max_{1 \le k \le n} |x_k - y_k| & \text{if } p = \infty, \end{cases}$$

where $x := (x_1, \ldots, x_n)$, and $y := (y_1, \ldots, y_n)$, designated by $l_p^{(n)}$ (real or complex, respectively).

Remarks 2.2.

– For $n = 1$, all these metrics coincide with $\rho(x,y) = |x - y|$.
– For $n = 2, 3$, and $p = 2$, we have the usual *Euclidean distance*.
– $(\mathbb{C}, \rho) = (\mathbb{R}^2, \rho_2)$.

4. Let $1 \le p \le \infty$. The set l_p of all real- or complex-termed sequences $(x_k)_{k \in \mathbb{N}}$ satisfying

$$\sum_{k=1}^{\infty} |x_k|^p < \infty \quad (1 \le p < \infty),$$

$$\sup_{k \in \mathbb{N}} |x_k| < \infty \quad (p = \infty)$$

(*p-summable/bounded* sequences, respectively) is a *metric space* relative to *p-metric*

$$\rho_p(x,y) = \begin{cases} [\sum_{k=1}^{\infty} |x_k - y_k|^p]^{1/p} & \text{if } 1 \le p < \infty, \\ \sup_{k \in \mathbb{N}} |x_k - y_k| & \text{if } p = \infty, \end{cases}$$

where $x := (x_k)_{k \in \mathbb{N}}$ and $y := (y_k)_{k \in \mathbb{N}} \in l_p$.

Remark 2.3. When contextually important to distinguish between the real and complex cases, we use the notations

$$l_p^{(n)}(\mathbb{R}), \; l_p(\mathbb{R}) \text{ and } l_p^{(n)}(\mathbb{C}), \; l_p(\mathbb{C}),$$

and refer to the corresponding spaces as *real* and *complex*, respectively.

Exercise 2.2. Verify Examples 2.1, 1, 2 and 3, 4 for $p = 1$ and $p = \infty$.

Remark 2.4. While verifying Examples 2.1, 3 and 4 for $p = 1$ and $p = \infty$ is straightforward, proving the *triangle inequality* for $1 < p < \infty$ requires *Minkowski's*[1] inequality.

---

1 Hermann Minkowski (1864–1909).

2.2 Hölder's and Minkowski's Inequalities

Here, we are to prove the celebrated *Hölder's*[2] and *Minkowski's inequalities* for n-tuples ($n \in \mathbb{N}$) and sequences. We use *Hölder's inequality* to prove *Minkowski's inequality* for the case of n-tuples. In its turn, to prove *Hölder's inequality* for n-tuples, we rely upon *Young's*[3] *inequality*, and hence, the latter is to be proved first. For the sequential case, *Hölder's* and *Minkowski's inequalities* are proved independently based on their analogues for n-tuples.

2.2.1 Conjugate Indices

Definition 2.2 (Conjugate Indices). We call $1 \le p, q \le \infty$ *conjugate indices* if they are related as follows:

$$\frac{1}{p} + \frac{1}{q} = 1 \text{ for } 1 < p, q < \infty,$$

$$q = \infty \text{ for } p = 1,$$

$$q = 1 \text{ for } p = \infty.$$

Examples 2.2. In particular, $p = 2$ and $q = 2$ are conjugate; so also are $p = 3$ and $q = 3/2$.

Remark 2.5. Thus, for $1 < p, q < \infty$,

$$q = \frac{p}{p-1} = 1 + \frac{1}{p-1} \rightarrow \begin{cases} \infty, & p \rightarrow 1+, \\ 1, & p \rightarrow \infty, \end{cases}$$

with $q > 2$ if $1 < p < 2$ and $1 < q < 2$ if $p > 2$, and the following relationships hold:

$$p + q = pq,$$
$$pq - p - q + 1 = 1 \Rightarrow (p-1)(q-1) = 1,$$
$$(p-1)q = p \Rightarrow p - 1 = \frac{p}{q},$$
$$(q-1)p = q \Rightarrow q - 1 = \frac{q}{p}. \tag{2.1}$$

---

2 Otto Ludwig Hölder (1859–1937).
3 William Henry Young (1863–1942).

2.2.2 Young's Inequality

Theorem 2.1 (Young's Inequality). *Let $1 < p, q < \infty$ be conjugate indices. Then, for any $a, b \geq 0$,*

$$ab \leq \frac{a^p}{p} + \frac{b^q}{q}.$$

Proof. The inequality is obviously true if $a = 0$ or $b = 0$ and for $p = q = 2$.

Exercise 2.3. Verify.

Suppose that $a, b > 0$ and $1 < p < 2$, or $p > 2$. Recall that

$$(p - 1)(q - 1) = 1, \quad \text{and} \quad (p - 1)q = p.$$

(see (2.1)).

Comparing the areas in the following figure, which corresponds to the case of $p > 2$, the case of $1 < p < 2$ being symmetric, we conclude that

$$A \leq A_1 + A_2,$$

where A is the area of the rectangle $[0, a] \times [0, b]$, the equality being the case *iff* $b = a^{p-1} = a^{p/q}$.

Hence,

$$ab \leq \int_0^a x^{p-1}\, dx + \int_0^b y^{q-1}\, dy = \frac{a^p}{p} + \frac{b^q}{q}. \qquad \square$$

Remark 2.6. As Figure 2.1 shows, equality in *Young's inequality* (Theorem 2.1) holds iff $a^p = b^q$.

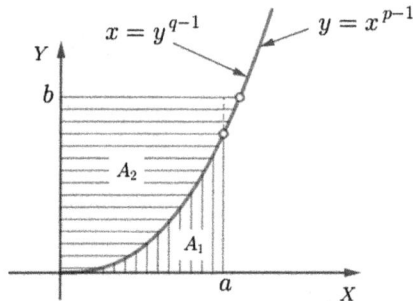

Figure 2.1: The case of $p > 2$.

2.2.3 The Case of *n*-Tuples

Definition 2.3 (*p*-Norm of an *n*-Tuple). Let $n \in \mathbb{N}$. For an *n*-tuple $x := (x_1, \ldots, x_n) \in \mathbb{C}^n$ $(1 \le p \le \infty)$, the *p-norm* of *x* is the distance of *x* from the *zero n-tuple* $0 := (0, \ldots, 0)$ in $l_p^{(n)}$:

$$\|x\|_p := \rho_p(x, 0) = \begin{cases} [\sum_{i=1}^n |x_i|^p]^{1/p} & \text{if } 1 \le p < \infty, \\ \max_{1 \le i \le n} |x_i| & \text{if } p = \infty. \end{cases}$$

Remarks 2.7.
- For an $x := (x_1, \ldots, x_n) \in \mathbb{C}^n$, $\|x\|_p = 0 \Leftrightarrow x = 0$.
- Observe that, for any $x := (x_1, \ldots, x_n), y := (y_1, \ldots, y_n) \in \mathbb{C}^n$,

$$\rho_p(x, y) = \|x - y\|_p,$$

where

$$x - y := (x_1 - y_1, \ldots, x_n - y_n).$$

Exercise 2.4. Verify.

Theorem 2.2 (Hölder's Inequality for *n*-Tuples). *Let $n \in \mathbb{N}$ and $1 \le p, q \le \infty$ be conjugate indices. Then, for any $x := (x_1, \ldots, x_n), y := (y_1, \ldots, y_n) \in \mathbb{C}^n$,*

$$\sum_{i=1}^n |x_i y_i| \le \|x\|_p \|y\|_q.$$

Proof. The symmetric cases of $p = 1$, $q = \infty$ and $p = \infty$, $q = 1$ are trivial.

Exercise 2.5. Verify.

Suppose that $1 < p, q < \infty$ and let $x := (x_1, \ldots, x_n), y := (y_1, \ldots, y_n) \in \mathbb{C}^n$ be arbitrary.

If

$$\|x\|_p = 0 \quad \text{or} \quad \|y\|_q = 0,$$

which (see Remarks 2.7) is equivalent to $x = 0$, or $y = 0$, respectively, *Hölder's inequality* is, obviously, true.

If

$$\|x\|_p \ne 0 \quad \text{and} \quad \|y\|_q \ne 0,$$

applying *Young's Inequality* (Theorem 2.1) to the nonnegative numbers

$$a_j = \frac{|x_j|}{\|x\|_p} \text{ and } b_j = \frac{|y_j|}{\|y\|_q}$$

for each $j = 1, \ldots, n$, we have the following:

$$\frac{|x_j y_j|}{\|x\|_p \|y\|_q} \le \frac{1}{p} \frac{|x_j|^p}{\sum_{i=1}^n |x_i|^p} + \frac{1}{q} \frac{|y_j|^q}{\sum_{i=1}^n |y_i|^q}, \quad j = 1, \ldots, n.$$

We obtain *Hölder's inequality* by adding the above n inequalities:

$$\frac{\sum_{j=1}^n |x_j y_j|}{\|x\|_p \|y\|_q} \le \frac{1}{p} \frac{\sum_{j=1}^n |x_j|^p}{\sum_{i=1}^n |x_i|^p} + \frac{1}{q} \frac{\sum_{j=1}^n |y_j|^q}{\sum_{i=1}^n |y_i|^q} = \frac{1}{p} + \frac{1}{q} = 1,$$

and multiplying through by $\|x\|_p \|y\|_q$. □

Theorem 2.3 (Minkowski's Inequality for n-Tuples). *Let* $1 \le p \le \infty$. *Then, for any* $(x_1, \ldots, x_n), (y_1, \ldots, y_n) \in \mathbb{C}^n$,

$$\|x + y\|_p \le \|x\|_p + \|y\|_p,$$

where

$$x + y := (x_1 + y_1, \ldots, x_n + y_n).$$

Proof. The cases of $p = 1$ and $p = \infty$ are trivial.

Exercise 2.6. Verify.

For an arbitrary $1 < p < \infty$, we have

$$\sum_{i=1}^n |x_i + y_i|^p = \sum_{i=1}^n |x_i + y_i|^{p-1} |x_i + y_i| \qquad \text{since} \quad |x_i + y_i| \le |x_i| + |y_i|, \ i = 1, \ldots, n;$$

$$\le \sum_{i=1}^n |x_i + y_i|^{p-1} |x_i| + \sum_{i=1}^n |x_i + y_i|^{p-1} |y_i|$$

by *Hölder's Inequality for n-Tuples* (Theorem 2.2);

$$\le \left[\sum_{i=1}^n |x_i + y_i|^{(p-1)q} \right]^{1/q} \left[\sum_{i=1}^n |x_i|^p \right]^{1/p} + \left[\sum_{i=1}^n |x_i + y_i|^{(p-1)q} \right]^{1/q} \left[\sum_{i=1}^n |y_i|^p \right]^{1/p}$$

since $(p-1)q = p$ (see (2.1));

$$= \left[\sum_{i=1}^n |x_i + y_i|^p \right]^{1/q} \left[\sum_{i=1}^n |x_i|^p \right]^{1/p} + \left[\sum_{i=1}^n |x_i + y_i|^p \right]^{1/q} \left[\sum_{i=1}^n |y_i|^p \right]^{1/p}$$

$$= \left[\sum_{i=1}^n |x_i + y_i|^p \right]^{1/q} \left(\left[\sum_{i=1}^n |x_i|^p \right]^{1/p} + \left[\sum_{i=1}^n |y_i|^p \right]^{1/p} \right).$$

Considering that, for $\sum_{i=1}^n |x_i + y_i|^p = 0$, *Minkowski's inequality* trivially holds, suppose that $\sum_{i=1}^n |x_i + y_i|^p > 0$. Then, dividing through by $[\sum_{i=1}^n |x_i + y_i|^p]^{1/q}$, we arrive at

$$\left[\sum_{i=1}^{n} |x_i + y_i|^p\right]^{1-1/q} \le \left[\sum_{i=1}^{n} |x_i|^p\right]^{1/p} + \left[\sum_{i=1}^{n} |y_i|^p\right]^{1/p},$$

which, in view of $1 - \frac{1}{q} = \frac{1}{p}$, implies that, in this case, *Minkowski's inequality* also holds, and completes the proof. □

2.2.4 Sequential Case

Definition 2.4 (*p*-Norm of a Sequence). For $x := (x_k)_{k \in \mathbb{N}} \in l_p$ ($1 \le p \le \infty$), the *norm* of x is the distance of x from the *zero sequence* $0 := (0, 0, 0, \dots)$ in l_p:

$$\|x\|_p := \rho_p(x, 0) = \begin{cases} [\sum_{k=1}^{\infty} |x_k|^p]^{1/p} & \text{if } 1 \le p < \infty, \\ \sup_{k \in \mathbb{N}} |x_k| & \text{if } p = \infty. \end{cases}$$

Remarks 2.8.
- For an $x := (x_k)_{k \in \mathbb{N}} \in l_p$, $\|x\|_p = 0 \Leftrightarrow x = 0$.
- Observe that, for any $x := (x_k)_{k \in \mathbb{N}}, y := (y_k)_{k \in \mathbb{N}} \in l_p$,

$$\rho_p(x, y) = \|x - y\|_p,$$

where

$$x - y := (x_k - y_k)_{k \in \mathbb{N}}.$$

Exercise 2.7. Verify.

Theorem 2.4 (Minkowski's Inequality for Sequences). *Let $1 \le p \le \infty$. Then, for any* $x := (x_k)_{k \in \mathbb{N}}, y := (y_k)_{k \in \mathbb{N}} \in l_p$, $x + y := (x_k + y_k)_{k \in \mathbb{N}} \in l_p$, *and*

$$\|x + y\|_p \le \|x\|_p + \|y\|_p.$$

Proof. The cases of $p = 1$ and $p = \infty$ are trivial.

Exercise 2.8. Verify.

For an arbitrary $1 < p < \infty$ and any $n \in \mathbb{N}$, by *Minkowski's Inequality for n-Tuples* (Theorem 2.3),

$$\left[\sum_{k=1}^{n} |x_k + y_k|^p\right]^{1/p} \le \left[\sum_{k=1}^{n} |x_k|^p\right]^{1/p} + \left[\sum_{k=1}^{n} |y_k|^p\right]^{1/p} \le \left[\sum_{k=1}^{\infty} |x_k|^p\right]^{1/p} + \left[\sum_{k=1}^{\infty} |y_k|^p\right]^{1/p}.$$

Passing to the limit as $n \to \infty$, we infer both the convergence for the series $\sum_{k=1}^{\infty} |x_k + y_k|^p$, i.e., the fact that $x + y \in l_p$, and the desired inequality. □

Exercise 2.9. Applying *Minkowski's inequality* (Theorems 2.3 and 2.4), verify Examples 2.1, 3, and 4, for $1 < p < \infty$.

Theorem 2.5 (Hölder's Inequality for Sequences). *Let* $1 \leq p, q \leq \infty$ *be conjugate indices. Then, for any* $x := (x_k)_{k \in \mathbb{N}} \in l_p$ *and any* $y := (x_k)_{k \in \mathbb{N}} \in l_q$, *the product sequence* $(x_k y_k)_{k \in \mathbb{N}} \in l_1$, *and*

$$\sum_{k=1}^{\infty} |x_k y_k| \leq \|x\|_p \|y\|_q.$$

Exercise 2.10. Prove based on *Hölder's Inequality for n-Tuples* (Theorem 2.2) similarly to proving *Minkowski's Inequality for Sequences* (Theorem 2.4).

Remark 2.9. The important special cases of *Hölder's inequality* with $p = q = 2$:

$$\sum_{k=1}^{n} |x_k y_k| \leq \left[\sum_{k=1}^{n} |x_k|^2 \right]^{1/2} \left[\sum_{k=1}^{n} |y_k|^2 \right]^{1/2} \quad (n \in \mathbb{N}),$$

$$\sum_{k=1}^{\infty} |x_k y_k| \leq \left[\sum_{k=1}^{\infty} |x_k|^2 \right]^{1/2} \left[\sum_{k=1}^{\infty} |y_k|^2 \right]^{1/2} \tag{2.2}$$

are known as the *Cauchy*[4]*–Schwarz*[5] *inequalities* (2.2) for *n*-tuples and sequences, respectively.

2.3 Subspaces of a Metric Space

Definition 2.5 (Subspace of a Metric Space). If (X, ρ) is a *metric space* and $Y \subseteq X$, then the restriction of the metric $\rho(\cdot, \cdot)$ to $Y \times Y$ is a metric on Y, and the metric space (Y, ρ) is called a *subspace* of (X, ρ).

Examples 2.3.
1. Any nonempty subset of $\mathbb{R}^n$ or $\mathbb{C}^n$ ($n \in \mathbb{N}$) is a metric space relative to *p*-metric $(1 \leq p \leq \infty)$.
2. The sets of real- or complex-termed sequences

$$c_{00} := \{ x := (x_k)_{k \in \mathbb{N}} \mid \exists N \in \mathbb{N} \, \forall k \geq N : x_k = 0 \} \quad \text{(eventually zero sequences)},$$

$$c_0 := \left\{ x := (x_k)_{k \in \mathbb{N}} \mid \lim_{k \to \infty} x_k = 0 \right\} \quad \text{(vanishing sequences)},$$

$$c := \left\{ x := (x_k)_{k \in \mathbb{N}} \mid \lim_{k \to \infty} x_k \in \mathbb{R} \text{ (or } \mathbb{C}) \right\} \quad \text{(convergent sequences)}$$

---

4 Augustin-Louis Cauchy (1789–1857).
5 Karl Hermann Amandus Schwarz (1843–1921).

endowed with the *supremum metric*

$$x := (x_k)_{k \in \mathbb{N}}, y := (y_k)_{k \in \mathbb{N}} \mapsto \rho_\infty(x, y) := \sup_{k \in \mathbb{N}} |x_k - y_k|$$

are *subspaces* of the space of *bounded sequences* l_∞ due to the set-theoretic inclusions

$$c_{00} \subset l_p \subset l_q \subset c_0 \subset c \subset l_\infty,$$

where $1 \le p < q < \infty$.

Exercise 2.11. Verify the inclusions and show that they are *proper*.

2.4 Function Spaces

More examples of metric spaces are found in the realm of functions sharing a certain common property, such as boundedness or continuity.

Examples 2.4.

1. Let T be a *nonempty set*. The set $M(T)$ of all real- or complex-valued functions *bounded* on T, i. e., all functions $f : T \to \mathbb{R}$ (or $\mathbb{C}$) such that

$$\sup_{t \in T} |f(t)| < \infty,$$

   is a *metric space* relative to the *supremum metric* (or *uniform metric*)

   $$M(T) \ni f, g \mapsto \rho_\infty(f, g) := \sup_{t \in T} |f(t) - g(t)|.$$

   Remark 2.10. The spaces $l_\infty^{(n)}$ ($n \in \mathbb{N}$) and l_∞ are particular cases of $M(T)$ with $T = \{1, \ldots, n\}$ and $T = \mathbb{N}$, respectively.

2. The set $C[a, b]$ of all real- or complex-valued functions *continuous* on an interval $[a, b]$ ($-\infty < a < b < \infty$) is a *metric space* relative to the *maximum metric* (or *uniform metric*)

   $$C[a, b] \ni f, g \mapsto \rho_\infty(f, g) = \max_{a \le t \le b} |f(t) - g(t)|$$

   as a *subspace* of $M[a, b]$.

   Exercise 2.12. Answer the following questions:
   (a) Why can max be used instead of sup for $C[a, b]$?
   (b) Can sup be replaced with max for the *uniform metric* on $M[a, b]$?

3. For any $-\infty < a < b < \infty$, the set P of all *polynomials* with real/complex coefficients and the set P_n of all *polynomials* of *degree at most n* ($n \in \mathbb{Z}_+$) are metric spaces as *subspaces* of $(C[a,b], \rho_\infty)$ due to the following (proper) inclusions:

$$P_m \subset P_n \subset P \subset C[a,b] \subset M[a,b],$$

where $0 \le m < n$.

4. The set $C[a,b]$ ($-\infty < a < b < \infty$) is a *metric space* relative to the *integral metric*

$$C[a,b] \ni f,g \mapsto \rho_1(f,g) = \int_a^b |f(t) - g(t)| \, dt.$$

However, the extension of the latter to the wider set $R[a,b]$ of real- or complex-valued functions *Riemann*[6] *integrable* on $[a,b]$ is only a *pseudometric*.

Exercise 2.13. Verify both statements.

5. The set $BV[a,b]$ ($-\infty < a < b < \infty$) of real- or complex-valued functions of *bounded variation* on $[a,b]$ ($-\infty < a < b < \infty$), i.e., all functions $f : [a,b] \to \mathbb{R}z$ (or $\mathbb{C}$) such that the *total variation* of f over $[a,b]$:

$$V_a^b(f) := \sup_P \sum_{k=1}^n |f(t_k) - f(t_{k-1})| < \infty,$$

where the *supremum* is taken over all partitions $P := \{a = t_0 < t_1 < \cdots < t_n = b\}$ ($n \in \mathbb{N}$) of $[a,b]$ (see, e. g., [46]), is a *metric space* relative to the *total-variation metric*

$$BV[a,b] \ni f,g \mapsto \rho(f,g) := |f(a) - g(a)| + V_a^b(f - g).$$

Exercise 2.14.

(a) Verify.

(b) For which functions in $BV[a,b]$ is $V_a^b(f) = 0$?

(c) Show that $d(f,g) := V_a^b(f - g)$, $f,g \in BV[a,b]$, is only a *pseudometric* on $BV[a,b]$.

Remark 2.11. When contextually important to distinguish between the real and complex cases, we can use the notations

$$M(T, \mathbb{R}), \; C([a,b], \mathbb{R}), \; BV([a,b], \mathbb{R}) \text{ and } M(T, \mathbb{C}), \; C([a,b], \mathbb{C}), \; BV([a,b], \mathbb{C}),$$

and refer to the corresponding spaces as *real* and *complex*, respectively.

---

6 Bernhard Riemann (1826–1866).

2.5 Further Properties of Metric

Theorem 2.6 (Generalized Triangle Inequality). *In a metric space* (X, ρ), *for any finite collection of points* $\{x_1, \ldots, x_n\} \subseteq X$ ($n \in \mathbb{N}$, $n \geq 3$),

$$\rho(x_1, x_n) \leq \rho(x_1, x_2) + \rho(x_2, x_3) + \cdots + \rho(x_{n-1}, x_n).$$

Exercise 2.15. Prove by induction.

Theorem 2.7 (Inverse Triangle Inequality). *In a metric space* (X, ρ), *for arbitrary* $x, y, z \in X$,

$$\left| \rho(x, y) - \rho(y, z) \right| \leq \rho(x, z).$$

Proof. Let $x, y, z \in X$ be arbitrary.

On one hand, by the *triangle inequality* and *symmetry*,

$$\rho(x, y) \leq \rho(x, z) + \rho(z, y) = \rho(x, z) + \rho(y, z),$$

which implies

$$\rho(x, y) - \rho(y, z) \leq \rho(x, z). \tag{2.3}$$

On the other hand, in the same manner,

$$\rho(y, z) \leq \rho(y, x) + \rho(x, z) = \rho(x, y) + \rho(x, z),$$

which implies

$$\rho(y, z) - \rho(x, y) \leq \rho(x, z). \tag{2.4}$$

Jointly, inequalities (2.3) and (2.4) are equivalent to the desired one. □

Theorem 2.8 (Quadrilateral Inequality). *In a metric space* (X, ρ), *for arbitrary* $x, y, u, v \in X$,

$$\left| \rho(x, y) - \rho(u, v) \right| \leq \rho(x, u) + \rho(y, v).$$

Proof. For any $x, y, u, v \in X$,

$$\left| \rho(x, y) - \rho(u, v) \right| = \left| \rho(x, y) - \rho(y, u) + \rho(y, u) - \rho(u, v) \right|$$
$$\leq \left| \rho(x, y) - \rho(y, u) \right| + \left| \rho(y, u) - \rho(u, v) \right|$$

by the *Inverse Triangle Inequality* (Theorem 2.7);

$$\leq \rho(x, u) + \rho(y, v). \qquad \square$$

Remark 2.12. With only the *symmetry* axiom and the *triangle inequality* used in the proofs of this section's statements, the latter, obviously, hold true for a *pseudometric*.

2.6 Convergence and Continuity

The notion of metric brings to life the important concepts of *limit* and *continuity*.

2.6.1 Convergence

Definition 2.6 (Limit and Convergence of a Sequence). A sequence of points $(x_n)_{n \in \mathbb{N}}$ (another notation is $\{x_n\}_{n=1}^{\infty}$) in a *metric space* (X, ρ) is said to *converge* (to be *convergent*) to a point $x \in X$ if

$$\forall \varepsilon > 0 \; \exists N \in \mathbb{N} \; \forall n \geq N : \rho(x_n, x) < \varepsilon,$$

i. e.,

$$\lim_{n \to \infty} \rho(x_n, x) = 0 \; (\rho(x_n, x) \to 0, \; n \to \infty).$$

We write in this case

$$\lim_{n \to \infty} x_n = x \quad \text{or} \quad x_n \to x, \; n \to \infty,$$

and say that x is the *limit* of $(x_n)_{n \in \mathbb{N}}$.

A sequence $(x_n)_{n \in \mathbb{N}}$ in a *metric space* (X, ρ) is called *convergent* if it converges to some $x \in X$, and *divergent* otherwise.

Examples 2.5.
1. A sequence $(x_n)_{n \in \mathbb{N}}$ is convergent in a discrete space (X, ρ_d) *iff* it is *eventually constant*, i. e.,

$$\exists N \in \mathbb{N} \; \forall n \geq N : x_n = x_N.$$

2. For $\mathbb{R}$ or $\mathbb{C}$ with the regular metric, the above definitions coincide with the familiar ones from classical analysis.
3. Convergence of a sequence in the space $l_p^{(n)}$ ($n \in \mathbb{N}$, $n \geq 2$ and $1 \leq p \leq \infty$) is equivalent to *componentwise convergence*, i. e.,

$$(x_1^{(k)}, \ldots, x_n^{(k)}) =: x^{(k)} \to x := (x_1, \ldots, x_n), \; k \to \infty,$$

iff

$$\forall i = 1, \ldots, n : x_i^{(k)} \to x_i, \; k \to \infty.$$

For instance, in $l_p^{(2)}$ ($1 \leq p \leq \infty$), the sequence $((1/n, 1-1/n))_{n \in \mathbb{N}}$ converges to $(0, 1)$, and the sequence $((1/n, (-1)^n))_{n \in \mathbb{N}}$ *diverges*.

4. Convergence of a sequence in the space l_p $(1 \leq p \leq \infty)$:

$$\left(x_n^{(k)}\right)_{n \in \mathbb{N}} =: x^{(k)} \to x := (x_n)_{n \in \mathbb{N}}, \; k \to \infty,$$

implies *termwise convergence*, i. e.,

$$\forall n \in \mathbb{N} : x_n^{(k)} \to x_n, \; k \to \infty.$$

The converse is *not* true. Indeed, in the space l_p $(1 \leq p \leq \infty)$, the sequence $(e_k :=$ $(\delta_{kn})_{n \in \mathbb{N}})_{k \in \mathbb{N}}$, where δ_{nk} is the *Kronecker*[7] *delta*, converges to the zero sequence $0 := (0, 0, 0, \dots)$ *termwise* but *diverges*.

Remark 2.13. The same example works in the spaces (c_{00}, ρ_∞) and (c_0, ρ_∞) (cf. Section 2.18, Problem 9 and 10).

5. Convergence in $(M(T), \rho_\infty)$ (see Examples 2.4) is the *uniform convergence* on T, i. e.,

$$f_n \to f, \; n \to \infty, \text{ in } (M(T), \rho_\infty)$$

iff

$$\forall \varepsilon > 0 \; \exists N = N(\varepsilon) \in \mathbb{N} \; \forall n \geq N \; \forall t \in T : \left|f_n(t) - f(t)\right| < \varepsilon.$$

Uniform convergence on T implies *pointwise convergence* on T:

$$\forall \varepsilon > 0 \; \forall t \in T \; \exists N = N(\varepsilon, t) \in \mathbb{N} \; \forall n \geq N : \left|f_n(t) - f(t)\right| < \varepsilon,$$

i. e.,

$$\forall t \in T : f_n(t) \to f(t), \; n \to \infty.$$

The same is true for l_∞ as a particular case of $M(T)$ with $T = \mathbb{N}$, in which case *"pointwise"* is *"termwise"*, and for $(C[a, b], \rho_\infty)$ $(-\infty < a < b < \infty)$ as a subspace of $(M[a, b], \rho_\infty)$.

Exercise 2.16.
(a) Verify each statement and give corresponding examples.
(b) Give an example showing that a function sequence $(f_n)_{n \in \mathbb{N}}$ in $C[a, b]$ pointwise convergent on $[a, b]$ need not converge on $[a, b]$ uniformly.

Remark 2.14. The limit of a sequence in a metric space need not exist. However, as the following statement asserts, whenever existent, the limit is *unique*.

---

7 Leopold Kronecker (1823–1891).

Theorem 2.9 (Uniqueness of Limit). *The limit of a convergent sequence $(x_n)_{n\in\mathbb{N}}$ in a metric space (X,ρ) is unique.*

Exercise 2.17. Prove.

Hint. Use the *triangle inequality*.

The following is a useful characterization of convergence.

Theorem 2.10 (Characterization of Convergence). *A sequence $(x_n)_{n\in\mathbb{N}}$ in a metric space (X,ρ) converges to a point $x \in X$ iff every subsequence $(x_{n(k)})_{k\in\mathbb{N}}$ of $(x_n)_{n\in\mathbb{N}}$ contains a subsequence $(x_{n(k(j))})_{j\in\mathbb{N}}$ such that*

$$x_{n(k(j))} \to x, j \to \infty.$$

Exercise 2.18. Prove.

Hint. Prove the *"if"* part *by contrapositive*.

2.6.2 Continuity, Uniform Continuity, and Lipschitz Continuity

Here, we introduce three notions of *continuity* in the order of increasing strength: regular, *uniform*, and in the *Lipschitz*[8] sense.

Definition 2.7 (Continuity of a Function). Let (X,ρ) and (Y,σ) be metric spaces. A function $f : X \to Y$ is called *continuous* at a point $x_0 \in X$ if

$$\forall \varepsilon > 0 \, \exists \delta > 0 \, \forall x \in X \text{ with } \rho(x,x_0) < \delta : \sigma(f(x),f(x_0)) < \varepsilon.$$

A function $f : X \to Y$ is called *continuous* on X if it is continuous at every point of X. The set of all such functions is designated as $C(X,Y)$, and we write $f \in C(X,Y)$.

Remarks 2.15.
- When X and Y are subsets of $\mathbb{R}$ or $\mathbb{C}$ with the regular distance, we obtain the familiar (ε, δ)-definitions from classical analysis.
- When $Y = \mathbb{R}$ or $Y = \mathbb{C}$, the shorter notation $C(X)$ is often used.

It is convenient to characterize continuity in terms of sequences.

Theorem 2.11 (Sequential Characterization of Local Continuity). *Let (X,ρ) and (Y,σ) be metric spaces. A function $f : X \to Y$ is continuous at a point $x_0 \in X$ iff, for each sequence $(x_n)_{n\in\mathbb{N}}$ in X such that*

$$\lim_{n\to\infty} x_n = x_0 \text{ in } (X,\rho),$$

---

[8] Sigismund Lipschitz (1832–1903).

we have

$$\lim_{n\to\infty} f(x_n) = f(x_0) \text{ in } (Y, \sigma).$$

Exercise 2.19. Prove.

Hint. Prove the *"if"* part *by contrapositive.*

Using the *Sequential Characterization of Local Continuity* (Theorem 2.11), one can easily prove the following two theorems:

Theorem 2.12 (Properties of Numeric Continuous Functions). *Let (X, ρ) be a metric space and let $Y = \mathbb{R}$ or $Y = \mathbb{C}$ with the regular distance.*
If f and g are continuous at a point $x_0 \in X$, then
(1) *for any $c \in \mathbb{R}$ (or $c \in \mathbb{C}$), cf is continuous at x_0,*
(2) *$f + g$ is continuous at x_0,*
(3) *$f \cdot g$ is continuous at x_0,*
(4) *provided $g(x_0) \neq 0$, $\frac{f}{g}$ is continuous at x_0.*

Theorem 2.13 (Continuity of Composition). *Let (X, ρ), (Y, σ), and (Z, τ), $f : X \to Y$ and $g : Y \to Z$.*
If, for some $x_0 \in X$, f is continuous at x_0 and g is continuous at $y_0 = f(x_0)$, then the composition $(g \circ f)(x) := g(f(x))$ is continuous at x_0.

Exercise 2.20. Prove Theorems 2.12 and 2.13 using the *sequential approach.*

Remark 2.16. The statements of Theorems 2.12 and 2.13 are naturally carried over to functions continuous on the whole space (X, ρ).

Definition 2.8 (Uniform Continuity). Let (X, ρ) and (Y, σ) be metric spaces. A function $f : X \to Y$ is said to be *uniformly continuous* on X if

$$\forall \varepsilon > 0 \; \exists \delta > 0 \; \forall x', x'' \in X \text{ with } \rho(x', x'') < \delta : \; \sigma(f(x'), f(x'')) < \varepsilon.$$

Remark 2.17. Each uniformly continuous function is continuous.

Exercise 2.21. Explain.

However, as the following example shows, a continuous function need not be uniformly continuous.

Example 2.6. For $X = Y := (0, \infty)$, $f(x) := x$ is uniformly continuous and $g(x) := \frac{1}{x}$ is continuous but not uniformly.

Exercise 2.22. Verify.

Definition 2.9 (Lipschitz Continuity). Let (X, ρ) and (Y, σ) be metric spaces. A function $f : X \to Y$ is said to be *Lipschitz continuous* (or *Lipschitzian*) on X with *Lipschitz con-*

stant L if

$$\exists L \geq 0 \, \forall x', x'' \in X : \; \sigma(f(x'), f(x'')) \leq L\rho(x', x'').$$

Example 2.7. On $\mathbb{R}$, the function $f(x) := mx$, $x \in X$, with an arbitrary $m \in \mathbb{R}$, is *Lipschitzian* since

$$|f(x) - f(y)| = |mx - my| = |m||x - y|, \; x, y \in X.$$

Remarks 2.18.

-   The smallest Lipschitz constant is called the *best Lipschitz constant*.

    Exercise 2.23. Explain why the *best Lipschitz constant* is well defined.

    Thus, in the prior example, m is the *best Lipschitz constant*.
-   A constant function is Lipschitz continuous with the best Lipschitz constant $L = 0$.
-   By the *Mean Value Theorem*, a real- or complex-valued differentiable function f on an interval $I \subseteq \mathbb{R}$ is Lipschitz continuous on I iff its derivative f' is bounded on I.

    Exercise 2.24. Explain.

    In particular, all functions in $C^1[a, b]$ ($-\infty < a < b < \infty$), i. e., *continuously differentiable* on $[a, b]$, are Lipschitz continuous on $[a, b]$.
-   Each Lipschitz continuous function is uniformly continuous.

    Exercise 2.25. Explain.

    However, as the following example shows, a uniformly continuous function need not be Lipschitz continuous.

Example 2.8. For $X = Y := [0, 1]$ with the regular distance, the function $f(x) := \sqrt{x}$ is uniformly continuous on $[0, 1]$, as follows from the *Heine–Cantor Uniform Continuity Theorem* (Theorem 2.56), but is *not* Lipschitz continuous on $[0, 1]$.

Exercise 2.26. Verify.

2.7 Balls, Separation, and Boundedness

The geometric concepts of *balls* and *spheres*, generalizing their familiar counterparts, are rather handy as is a generalized notion of *boundedness*.

Definition 2.10 (Balls and Spheres). Let (X, ρ) be a *metric space* and $r \geq 0$.

-   The *open ball* of radius r centered at a point $x_0 \in X$ is the set

    $$B(x_0, r) := \{x \in X \mid \rho(x, x_0) < r\}.$$

-   The *closed ball* of radius r centered at a point $x_0 \in X$ is the set

    $$\overline{B}(x_0, r) := \{x \in X \mid \rho(x, x_0) \leq r\}.$$

- The *sphere* of radius r centered at a point $x_0 \in X$ is the set

$$S(x_0, r) := \{x \in X \mid \rho(x, x_0) = r\} = \overline{B}(x_0, r) \setminus B(x_0, r).$$

Remarks 2.19.
- When contextually important to indicate which space the balls/spheres are considered in, the letter designating the space in question is added as a subscript. E. g., for (X, ρ), we use the notations

$$B_X(x_0, r), \ \overline{B}_X(x_0, r), \text{ and } S_X(x_0, r), \ x_0 \in X, r \geq 0.$$

- As is easily seen, for an arbitrary $x_0 \in X$,

$$B(x_0, 0) = \emptyset \text{ and } \overline{B}(x_0, 0) = S(x_0, 0) = \{x_0\} \quad (\textit{trivial cases}).$$

Exercise 2.27.
(a) Explain the latter.
(b) Describe balls and spheres in $\mathbb{R}$ and $\mathbb{C}$ with the regular distance and give some examples.
(c) Sketch the *unit sphere* $S(0, 1)$ in $(\mathbb{R}^2, \rho_1)$, $(\mathbb{R}^2, \rho_2)$, and $(\mathbb{R}^2, \rho_\infty)$.
(d) Describe balls and spheres in $(C[a, b], \rho_\infty)$.
(e) Let (X, ρ_d) be a *discrete* metric space and $x_0 \in X$ be arbitrary. Describe $B(x_0, r)$, $\overline{B}(x_0, r)$, and $S(x_0, r)$ for different values of $r \geq 0$.

Proposition 2.1 (Separation Property). *Let (X, ρ) be metric space. Then*

$$\forall x, y \in X, \ x \neq y \ \exists r > 0 : \ B(x, r) \cap B(y, r) = \emptyset,$$

i. e., distinct points in a metric space can be separated by disjoint balls.

Exercise 2.28. Prove.

The definitions of convergence and continuity can be naturally reformulated in terms of balls.

Definition 2.11 (Equivalent Definitions of Convergence and Continuity). A sequence of points $(x_n)_{n \in \mathbb{N}}$ in a *metric space* (X, ρ) is said to *converge* (to be *convergent*) to a point $x \in X$ if

$$\forall \varepsilon > 0 \ \exists N \in \mathbb{N} \ \forall n \geq N : \ x_n \in B(x_0, \varepsilon),$$

in which case we say that the sequence $(x_n)_{n \in \mathbb{N}}$ is *eventually* in the ε-ball $B(x, \varepsilon)$.
Let (X, ρ) and (Y, σ) be metric spaces. A function $f : X \to Y$ is called *continuous* at a point $x_0 \in X$ if

$$\forall \varepsilon > 0 \ \exists \delta > 0 : \ f(B_X(x_0, \delta)) \subseteq B_Y(f(x_0), \varepsilon).$$

Definition 2.12 (Bounded Set). Let (X, ρ) be a *metric space*. A nonempty set $A \subseteq X$ is called *bounded* if

$$\operatorname{diam}(A) := \sup_{x,y \in A} \rho(x,y) < \infty.$$

The number $\operatorname{diam}(A)$ is called the *diameter* of A.

Remark 2.20. The *empty set* $\emptyset$ is regarded as bounded with $\operatorname{diam}(\emptyset) := 0$.

Examples 2.9.
1. In a metric space (X, ρ), an open/closed ball of radius $r > 0$ is a bounded set of diameter *at most 2r*.
2. In $(\mathbb{R}, \rho)$, the sets $(0,1]$, $\{1/n\}_{n \in \mathbb{N}}$ are bounded, and the sets $(-\infty, 1)$, $\{n^2\}_{n \in \mathbb{N}}$ are not.
3. In l_∞, the set $\{(x_n)_{n \in \mathbb{N}} \mid |x_n| \le 1,\ n \in \mathbb{N}\}$ is bounded, and it is not in l_p $(1 \le p < \infty)$.
4. In $(C[0,1], \rho_\infty)$, the set $\{t^n\}_{n \in \mathbb{Z}_+}$ is bounded, and it is not in $(C[0,2], \rho_\infty)$.

Exercise 2.29.
(a) Verify.
(b) Show that a set A is bounded *iff* it is contained in some (closed) ball, i. e.,

$$\exists x \in X \ \exists r \ge 0 : \ A \subseteq \bar{B}(x, r).$$

(c) Describe all bounded sets in a *discrete* metric space (X, ρ_d).
(d) Give an example of a metric space (X, ρ), in which, for a ball $\bar{B}(x, r)$ with some $x \in X$ and $r > 0$,

$$\operatorname{diam}(\bar{B}(x,r)) < 2r.$$

Theorem 2.14 (Properties of Bounded Sets). *The bounded sets in a metric space (X, ρ) have the following properties:*
(1) *a subset of a bounded set is bounded;*
(2) *an arbitrary intersection of bounded sets is bounded;*
(3) *a finite union of bounded sets is bounded.*

Exercise 2.30.
(a) Prove.
(b) Give an example showing that an *infinite union* of bounded sets need not be bounded.

Definition 2.13 (Bounded Function). Let T be a nonempty set and (X, ρ) be a *metric space*. A function $f : T \to X$ is called *bounded* if the *set of its values* $f(T)$ is bounded in (X, ρ).

Remark 2.21. For $T = \mathbb{N}$, as a particular case, we obtain the definition of a *bounded sequence*.

2.8 Interior Points, Interiors, Open Sets

Now, we are ready to define *openness* and *closedness* for sets in metric spaces.

Definition 2.14 (Interior Point). Let (X, ρ) be a metric space. A point $x \in X$ is called an *interior point* of a nonempty set $A \subseteq X$ if A contains a nontrivial open ball centered at x, i. e.,

$$\exists r > 0 : B(x, r) \subseteq A.$$

Examples 2.10.
1. In an arbitrary metric space (X, ρ), any point $x \in X$ is, obviously, an interior point of the ball $B(x, r)$ or $\overline{B}(x, r)$ for each $r > 0$.
2. For the set $[0, 1)$ in $\mathbb{R}$ with the regular distance, the points $0 < x < 1$ are interior, and the point $x = 0$ is not.
3. A singleton $\{x\}$ in $\mathbb{R}$ with the regular distance has no interior points.

Exercise 2.31. Verify.

Definition 2.15 (Interior of a Set). The *interior* of a nonempty set A in a metric space (X, ρ) is the set of all interior points of A.

Notation. $\text{int}(A)$.

Examples 2.11.
1. In $\mathbb{R}$ with the regular distance,
    - $\text{int}((0, 1)) = \text{int}([0, 1)) = \text{int}((0, 1]) = \text{int}([0, 1]) = \text{int}([0, 1] \cup \{2\}) = (0, 1)$,
    - $\text{int}(\mathbb{R}) = \mathbb{R}$,
    - $\text{int}(\mathbb{N}) = \text{int}(\mathbb{Z}) = \text{int}(\mathbb{Q}) = \text{int}(\mathbb{Q}^c) = \emptyset$.
2. In the *discrete* metric space (X, ρ_d), for an arbitrary $x \in X$,

$$\text{int}(\{x\}) = \{x\}$$

since, for any $0 < r \le 1$, $B(x, r) = \{x\}$.

Remark 2.22. Thus, we have always the inclusion

$$\text{int}(A) \subseteq A,$$

the prior examples showing that the inclusion can be *proper* and that $\text{int}(A)$ can be *empty*.

Exercise 2.32. Give some more examples.

Definition 2.16 (Open Set). A *nonempty set A* in a metric space (X, ρ) is called *open* if each point of A is its *interior point*, i. e., $A = \text{int}(A)$.

Remark 2.23. The *empty set* $\emptyset$ is regarded as open, and the whole space (X, ρ) is trivially open in itself, i. e., X is an open set in (X, ρ).

Exercise 2.33.
(a) Verify that, in $\mathbb{R}$ with the regular distance, the intervals of the form (a, ∞), $(-\infty, b)$, and (a, b) $(-\infty < a < b < \infty)$ are open sets.
(b) Prove that, in an arbitrary metric space (X, ρ), any *open ball* $B(x_0, r)$ $(x_0 \in X, r \geq 0)$ is an *open set*.
(c) Describe all open sets in a *discrete* metric space (X, ρ_d).

The concept of openness in a metric space can be characterized sequentially.

Theorem 2.15 (Sequential Characterizations of Open Sets). *Let (X, ρ) be a metric space. A set $A \subseteq X$ is open in (X, ρ) iff, for any sequence $(x_n)_{n \in \mathbb{N}}$ convergent to a point $x \in A$ in (X, ρ), $(x_n)_{n \in \mathbb{N}}$ is eventually in A, i. e.,*

$$\exists N \in \mathbb{N} \, \forall n \geq N : x_n \in A.$$

Exercise 2.34. Prove.

Hint. Prove the *"if"* part *by contrapositive*.

Theorem 2.16 (Properties of Open Sets). *The open sets in a metric space (X, ρ) have the following properties:*
(1) *$\emptyset$ and X are open sets;*
(2) *an arbitrary union of open sets is open;*
(3) *a finite intersection of open sets is open.*

Exercise 2.35.
(a) Prove.
(b) Give an example showing that an *infinite intersection* of open sets need not be open.

Definition 2.17 (Metric Topology). The collection $\mathscr{G}$ of all open sets in a metric space (X, ρ) is called the *metric topology* generated by the metric ρ.

2.9 Limit and Isolated Points, Closures, Closed Sets

Definition 2.18 (Limit Point/Derived Set). Let (X, ρ) be a metric space. A point $x \in X$ is called a *limit point* (also an *accumulation point* or a *cluster point*) of a set A in X if any open ball centered at x contains a point of A *distinct* from x, i. e.,

$$\forall r > 0 : B(x,r) \cap (A \setminus \{x\}) \neq \emptyset.$$

The set A' of all limit points of A is called the *derived set* of A.

Example 2.12. In $\mathbb{R}$ with the regular distance,
- $\mathbb{N}' = \mathbb{Z}' = \emptyset$,
- $\{1/n\}'_{n \in \mathbb{N}} = \{0\}$,
- $(0,1)' = [0,1)' = (0,1]' = [0,1]' = ([0,1] \cup \{2\})' = [0,1]$,
- $\mathbb{Q}' = (\mathbb{Q}^c)' = \mathbb{R}$.

Exercise 2.36. Verify.

Remarks 2.24.
- As the prior example shows, a *limit point* x of a set A need not belong to A. It may even happen that none of them does, i. e.,

$$A' \subseteq A^c.$$

- Each open ball centered at a limit point x of a set A in a metric space (X, ρ) contains *infinitely many* points of A distinct from x.
  Thus, to have a *limit point*, a set A in a metric space (X, ρ) must necessarily be *infinite*. However, as the prior example shows, an infinite set need not have limit points.

Exercise 2.37.
(a) Verify.
(b) Describe the situation in a *discrete* metric space (X, ρ_d).
(c) Give examples showing that an *interior point* of a set need not be its *limit point*, and vice versa.

Limit points can be characterized sequentially as follows:

Theorem 2.17 (Sequential Characterization of Limit Points). *Let* (X, ρ) *be a metric space. A point* $x \in X$ *is a limit point of a set* $A \subseteq X$ *iff* A *contains a sequence of points* $(x_n)_{n \in \mathbb{N}}$ *distinct from* x *convergent to* x, i. e.,

$$x \in A' \iff \exists (x_n)_{n \in \mathbb{N}} \subseteq A \setminus \{x\} : x_n \to x, \, n \to \infty.$$

Exercise 2.38. Prove.

Definition 2.19 (Isolated Point). Let (X, ρ) be a *metric space*. A point x of a set A in X, which is not its limit point, is called an *isolated point* of A, i. e., there is an open ball centered at x containing no points of A other than x.

Remark 2.25. Thus, the set of all isolated points of A is $A \setminus A'$.

Example 2.13. In $\mathbb{R}$ with the regular distance,
-   each point of the sets $\mathbb{N}$, $\mathbb{Z}$, and $\{1/n\}_{n\in\mathbb{N}}$ is isolated,
-   no point of the sets $(0,1)$, $[0,1)$, $(0,1]$, $[0,1]$, $\mathbb{Q}$, and $\mathbb{Q}^c$ is isolated,
-   only the point 2 of the set $[0,1] \cup \{2\}$ is isolated.

Exercise 2.39. Show that a function f from a metric space (X,ρ) to a metric space (Y,σ) is *continuous* at all isolated points of X, if any.

Definition 2.20 (Closure of a Set). The *closure $\overline{A}$* of a set A in a metric space (X,ρ) is the set consisting of all points, which are either the points of A, or the limit points of A, i. e.,

$$\overline{A} := A \cup A'.$$

Example 2.14. In $\mathbb{R}$ with the regular distance,
-   $\overline{\mathbb{N}} = \mathbb{N}$ and $\overline{\mathbb{Z}} = \mathbb{Z}$,
-   $\overline{\{1/n\}_{n\in\mathbb{N}}} = \{1/n\}_{n\in\mathbb{N}} \cup \{0\}$,
-   $\overline{(0,1)} = \overline{[0,1)} = \overline{(0,1]} = \overline{[0,1]} = [0,1]$,
-   $\overline{\mathbb{Q}} = \overline{\mathbb{Q}^c} = \mathbb{R}$.

Exercise 2.40. Verify (see Example 2.12).

Remarks 2.26.
-   Obviously, $\emptyset' = \emptyset$, and hence, $\overline{\emptyset} = \emptyset$.
-   We always have the inclusion

$$A \subseteq \overline{A},$$

which, as the prior example shows, can be *proper*.
-   A point $x \in \overline{A}$ *iff* every nontrivial open ball centered at x contains a point of A (not necessarily distinct from x).

Exercise 2.41. Verify.

By the definition and the *Sequential Characterization of Limit Points* (Theorem 2.17), we obtain the following:

Theorem 2.18 (Sequential Characterization of Closures). *Let (X,ρ) be a metric space. For a set $A \subseteq X$,*

$$x \in \overline{A} \iff \exists (x_n)_{n\in\mathbb{N}} \subseteq A : x_n \to x, \, n \to \infty.$$

Exercise 2.42.
(a) Prove.

(b) Is it true that, in an arbitrary metric space (X, ρ), for each $x \in X$ and any $r > 0$,

$$\overline{B(x, r)} = \overline{B}(x, r)?$$

(c) Let A be a set in a metric space (X, ρ). Prove that, for each *open set* O in (X, ρ),

$$O \cap \overline{A} \neq \emptyset \iff O \cap A \neq \emptyset.$$

Definition 2.21 (Closed Set). Let (X, ρ) be a metric space. A set A in X is called *closed* if it contains all its limit points, i. e., $A' \subseteq A$, and hence, $A = \overline{A}$.

Remarks 2.27.
– Any metric space (X, ρ) is trivially closed in itself, i. e., X is a closed set in (X, ρ).
– Also closed are the sets with no limit points, in particular, *finite sets*, including the empty set $\emptyset$.
– A set in a metric space (X, ρ), which is simultaneously closed and open is called *clopen*. There are always at least two (*trivial*) clopen sets $\emptyset$ and X, and there can also exist nontrivial ones.

Exercise 2.43.
(a) Verify that in $\mathbb{R}$ with the regular distance, the intervals of the form $[a, \infty)$, $(-\infty, b]$, and $[a, b]$ $(-\infty < a < b < \infty)$ are closed sets.
(b) Verify that the sets $(0, 1)$ and $\{2\}$ are clopen in the metric space $(0, 1) \cup \{2\}$ with the regular distance.
(c) Describe all closed sets in a *discrete* metric space (X, ρ_d).

Theorem 2.19 (Characterizations of Closed Sets). *For an arbitrary nonempty set $A \subseteq X$ in a metric space (X, ρ), the following statements are equivalent.*
1. *Set A is closed in (X, ρ).*
2. *The complement A^c of set A is an open set in (X, ρ).*
3. *(Sequential Characterization) For any sequence $(x_n)_{n \in \mathbb{N}}$ in A convergent in (X, ρ), $\lim_{n \to \infty} x_n \in A$, i. e., the set A contains the limits of all its convergent sequences.*

Exercise 2.44.
(a) Prove.
(b) Show in *two different ways* that, in an arbitrary *metric space* (X, ρ), any *closed ball* $\overline{B}(x, r)$ $(x \in X, r \geq 0)$ is a *closed set*.

The properties of the closed sets follow from the properties of the open sets via *de Morgan's laws* (see Section 1.1.1), considering the fact that the closed and open sets are complementary.

Theorem 2.20 (Properties of Closed Sets). *The closed sets in a metric space (X, ρ) have the following properties:*

(1) ∅ *and X are closed sets;*
(2) *an arbitrary intersection of closed sets is closed;*
(3) *a finite union of closed sets is closed.*

Exercise 2.45.
(a) Prove.
(b) Give an example showing that an *infinite union* of closed sets need not be closed.

2.10 Exterior and Boundary

Relative to a set A in a metric space (X, ρ), each point of the space belongs to one of the three pairwise disjoint sets: *interior*, *exterior*, or *boundary* of A. The interior points having been defined above (see Definition 2.14), it remains pending to define the exterior and boundary ones.

Definition 2.22 (Exterior and Boundary Points). Let A be a set in a metric space (X, ρ).
– We say that $x \in X$ is an *exterior point* of A if it is an interior point of the complement A^c, i. e., there is an open ball centered at x and contained in A^c.
  All exterior points of a set A form its *exterior* $\text{ext}(A) := \text{int}(A^c)$.
– We say that $x \in X$ is a *boundary point* of A if it is neither interior nor exterior point of A, i. e., every open ball centered at x contains both a point of A and a point of A^c.
  All boundary points of a set A form its *boundary* ∂A.

Remarks 2.28.
– Thus, $x \in X$ is a *boundary point* of A *iff* any open ball centered at x contains both a point of A, and a point of A^c.
– By definition, for each set A in a metric space (X, ρ), $\text{int}(A)$, $\text{ext}(A)$, and ∂A form a *partition* of X, i. e., are *pairwise disjoint*, and

$$\text{int}(A) \cup \text{ext}(A) \cup \partial A = X.$$

– Since $\text{int}(A^c) = \text{ext}(A)$, and $\text{ext}(A^c) = \text{int}((A^c)^c) = \text{int}(A)$, we conclude that $\partial A^c = \partial A$.

Example 2.15. In $\mathbb{R}$ with the regular distance,
– $\text{int}(\mathbb{Z}) = \emptyset$, $\text{ext}(\mathbb{Z}) = \mathbb{Z}^c$, $\partial \mathbb{Z} = \mathbb{Z}$,
– $\text{int}([0, 1]) = (0, 1)$, $\text{ext}([0, 1]) = [0, 1]^c$, $\partial[0, 1] = \{0, 1\}$,
– $\text{int}(\mathbb{Q}) = \text{ext}(\mathbb{Q}) = \emptyset$, $\partial \mathbb{Q} = \mathbb{R}$,
– $\text{int}(\mathbb{R}) = \mathbb{R}$, $\text{ext}(\mathbb{R}) = \partial \mathbb{R} = \emptyset$.

Exercise 2.46.
(a) In $\mathbb{R}$ with the regular distance, determine $\text{int}(\mathbb{Q}^c)$, $\text{ext}(\mathbb{Q}^c)$, and $\partial \mathbb{Q}^c$.

(b) In an arbitrary metric space (X, ρ), determine $\mathrm{int}(\emptyset)$, $\mathrm{ext}(\emptyset)$, $\partial \emptyset$ and $\mathrm{int}(X)$, $\mathrm{ext}(X)$, ∂X.

(c) Determine $\mathrm{int}(A)$, $\mathrm{ext}(A)$, and ∂A of a nonempty proper subset A of a *discrete* metric space (X, ρ_d).

2.11 Dense Sets and Separability

Here, we consider the notions of the *denseness* of a set in a metric space, i. e., of a set's points being found arbitrarily close to all points of the space, and of the *separability* of a metric space, i. e., of a space containing a *countable* such a set (see Section 1.1.2).

Definition 2.23 (Dense Set). A set A in a metric space (X, ρ) is called *dense* if $\overline{A} = X$.

Remark 2.29. Thus, a set A is *dense* in a metric space (X, ρ) *iff* an arbitrary nontrivial open ball contains a point of A (see Remarks 2.26).

Example 2.16. The sets $\mathbb{Q}$ and $\mathbb{Q}^c$ are dense in $\mathbb{R}$.

Exercise 2.47. Verify.

From the *Sequential Characterization of Closure* (Theorem 2.18), follows

Theorem 2.21 (Sequential Characterization of Dense Sets). *A set A is dense in a metric space (X, ρ) iff*

$$\forall x \in X \; \exists \, (x_n)_{n \in \mathbb{N}} \subseteq A : \; x_n \to x, \; n \to \infty.$$

Definition 2.24 (Separable Metric Space). A metric space (X, ρ) containing a *countable dense subset* is a called *separable*.

Remark 2.30. Any countable metric space is, obviously, separable. However, as the following examples show, a metric space need not be countable to be separable.

Examples 2.17.
1. The spaces $l_p^{(n)}$ are *separable* for $(n \in \mathbb{N}, 1 \le p \le \infty)$, which includes the cases of $\mathbb{R}$ and $\mathbb{C}$ with the regular distances.
   Indeed, in this setting, one can consider as a countable dense set that of all ordered n-tuples with (real/complex) rational components.
2. The spaces l_p are *separable* for $1 \le p < \infty$.
   Indeed, as a countable dense set here, one can consider that of all eventually zero sequences with (real/complex) rational terms.
3. The space $(C[a, b], \rho_\infty)$ is separable, which follows from the classical *Weierstrass Approximation Theorem* (see, e. g., [45]) when we consider as a countable dense set that of all polynomials with rational coefficients.

4. More examples of separable metric spaces can be obtained as subspaces, Cartesian products, and continuous images of such spaces (see Propositions 2.19, 2.20, and 2.21, Section 2.18, Problems 19, 20, and 21, respectively).
5. The space l_∞ is *not separable*.

Exercise 2.48.
(a) Verify 1–3 using the *Properties of Countable Sets* (Theorem 1.3).
(b) Prove 5.

> **Hint.** In l_∞, consider the *uncountable* set B of all *binary sequences*, i. e., the sequences, whose only entries are 0 or 1 (see the *Uncountable Sets Proposition* (Proposition 1.1)). The balls of the *uncountable* collection
>
> $$\{B(x, 1/2) \mid x \in B\}$$
>
> are *pairwise disjoint*. Explain the latter and, assuming that there exists a countable dense subset in l_∞, arrive at a *contradiction*.

(c) Prove that the spaces $(C[a, b], \rho_p)$ $(1 \le p < \infty)$ (see Section 2.18, Problem 4) are separable.
(d) In which case is a *discrete* metric space (X, ρ_d) separable?

2.12 Equivalent Metrics, Homeomorphisms and Isometries

2.12.1 Equivalent Metrics

Definition 2.25 (Equivalent Metrics). Two metrics ρ_1 and ρ_2 on a nonempty set X are called *equivalent* if they generate the same *metric topology*.

Two metrics ρ_1 and ρ_2 on a nonempty set X are called *bi-Lipschitz equivalent* (or *bi-Lipschitzian*) if

$$\exists c, C > 0 \ \forall x, y \in X : \ c\rho_1(x, y) \le \rho_2(x, y) \le C\rho_1(x, y). \tag{2.5}$$

Remark 2.31. Both the equivalence and the bi-Lipschitz equivalence of metrics on a nonempty set X are *equivalence relations* (*reflexive, symmetric,* and *transitive*) on the set of all metrics on X.

Examples 2.18.
1. Any metric ρ on a nonempty set X is bi-Lipschitz equivalent to its constant multiple $c\rho$ $(c > 0)$ (see Section 2.18, Problem 3).
2. On $\mathbb{R}$, the standard metric ρ is equivalent to the generated by its *standard bounded metric*:

$$\mathbb{R} \ni x, y \mapsto d(x, y) := \min(\rho(x, y), 1)$$

(see Section 2.18, Problem 3), but the two metrics are *not* bi-Lipschitz equivalent.

3. All p-metrics $(1 \le p \le \infty)$ on $\mathbb{R}^n$ or $\mathbb{C}^n$ $(n \in \mathbb{N})$ are bi-Lipschitz equivalent.

Exercise 2.49. Verify the prior Remark 2.31 and Examples 2.18.

Hint. For 3, show that, on the n-space, for any $1 \le p < \infty$, ρ_p is bi-Lipschitz equivalent to ρ_∞; use Remark 2.31.

Remarks 2.32.
- Two metrics ρ_1 and ρ_2 on a nonempty set X are equivalent *iff*
  (i) for an arbitrary $x \in X$, each open ball centered at x in the space (X,ρ_1) contains an open ball centered at x in the space (X,ρ_2) and vice versa;
  (ii) any sequence $(x_n)_{n \in \mathbb{N}}$ convergent in the space (X,ρ_1) converges to the same limit in the space (X,ρ_2), and vice versa;
  (iii) the *identity mapping*

$$X \ni x \mapsto Ix := x \in X$$

  is *continuous* when considered as a mapping from (X,ρ_1) to (X,ρ_2), and vice versa.
- Two metrics ρ_1 and ρ_2 on a nonempty set X are bi-Lipschitz equivalent *iff* the *identity mapping*

$$X \ni x \mapsto Ix := x \in X$$

  is *Lipschitz continuous* when considered as a mapping from (X,ρ_1) to (X,ρ_2), and vice versa.
- If two metrics ρ_1 and ρ_2 on a nonempty set X are bi-Lipschitz equivalent, then they are equivalent, but, as Examples 2.18 demonstrate, not vice versa.

Exercise 2.50. Verify.

2.12.2 Homeomorphisms and Isometries

Definition 2.26 (Homeomorphism of Metric Spaces). A *homeomorphism* of a metric space (X,ρ) to a metric space (Y,σ) is a mapping

$$T : X \to Y,$$

which is *bijective* and *bicontinuous*, i. e., $T \in C(X,Y)$ and the inverse $T^{-1} \in C(Y,X)$.
  The space (X,ρ) is said to be *homeomorphic* to (Y,σ).

Remarks 2.33.
- The relation of *being homeomorphic to* is an *equivalence relation* on the set of all metric spaces, and thus, we can say that homeomorphism is *between* the spaces.

– Homeomorphic metric spaces are *topologically indistinguishable*, i. e., they have the same *topological properties*, such as *separability* or the existence of nontrivial *clopen* sets (*disconnectedness* (see, e. g., [52, 56])).

– Two metrics ρ_1 and ρ_2 on a nonempty set X are equivalent *iff* the *identity mapping*

$$X \ni x \mapsto Ix := x \in X$$

is a *homeomorphism* between the spaces (X, ρ_1) and (X, ρ_2) (see Remarks 2.32).

Exercise 2.51. Verify.

To show that two metric spaces are homeomorphic, one needs to specify a homeomorphism between them, whereas to show that they are not homeomorphic, one needs to indicate a topological property not shared by them.

Examples 2.19.

1. Any open bounded interval (a, b) $(-\infty < a < b < \infty)$ is homeomorphic to $(0, 1)$ via

$$f(x) := \frac{x - a}{b - a}.$$

2. An open interval of the form $(-\infty, b)$ in $\mathbb{R}$ is homeomorphic to $(-b, \infty)$ via

$$f(x) := -x.$$

3. An open interval of the form (a, ∞) in $\mathbb{R}$ is homeomorphic to $(1, \infty)$ via

$$f(x) := x - a + 1.$$

4. The interval $(1, \infty)$ is homeomorphic to $(0, 1)$ via

$$f(x) := \frac{1}{x}.$$

5. The interval $(0, 1)$ is homeomorphic to $\mathbb{R}$ via

$$f(x) := \cot(\pi x).$$

6. The interval $(0, 1)$ is *not* homeomorphic to the set $(0, 1) \cup \{2\}$ with the regular distance, the latter having nontrivial clopen sets (see Examples 2.43).

7. The spaces l_1 and l_∞ are *not* homeomorphic, the former being separable and the latter being nonseparable.

Remark 2.34. Thus, all open intervals in $\mathbb{R}$ (bounded or not) are homeomorphic.

The following is a very important particular case of a homeomorphism.

Definition 2.27 (Isometry of Metric Spaces). Let (X, ρ) and (Y, σ) be metric spaces.

An *isometry of X to Y* is a *one-to-one* (i. e., *injective*) mapping $T : X \to Y$, which is *distance preserving*, i. e.,

$$\sigma(Tx, Ty) = \rho(x, y), \ x, y \in X.$$

It is said to *isometrically embed X* in Y.

If an isometry $T : X \to Y$ is *onto* (i. e., *surjective*), it is called an *isometry between X and Y*, and the spaces are called *isometric*.

Remarks 2.35.
- The relation of *being isometric to* is an *equivalence relation* on the set of all metric spaces, and thus, we can say that isometry is *between* the spaces.
- An isometry between metric spaces (X, ρ) and (Y, σ) is, obviously, a homeomorphism between them, but not vice versa (see Examples 2.19).
- Isometric metric spaces are *metrically indistinguishable*.

Exercise 2.52. Identify isometries in Examples 2.19.

2.13 Completeness

Let us now deal with the fundamental concept of *completeness* of metric spaces underlying many important facts, such as the *Baire Category Theorem* (Theorem 2.32) and the *Banach Fixed-Point Theorem* (Theorem 2.33) (see Sections 2.14 and 2.15).

2.13.1 Cauchy/Fundamental Sequences

Definition 2.28 (Cauchy/Fundamental Sequence). A sequence $(x_n)_{n \in \mathbb{N}}$ in a metric space (X, ρ) is called a *Cauchy sequence*, or a *fundamental sequence*, if

$$\forall \, \varepsilon > 0 \; \exists N \in \mathbb{N} \; \forall m, n \geq N : \; \rho(x_m, x_n) < \varepsilon.$$

Remark 2.36. The latter is equivalent to

$$\rho(x_m, x_n) \to 0, \ m, n \to \infty,$$

or to

$$\sup_{k \in \mathbb{N}} \rho(x_{n+k}, x_n) \to 0, \ n \to \infty.$$

Examples 2.20.
1. In an arbitrary metric space (X, ρ), any *eventually constant* sequence $(x_n)_{n \in \mathbb{N}}$, i. e., such that

$$\exists N \in \mathbb{N} \, \forall n \geq N : x_n = x_N,$$

is fundamental, and only eventually constant sequences are fundamental in a *discrete* metric space (X, ρ_d).
2. The sequence $(1/n)_{n \in \mathbb{N}}$ is *fundamental* in $\mathbb{R}$, and the sequence $(n)_{n \in \mathbb{N}}$ is not.
3. The sequence $(x_n := (1, 1/2, \ldots, 1/n, 0, 0, \ldots))_{n \in \mathbb{N}}$ is *fundamental* in l_p $(1 < p \leq \infty)$, but not in l_1.
4. The sequence $(e_n := (\delta_{nk})_{k \in \mathbb{N}})_{n \in \mathbb{N}}$, where δ_{nk} is the *Kronecker delta*, is *not* fundamental in l_p $(1 \leq p \leq \infty)$.

Exercise 2.53. Verify.

Theorem 2.22 (Properties of Fundamental Sequences). *In a metric space (X, ρ),*
(1) *every fundamental sequence $(x_n)_{n \in \mathbb{N}}$ is bounded,*
(2) *every convergent sequence $(x_n)_{n \in \mathbb{N}}$ is fundamental,*
(3) *if a sequence $(x_n)_{n \in \mathbb{N}}$ is fundamental, then any sequence $(y_n)_{n \in \mathbb{N}}$ asymptotically equivalent to $(x_n)_{n \in \mathbb{N}}$, in the sense that*

$$\rho(x_n, y_n) \to 0, \ n \to \infty,$$

is also fundamental.

Exercise 2.54. Prove.

Remark 2.37. As the following examples demonstrate, a fundamental sequence need not converge.

Examples 2.21.
1. The sequence $(1/n)_{n \in \mathbb{N}}$ is fundamental and convergent to 0 in $\mathbb{R}$. However, in $\mathbb{R} \backslash \{0\}$ with the regular distance, it is still fundamental, but does not converge.
2. The sequence

$$(x_n := (1, 1/2, \ldots, 1/n, 0, 0, \ldots))_{n \in \mathbb{N}}$$

is fundamental, but divergent in (c_{00}, ρ_∞). However, in the wider space (c_0, ρ_∞), it converges to $x := (1/n)_{n \in \mathbb{N}}$.

Exercise 2.55. Verify.

Remark 2.38. By the *Sequential Characterization of Local Continuity* (Theorem 2.11), continuous functions preserve the convergence of sequences, i. e., continuous functions map convergent sequences to convergent sequences. The following statement shows that uniformly continuous functions preserve the fundamentality of sequences.

Proposition 2.2 (Fundamentality and Uniform Continuity). *Let (X,ρ) and (Y,σ) be metric spaces and a function $f : X \to Y$ be uniformly continuous on X. If $(x_n)_{n\in\mathbb{N}}$ is a fundamental sequence in the space (X,ρ), then the image sequence $(f(x_n))_{n\in\mathbb{N}}$ is a fundamental in the space (Y,σ), i. e., uniformly continuous functions map fundamental sequences to fundamental sequences.*

Exercise 2.56.
(a) Prove.
(b) Give an example showing that a continuous function need not preserve fundamentality.

2.13.2 Complete Metric Spaces

Definition 2.29 (Complete Metric Space). A metric space (X,ρ), in which every Cauchy/fundamental sequence converges, is called *complete*.

Examples 2.22.
1. The spaces $\mathbb{R}$ and $\mathbb{C}$ are *complete*, relative to the regular metric, as is known from analysis courses.
2. The spaces $\mathbb{R} \setminus \{0\}$, $(0,1)$, and $\mathbb{Q}$ are *incomplete* as subspaces of $\mathbb{R}$.
3. A *discrete* metric space (X,ρ_d) is *complete*.

Exercise 2.57. Verify 2 and 3.

4. **Theorem 2.23** (Completeness of the n-Space). *The (real or complex) space $l_p^{(n)}$ $(n \in \mathbb{N}, 1 \le p \le \infty)$ is complete.*

Exercise 2.58. Prove.

Hint. Considering the bi-Lipschitz equivalence of all p-metrics on the n-space (see Remarks 2.32) and Exercise 2.61, it suffices to show the completeness of the n-space relative to ρ_∞.

5. **Theorem 2.24** (Completeness of l_p $(1 \le p < \infty)$). *The (real or complex) space l_p $(1 \le p < \infty)$ is complete.*

Proof. Let $1 \le p < \infty$ and

$$x^{(n)} := (x_k^{(n)})_{k\in n}, \ n \in \mathbb{N},$$

be an arbitrary fundamental sequence in l_p.
Since

$$\forall \varepsilon > 0 \ \exists N \in \mathbb{N} \ \forall m,n \ge N : \rho_p(x_m,x_n) < \varepsilon,$$

for each $k \in \mathbb{N}$,

$$\left|x_k^{(m)} - x_k^{(n)}\right| \leq \left[\sum_{i=1}^{\infty} \left|x_i^{(m)} - x_i^{(n)}\right|^p\right]^{1/p} = \rho_p(x_m, x_n) < \varepsilon, \quad m, n \geq N,$$

which implies that, for every $k \in \mathbb{N}$, the numeric sequence $(x_k^{(n)})_{n \in \mathbb{N}}$ of the kth terms is *fundamental*, and hence, *converges*, i. e.,

$$\forall k \in \mathbb{N} \, \exists x_k \in \mathbb{R} \text{ (or } \mathbb{C}\text{)}: \ x_k^{(n)} \to x_k, \ n \to \infty.$$

Let us show that $x := (x_k)_{n \in \mathbb{N}} \in l_p$.
For any $K \in \mathbb{N}$ and arbitrary $m, n \geq N$,

$$\sum_{k=1}^{K} \left|x_k^{(m)} - x_k^{(n)}\right|^p \leq \sum_{k=1}^{\infty} \left|x_k^{(m)} - x_k^{(n)}\right|^p = \rho_p(x_m, x_n)^p < \varepsilon^p.$$

Whence, fixing arbitrary $K \in \mathbb{N}$ and $n \geq N$, and passing to the limit as $m \to \infty$, we obtain

$$\sum_{k=1}^{K} \left|x_k - x_k^{(n)}\right|^p \leq \varepsilon^p, \ K \in \mathbb{N}, n \geq N.$$

Now, for an arbitrary $n \geq N$, passing to the limit as $K \to \infty$, we arrive at

$$\sum_{k=1}^{\infty} \left|x_k - x_k^{(n)}\right|^p \leq \varepsilon^p, \ n \geq N, \tag{2.6}$$

which, in particular, implies that

$$x - x_N := \left(x_k - x_k^{(N)}\right)_{k \in \mathbb{N}} \in l_p,$$

and hence, by *Minkowski's Inequality for Sequences* (Theorem 2.4),

$$x = (x - x_N) + x_N \in l_p.$$

Whence, in view (2.6), we infer that

$$\rho_p(x, x_n) \leq \varepsilon, \ n \geq N,$$

which implies that

$$x_n \to x, \ n \to \infty,$$

in l_p and completes the proof. $\qquad\square$

6. **Theorem 2.25** (Completeness of $(M(T), \rho_\infty)$). *The (real or complex) space $(M(T), \rho_\infty)$ is complete.*

Proof. Let $(f_n)_{n \in \mathbb{N}}$ be an arbitrary fundamental sequence in $(M(T), \rho_\infty)$. Since

$$\forall \varepsilon > 0 \; \exists N \in \mathbb{N} \; \forall m, n \geq N : \rho_\infty(f_m, f_n) < \varepsilon,$$

for each $t \in T$,

$$|f_m(t) - f_n(t)| \leq \sup_{s \in T} |f_m(s) - f_n(s)| = \rho_\infty(f_m, f_n) < \varepsilon, \; m, n \geq N,$$

which implies that, for every $t \in T$, the numeric sequence $(f_n(t))_{n \in \mathbb{N}}$ of the values at t is *fundamental*, and hence, *converges*, i. e.,

$$\forall t \in T \; \exists f(t) \in \mathbb{R} \; (\text{or } \mathbb{C}) : f_n(t) \to f(t), \; n \to \infty.$$

Let us show that the function $T \ni t \mapsto f(t)$ belongs to $M(T)$, i. e., is *bounded* on T. Indeed, for any $t \in T$,

$$|f_m(t) - f_n(t)| < \varepsilon, \; m, n \geq N.$$

Whence, fixing arbitrary $t \in T$ and $n \geq N$, and passing to the limit as $m \to \infty$, we have

$$|f(t) - f_n(t)| \leq \varepsilon, \; t \in T, n \geq N,$$

i. e.,

$$\sup_{t \in T} |f(t) - f_n(t)| \leq \varepsilon, \; n \geq N. \tag{2.7}$$

Therefore,

$$\sup_{t \in T} |f(t)| \leq \sup_{t \in T} |f(t) - f_N(t)| + \sup_{t \in T} |f_N(t)| \leq \sup_{t \in T} |f_N(t)| + \varepsilon < \infty,$$

which implies that $f \in M(T)$.
Whence, in view (2.7), we infer that

$$\rho_\infty(f, f_n) \leq \varepsilon, \; n \geq N.$$

Consequently,

$$f_n \to f, \; n \to \infty,$$

in $(M(T), \rho_\infty)$, and completes the proof. $\qquad\qquad\qquad\qquad\qquad\square$

As a particular case, for $T := \mathbb{N}$, we obtain the following:

Corollary 2.1 (Completeness of l_∞). *The (real or complex) space l_∞ is complete.*

7. **Theorem 2.26** (Completeness of $(C[a,b],\rho_\infty)$ $(-\infty < a < b < \infty)$). *The (real or complex) space $(C[a,b],\rho_\infty)$ $(-\infty < a < b < \infty)$ is complete.*

Exercise 2.59. Prove. [25, Section 1.7, Exercise (i)]

8. The space P of all *polynomials* with real/complex coefficients is *incomplete* as a subspace of $(C[a,b],\rho_\infty)$.

Exercise 2.60. Give a corresponding *counterexample*.

9. **Proposition 2.3** (Incompleteness of $(C[a,b],\rho_p)$ $(1 \le p < \infty)$). *The space $(C[a,b], \rho_p)$ $(1 \le p < \infty)$ is incomplete.*

Proof. Let us fix a $c \in (a,b)$ (say, $c := (a+b)/2$), and consider the sequence (f_n) in $(C[a,b],\rho_p)$ defined for all $n \in \mathbb{N}$ sufficiently large so that $a < c - 1/n$ as follows:

$$
f_n(t) := \begin{cases} 0 & \text{for } a \le t \le c - 1/n, \\ [n(t - c + 1/n)]^{1/p} & \text{for } c - 1/n < t < c, \\ 1 & \text{for } c \le t \le b. \end{cases}
$$

The sequence (f_n) is *fundamental* in $(C[a,b],\rho_p)$. Indeed, for all sufficiently large $m, n \in \mathbb{N}$ with $m \le n$,

$$
\rho_p(f_m, f_n) := \left[\int_a^b |f_m(t) - f_n(t)|^p \, dt \right]^{1/p} = \left[\int_{c-1/m}^c |f_m(t) - f_n(t)|^p \, dt \right]^{1/p}
$$

by *Minkowski's inequality for functions* (see Section 2.18, Problem 4);

$$
\le \left[\int_{c-1/m}^c |f_m(t)|^p \, dt \right]^{1/p} + \left[\int_{c-1/m}^c |f_n(t)|^p \, dt \right]^{1/p}
$$

$$
= \left[\int_{c-1/m}^c m(t - c + 1/m) \, dt \right]^{1/p} + \left[\int_{c-1/n}^c n(t - c + 1/n) \, dt \right]^{1/p}
$$

$$
= \left(\frac{1}{2m} \right)^{1/p} + \left(\frac{1}{2n} \right)^{1/p} \to 0, \ m, n \to \infty.
$$

Assume that

$$
\exists f \in C[a,b] : f_n \to f, \ n \to \infty, \ \text{in } (C[a,b],\rho_p).
$$

Then

$$
\rho_p(f_n, f) = \left[\int_a^{c-1/n} |f(t)|^p \, dt + \int_{c-1/n}^c |f_n(t) - f(t)|^p \, dt + \int_c^b |1 - f(t)|^p \, dt \right]^{1/p}
$$

$$\rightarrow \int_a^c |f(t)|^p\, dt + \int_c^b |1 - f(t)|^p\, dt = 0,\ n \to \infty.$$

Whence, by the *continuity* of f on $[a, b]$, we infer that

$$f(t) = 0,\ t \in [a, c)\ \text{and}\ f(t) = 1,\ t \in (c, b],$$

which *contradicts* the continuity of f at $x = c$.

The obtained contradiction implies that the sequence (f_n) in $(C[a, b], \rho_p)$ does not converge in $(C[a, b], \rho_p)$. Hence, we conclude that the metric space $(C[a, b], \rho_p)$ ($1 \le p < \infty$) is *incomplete*. $\qquad\qquad\square$

Remark 2.39. The property of completeness is *isometrically invariant* (see Section 2.18, Problem 29) but is *not homeomorphically invariant*.

For instance, the complete space $\mathbb{R}$ is homeomorphic to the incomplete space $(0, 1)$ (see Examples 2.19 and 2.22).

A more sophisticated example is as follows: The set $X := \{1/n\}_{n \in \mathbb{N}}$ has the same *discrete* metric topology, relative to the regular metric ρ as relative to the discrete metric ρ_d, which implies that the two metrics are *equivalent* on X (see Section 2.12.1), i. e., the spaces (X, ρ) and (X, ρ_d) are *homeomorphic*, relative to the identity mapping

$$X \ni x \mapsto Ix := x \in X$$

(see Remarks 2.33). However, the former space is incomplete, whereas the latter is complete (see Examples 2.22).

Exercise 2.61.

(a) Explain.

(b) Prove that, if metrics ρ_1 and ρ_2 on a nonempty set X are *bi-Lipschitz equivalent*, the space (X, ρ_1) is complete *iff* the space (X, ρ_2) is complete.

2.13.3 Subspaces of Complete Metric Spaces

Theorem 2.27 (Characterization of Completeness). *For a subspace (Y, ρ) of a metric space (X, ρ) to be complete, it is necessary and, provided the space (X, ρ) is complete, sufficient that Y be a closed set in (X, ρ).*

Exercise 2.62. Prove.

Exercise 2.63. Apply the prior characterization to show the following:

(a) any *closed subset* of $\mathbb{R}$, in particular any *closed interval*, is a *complete* metric space, relative to the regular distance, i. e., as a subspace of the complete space $\mathbb{R}$ (see Examples 2.22);

(b) $\mathbb{R} \setminus \{0\}$, $(0,1)$, and $\mathbb{Q}$ are *incomplete* metric spaces, relative to the regular distance, i. e., as subspaces of the complete space $\mathbb{R}$;

(c) the spaces (c_0, ρ_∞) and (c, ρ_∞) are *complete* as closed subsets in the complete space l_∞ (see Examples 2.22);

(d) the space (c_{00}, ρ_∞) is *incomplete* as a nonclosed subset in the complete space (c_0, ρ_∞) (see Examples 2.22);

(e) the space $(C[a,b], \rho_\infty)$ $(-\infty < a < b < \infty)$ is *complete* as a closed subset in the complete space $(M[a,b], \rho_\infty)$ (see Examples 2.22);

(f) the space P of all *polynomials* with real/complex coefficients is *incomplete* as a nonclosed subset in the complete space $(C[a,b], \rho_\infty)$ (see Examples 2.22).

2.13.4 Nested Balls Theorem

The celebrated *Nested Intervals Theorem* (a characterization of the *completeness* of the *real numbers*) allows the following generalization:

Theorem 2.28 (Nested Balls Theorem). *A metric space (X, ρ) is complete iff, for any sequence of closed balls*

$$(B_n := \overline{B}(x_n, r_n))_{n \in \mathbb{N}}$$

such that

(1) $B_{n+1} \subseteq B_n$, $n \in \mathbb{N}$, *and*

(2) $r_n \to 0$, $n \to \infty$,

the intersection $\bigcap_{n=1}^\infty B_n$ is a singleton.

Proof. "*If*" part. Suppose that, for any sequence of closed balls

$$(B_n := \overline{B}(x_n, r_n))_{n \in \mathbb{N}},$$

satisfying conditions (1) and (2), the intersection $\bigcap_{n=1}^\infty B_n$ is a singleton and let $(x_n)_{n \in \mathbb{N}}$ be an arbitrary *fundamental sequence* in (X, ρ).

There exists a *subsequence* $(x_{n(k)})_{k \in \mathbb{N}}$ $(n(k) \in \mathbb{N}$, and $n(k) < n(k+1)$, $k \in \mathbb{N})$ such that

$$\forall k \in \mathbb{N} : \rho(x_m, x_{n(k)}) \le \frac{1}{2^{k+1}}, \quad m \ge n(k). \tag{2.8}$$

Exercise 2.64. Explain.

Consider the sequence of closed balls $(B_k := \overline{B}(x_{n(k)}, 1/2^k))_{k \in \mathbb{N}}$.
It is *nested*, i. e.,

$$B_{k+1} \subseteq B_k, \quad k \in \mathbb{N},$$

since, for any $y \in B_{k+1}$, by the *triangle inequality* and in view of (2.8),

$$\rho(y, x_{n(k)}) \le \rho(y, x_{n(k+1)}) + \rho(x_{n(k+1)}, x_{n(k)}) \le \frac{1}{2^{k+1}} + \frac{1}{2^{k+1}} = \frac{1}{2^k}, \ k \in \mathbb{N},$$

which implies that $y \in B_k$.

As for the *radii*, we have

$$1/2^k \to 0, \ k \to \infty.$$

Whence, by the premise, we infer that

$$\exists x \in X : \bigcap_{k=1}^{\infty} B_k = \{x\}.$$

Consequently,

$$0 \le \rho(x_{n(k)}, x) \le 1/2^k, \ k \in \mathbb{N},$$

which, by the *Squeeze Theorem*, implies that

$$x_{n(k)} \to x, \ k \to \infty, \text{ in } (X, \rho).$$

Since the *fundamental sequence* $(x_n)_{n \in \mathbb{N}}$ contains the *subsequence* $(x_{n(k)})_{n \in \mathbb{N}}$ *convergent* to x, by the *Fundamental Sequence with Convergent Subsequence Proposition* (Proposition 2.22, Section 2.18, Problem 25), $(x_n)_{n \in \mathbb{N}}$ also converges to x:

$$x_n \to x, \ n \to \infty, \text{ in } (X, \rho),$$

which proves the *completeness* of the space (X, ρ).

"*Only if*" part. Suppose that the metric space (X, ρ) is *complete* and let

$$(B_n := \overline{B}(x_n, r_n))_{n \in \mathbb{N}}$$

be an arbitrary sequence of closed balls, satisfying conditions (1) and (2).

Then the sequence of the centers $(x_n)_{n \in \mathbb{N}}$ is *fundamental* in (X, ρ). Indeed, for all $n, k \in \mathbb{N}$, since $B_{n+k} \subseteq B_n$,

$$0 \le \rho(x_{n+k}, x_n) \le r_n,$$

which, in view of $r_n \to 0, \ n \to \infty$, by the *Squeeze Theorem*, implies that

$$\sup_{k \in \mathbb{N}} \rho(x_{n+k}, x_n) \to 0, \ n \to \infty$$

(see Remark 2.36).

By the *completeness* of (X, ρ),

$$\exists x \in X : x_n \to x, \ n \to \infty, \ \text{in} \ (X, \rho).$$

Let us show that

$$\bigcap_{n=1}^{\infty} B_n = \{x\}.$$

Indeed, since

$$\forall n \in \mathbb{N} : x_m \in B_n, \ m \geq n,$$

in view of the *closedness* of B_n, by the *Sequential Characterization of Closed Sets* (Theorem 2.19),

$$\forall n \in \mathbb{N} : x = \lim_{m \to \infty} x_m \in B_n.$$

Hence, we have the inclusion

$$\{x\} \subseteq \bigcap_{n=1}^{\infty} B_n.$$

On the other hand, for any $y \in \bigcap_{n=1}^{\infty} B_n$, since $x, y \in B_n$ for each $n \in \mathbb{N}$, by the *triangle inequality*,

$$0 \leq \rho(x, y) \leq \rho(x, x_n) + \rho(x_n, y) \leq 2r_n,$$

which, in view of $r_n \to 0$, $n \to \infty$, by the *Squeeze Theorem*, implies that

$$\rho(x, y) = 0.$$

Hence, by the *separation axiom*, $x = y$.

Thus, the inverse inclusion

$$\{x\} \supseteq \bigcap_{n=1}^{\infty} B_n$$

holds as well, and we conclude that

$$\bigcap_{n=1}^{\infty} B_n = \{x\},$$

which completes the proof of the *"only if"* part and the entire statement. $\square$

Remark 2.40. Each of the *four* conditions in the *"only if"* part of the *Nested Balls Theorem* (the completeness of the space, the closedness of the balls, and conditions (1) and (2)) is essential and cannot be dropped.

Exercise 2.65. Verify by providing corresponding *counterexamples*.

A more general version of the *Nested Balls Theorem*, which can be proved by mimicking the proof of the latter, is the following:

Theorem 2.29 (Generalized Nested Balls Theorem). *A metric space (X, ρ) is complete iff for every sequence of nonempty closed sets $(B_n)_{n \in \mathbb{N}}$ such that*
(1) $B_{n+1} \subseteq B_n$, $n \in \mathbb{N}$, and
(2) $\operatorname{diam}(B_n) := \sup_{x,y \in B_n} \rho(x, y) \to 0$, $n \to \infty$,

the intersection $\bigcap_{n=1}^{\infty} B_n$ is a singleton.

2.14 Category and Baire Category Theorem

The *Nested Balls Theorem* (Theorem 2.28) is applied to prove the *Baire*[9] *Category Theorem* (Theorem 2.32), which is one of the most important facts about complete metric spaces, underlying a number of fundamental statements, such as the *Uniform Boundedness Principle* (Theorem 4.6) and the *Open Mapping Theorem* (Theorem 4.7) (see Sections 4.3 and 4.4).

2.14.1 Nowhere Denseness

First, we introduce and study *nowhere dense sets*, that means, sets in a metric space, which are, in a sense, "scarce".

Definition 2.30 (Nowhere Dense Set). A set A in a metric space (X, ρ) is called *nowhere dense* if the interior of its closure $\overline{A}$ is empty:

$$\operatorname{int}(\overline{A}) = \emptyset,$$

i. e., $\overline{A}$ contains no nontrivial open balls.

Remarks 2.41.
- A set A is nowhere dense in a metric space (X, ρ) iff its closure $\overline{A}$ is nowhere dense in (X, ρ) (see Section 2.18, Problem 16).
- For a closed set A in a metric space (X, ρ), since $A = \overline{A}$, its nowhere denseness is simply the emptiness of its interior:

$$\operatorname{int}(A) = \emptyset.$$

---

9 René-Louis Baire (1874–1932).

Examples 2.23.
1. The empty set $\emptyset$ is nowhere dense in any metric space (X, ρ), and only the empty set is nowhere dense in a *discrete* metric space (X, ρ_d).
2. Finite sets, the sets $\mathbb{Z}$ of all integers and $\{1/n\}_{n \in \mathbb{N}}$ are nowhere dense in $\mathbb{R}$.
3. The celebrated *Cantor set* (see, e. g., [33, 38, 40]) is a closed nowhere dense set in $\mathbb{R}$ as well as its two-dimensional analogue, the *Sierpinski*[10] *carpet*, in $\mathbb{R}^2$ with the Euclidean distance (i. e., in $l_2^{(2)}(\mathbb{R})$) (see, e. g., [52]).
4. A straight line is a closed nowhere dense set in $l_2^{(2)}(\mathbb{R})$.
5. An arbitrary *dense* set in a metric space (X, ρ) is *not* nowhere dense.
   In particular, the sets $\mathbb{Q}$ of the rationals and the set $\mathbb{Q}^c$ of the irrationals are *not* nowhere dense in $\mathbb{R}$.
6. However, a set need not be dense in a metric space need not be nowhere dense. Thus, a nondegenerate proper interval I in $\mathbb{R}$ is neither dense nor nowhere dense.

Exercise 2.66. Verify.

Remark 2.42. Being formulated entirely in terms of closure and interior, *nowhere denseness* is a *topological property*, i. e., it is preserved by a *homeomorphism*.

Proposition 2.4 (Characterization of Nowhere Denseness). *A set A is nowhere dense in a metric space (X, ρ) iff its exterior $\mathrm{ext}(A)$ is dense in (X, ρ).*

Proof. Let $A \subseteq X$ be arbitrary. Then, in view of

$$\mathrm{ext}(A) = \overline{A}^c$$

(see Section 2.18, Problem 18) and

$$\mathrm{ext}(\overline{A}^c) = \mathrm{int}(\overline{A})$$

(see Remarks 2.28), we have the following disjoint union representation:

$$X = \mathrm{ext}(\overline{A}^c) \cup \overline{A}^c = \mathrm{int}(\overline{A}) \cup \overline{\mathrm{ext}(A)},$$

which implies that

$$\mathrm{int}(\overline{A}) = \emptyset \iff \overline{\mathrm{ext}(A)} = X,$$

completing the proof. $\qquad\square$

Since for a closed set A in a metric space (X, ρ),

$$\mathrm{ext}(A) = A^c$$

(see Section 2.18, Problem 18), we obtain the following:

---

10 Waclaw Sierpinski (1882–1969).

Corollary 2.2 (Characterization of Nowhere Denseness of Closed Sets). *A closed set A in a metric space (X, ρ) is nowhere dense iff its complement A^c is dense in (X, ρ).*

Example 2.24. Thus, the complement of the *Cantor set* is an open dense set in $\mathbb{R}$.

Remark 2.43. The *"only if"* part of the prior corollary holds true regardless whether the set is closed or not, i. e., the complement of an arbitrary nowhere dense set is necessarily dense.

Exercise 2.67. Show.

Considering the mutually complementary nature of closed and open sets (see *Characterizations of Closed Sets* (Theorem 2.19)), we can equivalently reformulate the prior statement as follows:

Corollary 2.3 (Characterization of Denseness of Open Sets). *An open set A in a metric space (X, ρ) is dense iff its complement A^c is nowhere dense in (X, ρ).*

Remark 2.44. The requirement of the *openness* of the set in the *"only if"* part of the prior corollary is essential and cannot be dropped, i. e., the complement of a dense set, which is not open, need not be nowhere dense.

Exercise 2.68. Give a corresponding example.

Theorem 2.30 (Properties of Nowhere Dense Sets). *The nowhere dense sets in a metric space (X, ρ) have the following properties:*
(1) *a subset of a nowhere dense set is nowhere dense;*
(2) *an arbitrary intersection of nowhere dense sets is nowhere dense;*
(3) *a finite union of nowhere dense sets is nowhere dense.*

Exercise 2.69. Prove (see *Properties of Bounded Sets* (Theorem 2.14)).

Hint. To prove (3), first show that, for any sets A and B in (X, ρ),

$$\overline{A \cup B} = \overline{A} \cup \overline{B},$$

which, by *de Morgan's laws*, is equivalent to

$$\text{ext}(A \cup B) = \text{ext}(A) \cap \text{ext}(B)$$

(see Section 2.18, Problem 18). Then exploit the *denseness* jointly with *openness* (cf. *Finite Intersections of Open Dense Sets Proposition* (Proposition 2.24, Section 2.18, Problem 30)).

Remark 2.45. An *infinite* union of nowhere dense sets in a metric space (X, ρ) need not be nowhere dense.

For instance, any *singleton* is nowhere dense in $\mathbb{R}$ (see Examples 2.23). However, the set $\mathbb{Q}$ of all rationals, being a countably infinite union of singletons, is *dense* in $\mathbb{R}$.

2.14.2 Category

Here, based on the concept of *nowhere denseness*, we naturally distinguish two categories of sets of a metric space, the notion of *category* giving a sense of a set's "fullness" and being closely related to the notion of *completeness*.

Definition 2.31 (First-/Second-Category Set). A set A in a metric space (X, ρ) is said to be of the *first category* if it can be represented as a *countable union* of *nowhere dense* sets.

Otherwise, A is said to be of the *second category* in (X, ρ).

Examples 2.25.
1. Any *nowhere dense set* in a metric space (X, ρ) is a *first-category set*.
   In particular, the set $\mathbb{N}$ of all naturals is of the first category in $\mathbb{R}$.
2. A *dense set* in a metric space (X, ρ) can be of the first-category as well. For instance, the set $\mathbb{Q}$ of all rationals in $\mathbb{R}$ (see Remark 2.45).
3. The space (c_{00}, ρ_{∞}) is of the *first category* in itself (also in (c_0, ρ_{∞}), (c, ρ_{∞}), and l_{∞}).
   Indeed,

$$c_{00} = \bigcup_{n=1}^{\infty} U_n,$$

where

$$U_n := \{x := (x_1, \ldots, x_n, 0, 0, \ldots) \in c_{00}\}, \ n \in \mathbb{N}.$$

Each U_n is *closed*, being an *isometric embedding* of the *complete* space $l_{\infty}^{(n)}$ into (c_{00}, ρ_{∞}), and *nowhere dense* in (c_{00}, ρ_{∞}) (also in (c_0, ρ_{∞}), (c, ρ_{∞}), and l_{∞}) since

$$\forall n \in \mathbb{N} \ \forall x := (x_1, \ldots, x_n, 0, 0, \ldots) \in U_n \ \forall \varepsilon > 0$$
$$\exists y := (x_1, \ldots, x_n, \varepsilon/2, 0, 0 \ldots) \in c_{00} \setminus U_n : \ \rho_{\infty}(x, y) = \varepsilon/2 < \varepsilon,$$

which implies that $\mathrm{int}(U_n) = \emptyset$, $n \in \mathbb{N}$.
4. Whereas, by the *Baire Category Theorem* (Theorem 2.32), the complete spaces $\mathbb{R}$ and $\mathbb{C}$ are of the second category in themselves, the *isometric embedding*

$$\{(x, 0) \mid x \in \mathbb{R}\}$$

of $\mathbb{R}$ into $\mathbb{C}$ is of the first category (*nowhere dense*, to be precise) in $\mathbb{C}$ (see Examples 2.23).
5. Every nonempty set in a *discrete* metric space (X, ρ_d) is a *second-category set*.

Exercise 2.70. Explain 5.

Remark 2.46. Being formulated entirely in terms of nowhere denseness and union, *first category* is a *topological property*, i. e., it is preserved by a *homeomorphism*, and hence, so is *second category*.

Example 2.26. Thus, any open interval I in $\mathbb{R}$, being homeomorphic to $\mathbb{R}$ (see Remark 2.34), is of the second category in itself as a subspace of $\mathbb{R}$ and, since the interior of a set in I coincides with its interior in $\mathbb{R}$, is also of the second category in $\mathbb{R}$.

Theorem 2.31 (Properties of First-Category Sets). *The first-category sets in a metric space (X,ρ) have the following properties:*
(1) *a subset of a first-category set is a first-category set;*
(2) *an arbitrary intersection of first-category sets is a first-category set;*
(3) *a countable union of first-category sets is a first-category set.*

Exercise 2.71. Prove.

Hint. To prove (3), use the *Properties of Countable Sets* (Theorem 1.3).

We obtain the following:

Corollary 2.4 (Set With a Second-Category Subset). *A set A in a metric space (X,ρ) containing a second-category subset B is of the second category.*

Examples 2.27.
1. Any nontrivial (open or closed) interval I in $\mathbb{R}$ is of the *second category* in $\mathbb{R}$, and in itself as a subspace of $\mathbb{R}$, since it contains an open interval, which is a second-category set in $\mathbb{R}$ (see Example 2.26).
2. More generally, for the same reason, any set $A \subseteq \mathbb{R}$ with $\text{int}(A) \neq \emptyset$ is of the *second category* in $\mathbb{R}$, and in itself as a subspace of $\mathbb{R}$.

2.14.3 Baire Category Theorem

Now, we apply the *Nested Balls Theorem* (Theorem 2.28) to prove the celebrated *Baire Category Theorem* already referred to in the prior section (see Examples 2.25).

Theorem 2.32 (Baire Category Theorem). *A complete metric space (X,ρ) is of the second category in itself, i. e., X is of the second category in (X,ρ).*

Proof. Let us prove the statement *by contradiction* assuming that there exists a complete metric space (X,ρ) of the first category in itself, i. e., X can be represented as a countable union of nowhere dense sets:

$$X = \bigcup_{n=1}^{\infty} U_n.$$

Without loss of generality, we can regard all U_n, $n \in \mathbb{N}$, to be *closed*.

Exercise 2.72. Explain.

Being nowhere dense, the closed set U_1 is a *proper subset* of X, and hence, by the *Characterization of Nowhere Denseness of Closed Sets* (Corollary 2.2), its *open* complement U_1^c is *dense* in (X, ρ), and the more so, U_1^c is *nonempty*. Therefore,

$$\exists\, x_1 \in U_1^c \,\exists\, 0 < \varepsilon_1 < 1 : \overline{B}(x_1, \varepsilon_1) \subseteq U_1^c,$$

i. e.,

$$\overline{B}(x_1, \varepsilon_1) \cap U_1 = \emptyset.$$

Since the open ball $B(x_1, \varepsilon_1/2)$ is not contained in the closed nowhere dense set U_2,

$$\exists\, x_2 \in B(x_1, \varepsilon_1/2) \,\exists\, 0 < \varepsilon_2 < 1/2 : \overline{B}(x_2, \varepsilon_2) \cap U_2 = \emptyset$$

and

$$\overline{B}(x_2, \varepsilon_2) \subseteq B(x_1, \varepsilon_1/2).$$

Continuing inductively, we obtain a sequence of closed balls $\{\overline{B}(x_n, \varepsilon_n)\}_{n=1}^{\infty}$ such that

(1) $\overline{B}(x_{n+1}, \varepsilon_{n+1}) \subseteq \overline{B}(x_n, \varepsilon_n)$, $n \in \mathbb{N}$.
(2) $0 < \varepsilon_n < 1/2^{n-1}$, $n \in \mathbb{N}$.
(3) $\overline{B}(x_n, \varepsilon_n) \cap U_n = \emptyset$, $n \in \mathbb{N}$.

In view of the *completeness* of the space (X, ρ), by the *Nested Balls Theorem* (Theorem 2.28), (1) and (2) imply that

$$\bigcap_{n=1}^{\infty} \overline{B}(x_n, \varepsilon_n) = \{x\}$$

with some $x \in X$.

Since $x \in \overline{B}(x_n, \varepsilon_n)$ for each $n \in \mathbb{N}$, by (3), we conclude that

$$x \notin \bigcup_{n=1}^{\infty} U_n = X,$$

which is a *contradiction* proving the statement. $\qquad\square$

Examples 2.28.
1. By the *Baire Category Theorem*, the complete spaces $\mathbb{R}$ and $\mathbb{C}$ are of the second category in themselves (see Examples 2.25).
2. Any set $A \subseteq \mathbb{R}$ with $\mathrm{int}(A) \neq \emptyset$ is of the *second category* in $\mathbb{R}$, and in itself as a subspace of $\mathbb{R}$ (see Examples 2.27).
   However, a set need not have a nonempty interior to be of the second category. Indeed, as follows from the *Baire Category Theorem*, the set $\mathbb{Q}^c$ of all irrationals with $\mathrm{int}(\mathbb{Q}^c) = \emptyset$ is of the *second category* in $\mathbb{R}$, as well as the complement A^c of any first-category set A in a complete metric space (X, ρ) (see the *Complement of a First-Category Set Proposition* (Proposition 2.25, Section 2.18, Problem 32)).

3. By the *Baire Category Theorem* and the *Characterization of Completeness* (Theorem 2.27), the set of all integers $\mathbb{Z}$ and the *Cantor set* are of the *second category* in themselves as closed subspaces of the complete space $\mathbb{R}$, but are of the *first category* (*nowhere dense*, to be precise) in $\mathbb{R}$ (see Examples 2.23).

Remarks 2.47.
− The converse to the *Baire Category Theorem* is not true, i. e., there exist incomplete metric spaces of the second category in themselves.
  For instance, an open interval I in $\mathbb{R}$ is of the *second category* in itself (see Example 2.26), but is *incomplete* as a subspace of $\mathbb{R}$ (see the *Characterization of Completeness* (Theorem 2.27)).
− The proof of the *Baire Category Theorem* requires a weaker form of the *Axiom of Choice* (see Appendix A), the *Axiom of Dependent Choices* (see, e. g., [36, 50]).

Corollary 2.5 (Second-Category Properties of Complete Metric Spaces). *In a complete metric space* (X, ρ),
(1) *any representation of X as a countable union of its subsets*

$$X = \bigcup_{n=1}^{\infty} U_n$$

contains at least one subset, which is not nowhere dense, i. e.,

$$\exists N \in \mathbb{N} : \operatorname{int}(\overline{U_N}) \neq \emptyset;$$

(2) *any countable intersection $\bigcap_{n=1}^{\infty} U_n$ of open dense sets is nonempty and dense in* (X, ρ).

Proof. Here, we are to prove part (2) only. Proving part (1) is left to the reader as an exercise.
  Let $(U_n)_{n \in \mathbb{N}}$ be an arbitrary sequence of open dense sets in (X, ρ).
  Then, by the *Characterization of Denseness of Open Sets* (Corollary 2.3), $(U_n^c)_{n \in \mathbb{N}}$ is a sequence of nowhere dense closed sets.
  The assumption that

$$\bigcap_{n=1}^{\infty} U_n = \emptyset,$$

by *de Morgan's laws*, implies that

$$\bigcup_{n=1}^{\infty} U_n^c = \left(\bigcap_{n=1}^{\infty} U_n \right)^c = \emptyset^c = X,$$

which, by the *Baire Category Theorem* (Theorem 2.32), *contradicts* the completeness of the space (X,ρ), and hence,

$$\bigcap_{n=1}^{\infty} U_n \neq \emptyset.$$

Assume that the set $\bigcap_{n=1}^{\infty} U_n$ is *not* dense in (X,ρ). Therefore,

$$\exists x \in X \ \exists r > 0 : \overline{B}(x,r) \cap \bigcap_{n=1}^{\infty} U_n = \emptyset.$$

I. e.,

$$\overline{B}(x,r) \subseteq \left(\bigcap_{n=1}^{\infty} U_n\right)^c = \bigcup_{n=1}^{\infty} U_n^c.$$

The closed ball $\overline{B}(x,r)$, being a closed set in (X,ρ), by the *Characterization of Completeness* (Theorem 2.27), the space $(\overline{B}(x,r),\rho)$ is *complete* as a subspace of the complete space (X,ρ):

$$\overline{B}(x,r) = \bigcup_{n=1}^{\infty} V_n,$$

where, for each $n \in \mathbb{N}$, the set

$$V_n := \overline{B}(x,r) \cap U_n^c,$$

is *nowhere dense* in (X,ρ), and hence, in $(\overline{B}(x,r),\rho)$.

Exercise 2.73. Explain.

This, by the *Baire Category Theorem* (Theorem 2.32), *contradicts* the completeness of the space $(\overline{B}(x,r),\rho)$, and hence, proves that the set $\bigcap_{n=1}^{\infty} U_n$ is *dense* in (X,ρ). □

Exercise 2.74. Prove part (1).

Remarks 2.48.
- Part (2) of the prior corollary generalizes *Finite Intersections of Open Dense Sets Proposition* (Proposition 2.24, Section 2.18, Problem 30).
- As follows from the proof of part (2) of the prior corollary, any nontrivial closed/ open ball in a complete metric space (X,ρ) is a set of the second category. Hence, by the *Set With a Second-Category Subset Corollary* (Corollary 2.4), one can derive the following profound conclusion: Any set A in a complete metric space (X,ρ) with $\text{int}(A) \neq \emptyset$ is of the *second category* in (X,ρ) (cf. Examples 2.28).
- An uncountable intersection of open dense sets need not be dense.

Exercise 2.75. Give a corresponding example (see Section 2.18, Problem 30).

2.15 Banach Fixed-Point Theorem and Applications

The *Banach*[11] *Fixed-Point Theorem* (1922) or the *Banach Contraction Principle* (also *Contraction Mapping Principle*) is one of the central results of nonlinear functional analysis with multiple applications. In abstract setting, it utilizes the *method of successive approximations* introduced by Liouville[12] (1837) and developed by Picard[13] (1890) (see, e. g., [28, 1, 55, 14]).

2.15.1 Fixed Points

We start our discourse with introducing the notion of *fixed point*, which does not require metric structure.

Definition 2.32 (Fixed Point). Let X be a nonempty set and $T : X \to X$ be a mapping on X. A point $x \in X$ is called a *fixed point* for T if

$$Tx = x.$$

Exercise 2.76. Give an example of a set X and a mapping T on X such that
(a) T has *no* fixed points,
(b) T has a *unique* fixed point,
(c) *each* point in X is a fixed point for T.

Remarks 2.49.
–   For a function $f : I \to I$, where I is an interval of the real axis $\mathbb{R}$, finding fixed points is equivalent to finding the x-coordinates of the points of intersection of its graph with the line $y = x$.
–   As is easily seen, if $x \in X$ is a fixed point for a mapping $T : X \to X$, it is also a fixed point for T^n for all $n \in \mathbb{N}$, where T^n is understood as the result of consecutively applying T n times but not vice versa (see *Unique Fixed Point Proposition* (Proposition 2.26, Section 2.18, Problem 33)).

Exercise 2.77. Verify.

2.15.2 Contractions

To proceed, we also need the concept of a *contraction*.

---

11 Stefan Banach (1892–1945).
12 Joseph Liouville (1809–1882).
13 Émile Picard (1856–1941).

Definition 2.33 (Contraction). Let (X, ρ) be a metric space. A Lipschitz continuous mapping $T : X \to X$ with Lipschitz constant $0 \le L < 1$, i. e., such that

$$\exists L \in [0, 1) \; \forall x', x'' \in X : \; \rho(Tx', Tx'') \le L\rho(x', x''),$$

is called a *contraction* on (X, ρ) with *contraction constant L*.

Examples 2.29.
1. On $\mathbb{R}$, the function and $f(x) := mx$, $x \in X$, is a contraction *iff* $|m| < 1$. In particular, the zero function $f(x) := 0$ is a contraction on $\mathbb{R}$.
2. More generally, by the *Mean Value Theorem*, on an interval I of the real axis $\mathbb{R}$, a differentiable function $f : I \to I$ is a contraction *iff*

$$\exists L \in [0, 1) \; \forall x \in I : \; |f'(x)| \le L$$

(see Remarks 2.18).

Definition 2.34 (Weak Contraction). Let (X, ρ) be a metric space. We call a mapping $T : X \to X$ satisfying the condition

$$\rho(Tx, Ty) < \rho(x, y), \; x, y \in X, \; x \ne y,$$

a *weak contraction* (or a *shrinking mapping*) on (X, ρ).

Remark 2.50. Any contraction on a metric space is, clearly, a weak contraction, but, as the subsequent example shows, not vice versa.

Example 2.30. The function

$$f(x) := x + \frac{1}{x}$$

is a *weak contraction*, but *not* a contraction on the complete metric space $X := [1, \infty)$ (see Exercise 2.63).

Exercise 2.78. Verify.

Hint. Apply the *Mean Value Theorem* (see Examples 2.29).

Proposition 2.5 (Uniqueness of Fixed Point for Weak Contraction). *Every weak contraction, in particular every contraction, on a metric space (X, ρ) has at most one fixed point.*

Exercise 2.79.
(a) Prove.
(b) Give an example showing that a weak contraction need not have fixed points.

2.15.3 Banach Fixed-Point Theorem

Theorem 2.33 (Banach Fixed-Point Theorem). *A contraction T with contraction constant $0 \leq L < 1$ on a complete metric space (X, ρ) has a unique fixed point x^*, which can be found as the limit of the recursive sequence of successive approximations*

$$\{T^n x_0\}_{n=0}^{\infty},$$

relative to T for an arbitrary initial point $x_0 \in X$; the error of the nth successive approximation can be estimated by the inequality

$$\rho(x^*, T^n x_0) \leq \frac{L^n}{1-L} \rho(x_0, Tx_0), \quad n \in \mathbb{N}. \tag{2.9}$$

Proof. *Existence* part. Let $x_0 \in X$ be arbitrary. We show that the recursive sequence

$$\{T^n x_0\}_{n=0}^{\infty} \quad (T^0 := I, I \text{ is the } identity \text{ } mapping \text{ on } X)$$

of successive approximations relative to T is *fundamental* in (X, ρ).

Indeed, for any $n, p \in \mathbb{N}$,

$\rho(T^{n+p} x_0, T^n x_0)$ by the *Lipschitz condition* applied n times;

$\leq L\rho(T^{n+p-1} x_0, T^{n-1} x_0) \leq \cdots \leq L^n \rho(T^p x_0, x_0)$

       by the *generalized triangle inequality* (Theorem 2.6);

$\leq L^n [\rho(T^p x_0, T^{p-1} x_0) + \rho(T^{p-1} x_0, T^{p-2} x_0) + \cdots + \rho(Tx_0, x_0)]$

       similarly, by the *Lipschitz condition*;

$\leq L^n [L^{p-1} + L^{p-2} + \cdots + 1] \rho(Tx_0, x_0)$ considering that $0 \leq L < 1$;

$\leq L^n \left[\sum_{k=0}^{\infty} L^k \right] \rho(Tx_0, x_0)$

       by the sum of a convergent geometric series formula and the *symmetry axiom*;

$$= \frac{L^n}{1-L} \rho(x_0, Tx_0), \tag{2.10}$$

which, in view of $0 \leq L < 1$, implies that

$$\sup_{p \in \mathbb{N}} \rho(T^{n+p} x_0, T^n x_0) \leq \frac{L^n}{1-L} \rho(x_0, Tx_0) \to 0, \quad n \to \infty,$$

proving the fundamentality of the sequence $\{T^n x\}_{n=0}^{\infty}$ (see Remark 2.36).

By the *completeness* of the space (X, ρ), the sequence $\{T^n x_0\}_{n=0}^{\infty}$ converges in (X, ρ), i.e.,

$$\exists x^* \in X: \lim_{n \to \infty} T^n x_0 = x^*.$$

By the *continuity* of T on X,

$$Tx^* = T \lim_{n \to \infty} T^n x_0 = \lim_{n \to \infty} T^{n+1} x_0 = x^*.$$

That means, x^* is a fixed point of T.

Uniqueness part. The uniqueness of the fixed point for T follows from the *Uniqueness of Fixed Point for Weak Contraction Proposition* (Proposition 2.5).

Error estimate part. Fixing an arbitrary $n \in \mathbb{N}$ in (2.10) and letting $p \to \infty$, by the *Joint Continuity of Metric* (Theorem 2.15, Section 2.18, Problem 7), we obtain the *error estimate* given by (2.9). □

Remarks 2.51.

-   As is well put in [1], the *Banach Fixed-Point Theorem* meets all requirements for a useful mathematical statement: *existence, uniqueness, construction,* and *error estimate.*
-   The *Banach Fixed Point Theorem* has a *constructive nature* providing an algorithm for finding the fixed point as the limit of the sequence of *successive approximations* starting at an *arbitrary* initial point called the *method of successive approximations.*
-   Error estimate (2.9) in the *Banach Fixed-Point Theorem* affords a *stopping rule* for the iteration process to discontinue the iterations as soon as the desired approximation accuracy has been attained.
    The mapping T being a contraction with contraction constant $0 < L < 1$ on a complete metric space (X, ρ), the *stopping point* for a preassigned accuracy $\varepsilon > 0$ is the smallest $N \in \mathbb{Z}_+$ such that

$$\rho(x^*, T^N x_0) < \varepsilon.$$

By error estimate (2.9), this is attained when

$$\frac{L^N}{1-L} \rho(x_0, Tx_0) < \varepsilon,$$

which is equivalent to finding the smallest $N \in \mathbb{Z}_+$ such that

$$N > \frac{\ln \varepsilon + \ln(1-L) - \ln \rho(x_0, Tx_0)}{\ln L},$$

assuming that $\rho(x_0, Tx_0) > 0$ [55].
-   As the example of the zero function $f(x) := 0$, which is a contraction on the complete metric space $\mathbb{R}$ with contraction constant $L = 0$ and has the unique fixed point $x^* = 0$, shows, it is possible to arrive at the fixed point after a finite number iterations.
-   Each of the two conditions of the *Banach Fixed-Point Theorem* (the *completeness* of the space (X, ρ) and the *contractiveness* of the mapping $T : X \to X$) is essential and cannot be relaxed or dropped (see Section 2.18, Problem 35).

Exercise 2.80. Verify by providing corresponding counterexamples.

2.15.4 Applications

The *Banach Fixed-Point Theorem* (Theorem 2.33) is very instrumental when applied directly or indirectly to various equations, including integral and differential, in particular, to equations, such as

$$e^{-x} = x,$$

which cannot be solved exactly.

2.15.4.1 Equations of the Form $f(x) = x$

Theorem 2.34 (Equations of the Form $f(x) = x$). *If a function $f : I \rightarrow I$, where I is a closed interval of the real axis $\mathbb{R}$, is a contraction on I, then the equation*

$$f(x) = x$$

has a unique solution $x^* \in I$, which can be found as the limit of the sequence of successive approximations relative to f, starting at an arbitrary initial point $x_0 \in I$.

The statement follows directly from the *Banach Fixed-Point Theorem* (Theorem 2.33), considering the completeness of the closed interval I as a subspace of $\mathbb{R}$ (see Exercise 2.63).

Example 2.31. The prior theorem applies to the equation

$$e^{-x} = x$$

on the interval $[\frac{1}{e}, 1]$, but *not* on the interval $[0, 1]$. However, as is seen from the graphs, in this case, the method of successive approximations works for an arbitrary initial point $x_0 \in \mathbb{R}$.

Exercise 2.81.
(a) Verify.

    Hint. Apply the *Mean Value Theorem* (see Examples 2.29).

(b) Graph $y = e^{-x}$ and $y = x$, and construct several successive approximations corresponding to various values of the initial point $x_0 \in \mathbb{R}$.
(c) Is the function $f(x) = e^{-x}$ a contraction on $\mathbb{R}$? How about $f^2(x) := (f \circ f)(x) = e^{-e^{-x}}$?

2.15.4.2 Equations of the Form $F(x) = 0$
To an equation of the form

$$F(x) = 0,$$

the *Banach Fixed-Point Theorem* (Theorem 2.33) can be applied indirectly by replacing it with an equivalent one of the form

$$f(x) = x.$$

Theorem 2.35 (Equations of the Form $F(x) = 0$). *If a function $F \in C^1([a,b], \mathbb{R})$ ($-\infty < a < b < \infty$) satisfies the conditions*
1. $F(a) < 0 < F(b)$,
2. $F'(x) > 0, x \in [a, b]$,

then the equation

$$F(x) = 0$$

has a unique solution $x^* \in [a, b]$, which can be found as the limit of the sequence of successive approximations relative to the contractive function

$$[a, b] \ni x \mapsto f(x) := x - \frac{1}{M}F(x) \in [a, b], \qquad (2.11)$$

where

$$M > \max_{a \leq x \leq b} |f'(x)|,$$

starting at an arbitrary initial point $x_0 \in [a, b]$.

Proof. Considering the continuity of the derivative F' on $[a, b]$, by the *Extreme Value Theorem*, we infer from condition 2 that

$$\exists m, M > 0 : m < F'(x) < M, \ x \in [a, b].$$

Let us fix such m and M and show that the function

$$[a, b] \ni x \mapsto f(x) := x - \frac{1}{M}F(x) \in [a, b]$$

is a *contraction* on $[a, b]$.
    Indeed,

$$\forall x \in [a, b] : \ 0 = 1 - \frac{M}{M} < f'(x) = 1 - \frac{M}{M}F'(x) < 1 - \frac{m}{M} < 1. \qquad (2.12)$$

    Whence, by the *Monotonicity Test*, we infer that the function f is *strictly increasing* on $[a, b]$, which, in view of condition 1, implies that

$$a < a - \frac{1}{M}F(a) = f(a) < f(b) = b - \frac{1}{M}F(b) < b.$$

Hence,

$$\forall x \in [a, b] : \ a < f(a) \leq f(x) \leq f(b) < b,$$

i. e., $f : [a, b] \to [a, b]$.

Furthermore, (2.12) implies that f is a *contraction* on $[a, b]$ (see Examples 2.29). By the prior theorem, the equation

$$f(x) := x - \frac{1}{M} F(x) = x,$$

and hence, the original equation

$$F(x) = 0,$$

which, as is easily seen, is *equivalent* to it, has a unique solution $x^* \in [a, b]$, which can be found as the limit of the sequence of successive approximations, relative to the contractive function f, starting at an arbitrary initial point $x_0 \in [a, b]$. $\qquad \square$

Remarks 2.52.
– Observe that the *existence* of the solution to the equation

$$F(x) = 0$$

follows from condition 1 by the *Intermediate Value Theorem*, and its *uniqueness* follows from condition 2 by the *Monotonicity Test*. However, with the aforementioned theorems saying nothing about how to find the solution, the *Banach Fixed-Point Theorem* (Theorem 2.33) is instrumental in terms of providing an algorithm for approximating the solution by the sequence of successive approximations, relative to the contractive function f defined by (2.11), starting at an arbitrary initial point $x_0 \in [a, b]$.
– The analogue of the prior theorem holds true for a function $F \in C^1([a, b], \mathbb{R})$ ($-\infty < a < b < \infty$), satisfying the "symmetric" conditions:
  1. $F(a) > 0 > F(b)$,
  2. $F'(x) < 0, x \in [a, b]$.

Exercise 2.82. Explain.

Example 2.32. The prior theorem applies to the equation

$$x^2 - 2x = 0$$

on the interval $[3/2, 5/2]$, whose the unique solution $x = 2$ can be found via the *method of successive approximation* as described above.

Exercise 2.83.
(a) Verify.
(b) Can the solution $x = 0$ be found similarly? If yes, how can the interval containing $x = 0$ be chosen?

2.15.4.3 Linear Systems

The *Banach Fixed-Point Theorem* (Theorem 2.33) can also be applied in linear algebra to solve linear systems.

Theorem 2.36 (Linear Systems). *If an $n \times n$ ($n \in \mathbb{N}$) matrix $A := [a_{ij}]$ with real or complex entries satisfies the condition*

$$\lambda := \left[\sum_{i,j=1}^{n} |a_{ij} + \delta_{ij}|^2 \right]^{1/2} < 1, \tag{2.13}$$

where δ_{ij} is the Kronecker delta. Then, for any $b \in \mathbb{R}^n$ or $b \in \mathbb{C}^n$, the linear system

$$Ax = b$$

has a unique solution $x^* \in \mathbb{R}^n$ or $x^* \in \mathbb{C}^n$, which can be found as the limit of the sequence of successive approximations, relative to the contractive mapping

$$Tx := Ax + x - b$$

on the real or complex complete metric space $l_2^{(n)}$, respectively, starting at an arbitrary initial point $x_0 \in \mathbb{R}^n$ or $x_0 \in \mathbb{C}^n$; the error of the nth successive approximation can be estimated by the inequality

$$\rho_2(x^*, T^n x_0) \le \frac{\lambda^n}{1-\lambda} \rho_2(x_0, Tx_0), \quad n \in \mathbb{N}.$$

Proof. Let $\mathbb{F}$ stand for $\mathbb{R}$ or $\mathbb{C}$. As is easily seen, solving the linear system is *equivalent* to finding the fixed-points for the mapping

$$Tx := Ax + x - b$$

on the *complete* metric space $l_2^{(n)}$ (see the *Completeness of the n-Space Theorem* (Theorem 2.23)).

Let us show that T is a *contraction* with contraction constant λ defined by (2.13). Indeed, for any $x := (x_1, \ldots, x_n)$, $y := (y_1, \ldots, y_n) \in \mathbb{F}^n$,

$$\rho_2(Tx, Ty) = \left[\sum_{i=1}^{n} \left| \sum_{j=1}^{n} a_{ij} x_j + x_i - b_i - \sum_{j=1}^{n} a_{ij} y_j - y_i + b_i \right|^2 \right]^{1/2}$$

$$= \left[\sum_{i=1}^{n} \left| \sum_{j=1}^{n} (a_{ij} + \delta_{ij})(x_j - y_j) \right|^2 \right]^{1/2} \le \left[\sum_{i=1}^{n} \left[\sum_{j=1}^{n} |(a_{ij} + \delta_{ij})(x_j - y_j)| \right]^2 \right]^{1/2}$$

by the *Cauchy-Schwarz inequality* (see (2.2)) applied for each $i = 1, \ldots, n$;

$$\le \left[\sum_{i=1}^{n} \sum_{j=1}^{n} |a_{ij} + \delta_{ij}|^2 \sum_{j=1}^{n} |x_j - y_j|^2 \right]^{1/2}$$

$$= \left[\sum_{i,j=1}^{n} |a_{ij} + \delta_{ij}|^2 \right]^{1/2} \left[\sum_{j=1}^{n} |x_j - y_j|^2 \right]^{1/2} = \lambda \rho_2(x,y),$$

where, by condition (2.13), $\lambda \in [0,1)$.

By the *Banach Fixed-Point Theorem* (Theorem 2.33), the contractive mapping T has a unique *fixed point* $x^* \in \mathbb{F}^n$, which is also the unique solution of the original linear system. It can be found as the limit of the sequence of successive approximations, relative to T, starting at any point $x_0 \in \mathbb{F}^n$, the error of the nth successive approximation estimated by the inequality

$$\rho_2(x^*, T^n x_0) \le \frac{\lambda^n}{1-\lambda} \rho_2(x_0, Tx_0), \quad n \in \mathbb{N}. \qquad \square$$

Example 2.33. The matrix

$$A := \begin{bmatrix} -1/2 & 1/2 \\ 0 & -1/2 \end{bmatrix},$$

satisfies condition (2.13) of the prior theorem with

$$\lambda = \sqrt{\frac{3}{4}} = \frac{\sqrt{3}}{2} < 1$$

for any $b \in \mathbb{C}^2$. The linear system

$$Ax = b$$

has a unique solution, which is entirely consistent with the fact $\det A \neq 0$ (see, e. g., [34, 54]).

Remark 2.53. In general, by the prior theorem, we deduce the following interesting linear algebra fact: If an $n \times n$ ($n \in \mathbb{N}$) matrix $A := [a_{ij}]$ with real or complex entries satisfies condition (2.13), then necessarily $\det A \neq 0$ (see, e. g., [34, 54]).

2.15.4.4 Fredholm Integral Equations of the Second Kind

Here, we further extend the realm of applications of the *Banach Fixed-Point Theorem* (Theorem 2.33) to integral equations.

Definition 2.35 (Fredholm Integral Equation). Let $K(\cdot, \cdot)$ be a nonzero continuous function on a closed square $[a,b] \times [a,b]$ ($-\infty < a < b < \infty$), $y \in C[a,b]$, and $\lambda \neq 0$. The equations[14]

$$\int_a^b K(t,s)x(s)\,ds = y(t)$$

---

14 Erik Ivar Fredholm (1866–1927).

and

$$x(t) - \lambda \int_a^b K(t,s)x(s)\, ds = y(t) \qquad (2.14)$$

to be solved relative to the unknown function $x \in C[a,b]$ are, respectively, a *Fredholm integral equation of the first kind* and a *Fredholm integral equation of the second kind*.

Theorem 2.37 (Existence and Uniqueness Theorem for Fredholm Integral Equations of the Second Kind). *Let $K(\cdot,\cdot)$ be a nonzero continuous function on a closed square $[a,b] \times [a,b]$ ($-\infty < a < b < \infty$),*

$$M := \max_{(t,s)\in[a,b]^2} |K(t,s)|,$$

and $\lambda \neq 0$ be such that

$$|\lambda| < \frac{1}{M(b-a)}.$$

Then, for any $y \in C[a,b]$, the Fredholm integral equation of the second kind given by (2.14) has a unique solution $x^* \in C[a,b]$, which can be found as the limit of the sequence of successive approximations, relative to the contractive mapping

$$[Tx](t) := \lambda \int_a^b K(t,s)x(s)\, ds + y(t), \ t \in [a,b], \qquad (2.15)$$

on the complete metric space $(C[a,b],\rho_\infty)$, starting at an arbitrary initial function $x_0 \in C[a,b]$.

Proof. Observe that, by the *Weierstrass Extreme Value Theorem* (Theorem 2.55), the number

$$M := \max_{(t,s)\in[a,b]^2} |K(t,s)|$$

is well defined.

Exercise 2.84. Complete the proof by showing that the mapping defined by (2.15) is a *contraction* on the complete space $(C[a,b],\rho_\infty)$, and then applying the *Banach Fixed-Point Theorem* (Theorem 2.33). □

2.15.4.5 Picard's Existence and Uniqueness Theorem
The following central result of the theory of differential equations can also be proved based on the *Banach Fixed-Point Theorem* (Theorem 2.33).

Theorem 2.38 (Picard's Existence and Uniqueness Theorem). *Let a real-valued function f of two real variables be*

1. *continuous on a closed rectangle*

$$R := [x_0 - a, x_0 + a] \times [y_0 - b, y_0 + b] \subset \mathbb{R}^2,$$

with some $(x_0, y_0) \in \mathbb{R}^2$ *and* $0 < a, b < \infty$, *and*

2. *Lipschitz continuous in y on R with Lipschitz constant* $L > 0$, *i. e.,*

$$\exists L > 0 \; \forall (x, y_1), (x, y_2) \in R : \; |f(x, y_1) - f(x, y_2)| \le L|y_1 - y_2|.$$

Then there exists an $h \in (0, a]$ *such that the initial value problem*

$$\frac{dy}{dx} = f(x, y), \; y(x_0) = y_0, \tag{2.16}$$

has a unique solution on the interval $[x_0 - h, x_0 + h]$, *which can be found as the limit of the sequence of successive approximations, relative to the contractive mapping*

$$[Ty](x) := y_0 + \int_{x_0}^{x} f(t, y(t)) \, dt, \; x \in [x_0 - h, x_0 + h], \tag{2.17}$$

on a closed subspace B of the complete metric space $(C[x_0 - h, x_0 + h], \rho_\infty)$, *starting at the initial function* $y_0(x) \equiv y_0 \in B$.

Proof. Observe that, due to condition 1, by the *Fundamental Theorem of Calculus*, the initial value problem (2.16) on an interval $[x_0 - h, x_0 + h]$ with an arbitrary $0 < h \le a$ is *equivalent* to the integral equation

$$y(x) = y_0 + \int_{x_0}^{x} f(t, y(t)) \, dt \tag{2.18}$$

relative to $y \in C[x_0 - h, x_0 + h]$.

Exercise 2.85. Explain.

In view of condition 1, by the *Weierstrass Extreme Value Theorem* (Theorem 2.55),

$$\exists M > 0 : \; \max_{(x,y) \in R} |f(x, y)| \le M.$$

Let us fix such an M and a number h so that

$$0 < h \le \min\left[a, \frac{b}{M}\right] \quad \text{and} \quad h < \frac{1}{L}. \tag{2.19}$$

The set

$$B := \{y \in C[x_0 - h, x_0 + h] \mid \forall x \in [x_0 - h, x_0 + h] : \; y_0 - b \le y(x) \le y_0 + b, \; y(x_0) = y_0\}$$

is a *closed subset* of the complete space $(C[x_0 - h, x_0 + h], \rho_\infty)$.

Exercise 2.86. Verify.

Hence, by the *Characterization of Completeness* (Theorem 2.27), B is a *complete* metric space relative to the *maximum metric* ρ_∞ (see Examples 2.4).

Now, utilizing condition 2 and estimates (2.19), let us show that the mapping T defined by (2.17) is a *contraction* on the *complete* metric space (B, ρ_∞).

The mapping $T : B \to B$ since, in view of (2.19), for any $y \in B$,

$$\forall x \in [x_0 - h, x_0 + h] : \left|[Ty](x) - y_0\right| = \left|\int_{x_0}^{x} f(t, y(t))\, dt\right| \leq \left|\int_{x_0}^{x} |f(t, y(t))|\, dt\right|$$

$$\leq M|x - x_0| \leq Mh \leq M\frac{b}{M} = b,$$

and

$$[Ty](x_0) = y_0.$$

The mapping T is also a *contraction* on B since, for any $y_1, y_2 \in B$,

$$\rho_\infty(Ty_1, Ty_2) = \max_{x_0 - h \leq x \leq x_0 + h} \left|[Ty_1](x) - [Ty_2](x)\right|$$

$$= \max_{x_0 - h \leq x \leq x_0 + h} \left|\int_{x_0}^{x} [f(t, y_1(t)) - f(t, y_2(t))]\, dt\right|$$

$$\leq \max_{x_0 - h \leq x \leq x_0 + h} \left|\int_{x_0}^{x} |f(t, y_1(t)) - f(t, y_2(t))|\, dt\right| \qquad \text{by condition 2;}$$

$$\leq \max_{x_0 - h \leq x \leq x_0 + h} \left|\int_{x_0}^{x} L|y_1(t) - y_2(t)|\, dt\right|$$

$$\leq \max_{x_0 - h \leq x \leq x_0 + h} \left[\max_{x_0 - h \leq t \leq x_0 + h} L|y_1(t) - y_2(t)||x - x_0|\right] = Lh\rho_\infty(y_1, y_2),$$

where, by (2.19), $0 < Lh < 1$.

By the *Banach Fixed-Point Theorem* (Theorem 2.33), the contraction T has a *unique fixed point* in B, which is the *unique solution* on the interval $[x_0 - h, x_0 + h]$ of integral equation (2.18), and hence, of the equivalent initial value problem given by (2.16); it can be found as the *uniform limit* on $[x_0 - h, x_0 + h]$ (see Examples 2.5) of the sequence of successive approximations

$$y_0(x) \equiv y_0, \ y_n(x) := [Ty_{n-1}](x) = y_0 + \int_{x_0}^{x} f(t, y_{n-1}(t))\, dt, \ n \in \mathbb{N}. \qquad \square$$

Remarks 2.54.

− *Picard's Existence and Uniqueness Theorem* applies when a real-valued function $f(\cdot, \cdot)$ and its partial derivative $f_y(\cdot, \cdot)$ relative to y are both continuous on a closed rectangle:

$$R := [x_0 - a, x_0 + a] \times [y_0 - b, y_0 + b],$$

with some $(x_0, y_0) \in \mathbb{R}^2$ and $0 < a, b < \infty$.

Exercise 2.87. Explain.

− The condition of *continuity* in x and y is essential for the *local existence* (see *Peano's Existence Theorem* (Theorem 2.60)), whereas the additional condition of *Lipschitz continuity* in y is essential for the *uniqueness* of the initial-value problem solutions (see Remark 2.82 and Example 2.46).

The following statement is an immediate corollary:

Theorem 2.39 (Picard's Global Existence and Uniqueness Theorem). *Let I be an open interval containing a point x_0 and a real-valued function f of two real variables be*
1. *continuous on the slab*

$$S := I \times \mathbb{R} \subseteq \mathbb{R}^2$$

and
2. *Lipschitz continuous in y on S with Lipschitz constant $L > 0$, i. e.,*

$$\exists L > 0 \,\forall\, (x, y_1), (x, y_2) \in S : \ |f(x, y_1) - f(x, y_2)| \le L|y_1 - y_2|.$$

Then, for any $x_0 \in I$ and $y_0 \in \mathbb{R}$, the initial value problem

$$\frac{dy}{dx} = f(x, y), \ y(x_0) = y_0,$$

has a unique solution on the entire interval I.

Picard's Existence and Uniqueness Theorem (Theorem 2.38) and *Picard's Global Existence and Uniqueness Theorem* (Theorem 2.39) can be generalized to systems of differential equations as follows:

Theorem 2.40 (Picard's Existence and Uniqueness Theorem for Systems of Differential Equations). *Let a vector function f of $n + 1$ ($n \in \mathbb{N}$) variables with values in $\mathbb{R}^n$ be*
1. *continuous on a region*

$$R := \{(x, y) \mid |x - x_0| \le a, \ \rho_p(y, y_0) \le b\} \subset \mathbb{R}^{n+1},$$

with some $x_0 \in \mathbb{R}$, $y_0 \in \mathbb{R}^n$, and $0 < a, b < \infty$, and

2. *Lipschitz continuous in y on R with Lipschitz constant L > 0, i. e.,*

$$\exists L > 0 \; \forall \, (x, y_1), (x, y_2) \in R : \; \rho_p(f(x, y_1), f(x, y_2)) \leq L\rho_p(y_1, y_2),$$

where ρ_p is p-metric on $\mathbb{R}^n$ ($1 \leq p \leq \infty$).

Then there exists an $h \in (0, a]$ such that the initial value problem

$$\frac{dy}{dx} = f(x, y), \; y(x_0) = y_0,$$

relative to the n-component unknown function $y(x) := (y_1(x), \ldots, y_n(x))$, has a unique solution on the interval $[x_0 - h, x_0 + h]$, which can be found as the uniform limit of the sequence of successive approximations

$$y_0(x) = y_0, \; y_k(x) := [Ty_{k-1}](x) = y_0 + \int_{x_0}^{x} f(t, y_{k-1}(t)) \, dt, \; k \in \mathbb{N},$$

on $[x_0 - h, x_0 + h]$, with differentiation, integration, and convergence understood in the componentwise sense.

Theorem 2.41 (Picard's Global Existence and Uniqueness Theorem for Systems of Differential Equations). *Let I be an open interval containing a point x_0 and a vector function f of $n + 1$ ($n \in \mathbb{N}$) variables with values in $\mathbb{R}^n$ be*
1. *continuous on the set*

$$S := I \times \mathbb{R}^n \subseteq \mathbb{R}^{n+1}$$

   and
2. *Lipschitz continuous in y on R with Lipschitz constant L > 0, i. e.,*

$$\exists L > 0 \; \forall \, (x, y_1), (x, y_2) \in R : \; \rho_p(f(x, y_1), f(x, y_2)) \leq L\rho_p(y_1, y_2),$$

where ρ_p is p-metric on $\mathbb{R}^n$ ($1 \leq p \leq \infty$).

Then, for any $x_0 \in I$ and $y_0 \in \mathbb{R}^n$, the initial value problem

$$\frac{dy}{dx} = f(x, y), \; y(x_0) = y_0,$$

relative to the n-component unknown function $y(x) := (y_1(x), \ldots, y_n(x))$, with differentiation understood in the componentwise sense, has a unique solution on the entire interval I.

Remark 2.55. Due to the bi-Lipschitz equivalence of all p-metrics on $\mathbb{R}^n$ (see Examples 2.18), the value of $1 \leq p \leq \infty$ in the two prior theorems is superfluous.

The statement that follows is an important particular case.

Theorem 2.42 (Global Existence and Uniqueness Theorem for Constant-Coefficient Homogeneous Linear Systems of Differential Equations). *Let $A := [a_{ij}]$ be a $n \times n$ $(n \in \mathbb{N})$ matrix with real or complex entries. Then, for any initial values $t_0 \in \mathbb{R}$ and $y_0 \in \mathbb{R}^n$ or $y_0 \in \mathbb{C}^n$, the initial value problem*

$$y'(t) = Ay(t),\ y(0) = y_0,$$

relative to the n-component unknown function $y(t) := (y_1(t), \dots, y_n(t))$, with differentiation understood in the componentwise sense, has a unique solution on $\mathbb{R}$ given by the exponential formula

$$y(t) = e^{tA}y_0 := \left(\sum_{k=0}^{\infty} \frac{t^k A^k}{k!} \right) y_0,\ t \in \mathbb{R}. \tag{2.20}$$

In the latter formula, $A^0 := I$ is the $n \times n$ identity matrix, the convergence of the matrix series understood in the entrywise sense.

Exercise 2.88. Prove via the termwise differentiation of the matrix series (see Section 5.5.4).

Hint. To demonstrate the Lipschitz continuity of

$$f(t, y) := Ay \in l_2^{(n)},\ t \in \mathbb{R}, y \in \mathbb{R}^n \text{ or } y \in \mathbb{C}^n,$$

in y, use reasoning based on the *Cauchy-Schwarz inequality* (see (2.2)) similar to that found in the proof of the *Linear Systems Theorem* (Theorem 2.36).

Examples 2.34.
1. The initial value problem

$$y'(t) = Ay(t),\ y(0) = y_0,$$

with

$$A := \begin{bmatrix} 0 & 0 \\ 0 & 2 \end{bmatrix},$$

in view of

$$A^n = \begin{bmatrix} 0^n & 0 \\ 0 & 2^n \end{bmatrix},\ n \in \mathbb{N},$$

has the unique solution

$$y(t) = \left(\sum_{n=0}^{\infty} \frac{t^n}{n!} \begin{bmatrix} 0 & 0 \\ 0 & 2 \end{bmatrix}^n \right) y_0 = \begin{bmatrix} e^{0t} & 0 \\ 0 & e^{2t} \end{bmatrix} y_0 = \begin{bmatrix} 1 & 0 \\ 0 & e^{2t} \end{bmatrix} y_0,\ t \in \mathbb{R},$$

on $\mathbb{R}$ for any $y_0 \in \mathbb{R}^2$ or $y_0 \in \mathbb{C}^2$.

2. The initial value problem

$$y'(t) = Ay(t), \ y(0) = y_0,$$

with

$$A := \begin{bmatrix} 0 & 1 \\ 0 & 0 \end{bmatrix},$$

in view of

$$A^n = \begin{bmatrix} 0 & 0 \\ 0 & 0 \end{bmatrix}, \ n = 2, 3, \ldots,$$

has the unique solution

$$y(t) = \left(\begin{bmatrix} 1 & 0 \\ 0 & 1 \end{bmatrix} + t \begin{bmatrix} 0 & 1 \\ 0 & 0 \end{bmatrix} \right) y_0 = \begin{bmatrix} 1 & t \\ 0 & 1 \end{bmatrix} y_0, \ t \in \mathbb{R},$$

on $\mathbb{R}$ for any $y_0 \in \mathbb{R}^2$ or $y_0 \in \mathbb{C}^2$.

3. The initial value problem

$$y'(t) = Ay(t), \ y(0) = y_0,$$

with

$$A := \begin{bmatrix} 0 & 1 \\ -1 & 0 \end{bmatrix},$$

in view of

$$A^{2n} = (-1)^n I, \text{ and } A^{2n-1} = (-1)^{n-1} A, \ n \in \mathbb{N}, \tag{2.21}$$

has the unique solution

$$y(t) = \left(\sum_{n=0}^{\infty} \frac{(-1)^n t^{2n}}{(2n)!} I + \sum_{n=1}^{\infty} \frac{(-1)^{n-1} t^{2n-1}}{(2n-1)!} A \right) y_0 = \begin{bmatrix} \cos t & \sin t \\ -\sin t & \cos t \end{bmatrix} y_0, \ t \in \mathbb{R},$$

on $\mathbb{R}$ for any $y_0 \in \mathbb{R}^2$ or $y_0 \in \mathbb{C}^2$.

Exercise 2.89. Verify (2.21).

2.16 Compactness

The notion of *compactness*, naturally emerging from our spatial intuition, is of utmost importance both in theory and applications. We study it and related concepts here.

2.16.1 Total Boundedness

2.16.1.1 Definitions and Examples

Total boundedness is a notion inherent to metric spaces that is stronger than *boundedness*, but weaker than *precompactness*.

Definition 2.36 (ε-Net). Let (X,ρ) be a metric space and $\varepsilon > 0$. A set $N_\varepsilon \subseteq X$ is called an ε-net for a set $A \subseteq X$ if A can be *covered* by the collection $\{B(x,\varepsilon) \mid x \in N_\varepsilon\}$ of all open ε-balls centered at the points of N_ε, i. e.,

$$A \subseteq \bigcup_{x \in N_\varepsilon} B(x,\varepsilon).$$

Examples 2.35.
1. For an arbitrary nonempty set A in a metric space (X,ρ) and any $\varepsilon > 0$, A is an ε-net for itself.
2. A *dense set A* in a *metric space (X,ρ)* is an ε-net for the entire X with any $\varepsilon > 0$. In particular, for any $\varepsilon > 0$, $\mathbb{Q}$ and $\mathbb{Q}^c$ are ε-nets for $\mathbb{R}$.
3. For any $\varepsilon > 0$, $\mathbb{Q}$ is an ε-net for $\mathbb{Q}^c$; $\mathbb{Q}^c$ is also an ε-net for $\mathbb{Q}$.
4. For any $n \in \mathbb{N}$ and $\varepsilon > 0$, the set

$$N_\varepsilon := \left\{ \left(\frac{\varepsilon k_1}{\sqrt{n}}, \ldots, \frac{\varepsilon k_n}{\sqrt{n}} \right) \middle| k_i \in \mathbb{Z}, \ i = 1, \ldots, n \right\}$$

is an ε-net for $\mathbb{R}^n$ with the *Euclidean metric*, i. e., for $l_2^{(n)}(\mathbb{R})$.

Exercise 2.90. Verify and make a drawing for $n = 2$.

Remark 2.56. Example 4 can be modified to furnish an ε-net for $\mathbb{R}^n$ endowed with any p-metric $(1 \le p \le \infty)$, i. e., for $l_p^{(n)}(\mathbb{R})$, and can be naturally stretched to the *complex n-space* $\mathbb{C}^n$ relative to any p-metrics $(1 \le p \le \infty)$. That implies to any space $l_p^{(n)}(\mathbb{C})$.

Exercise 2.91. Verify for $n = 2$, $p = 1$, and for $n = 2$, $p = \infty$.

Remark 2.57. As is seen from the prior examples, an ε-net N_ε for a set A in a metric space (X,ρ) need not consist of points of A. It may even happen that A and N_ε are *disjoint*.

Definition 2.37 (Total Boundedness). A set A in a metric space (X,ρ) is called *totally bounded* if, for any $\varepsilon > 0$, there exists a *finite* ε-net for A:

$$\forall \varepsilon > 0 \ \exists N \in \mathbb{N}, \ \exists \{x_1, \ldots, x_N\} \subseteq X : A \subseteq \bigcup_{n=1}^{N} B(x_n, \varepsilon),$$

i. e., A can be covered by a finite number of open balls with arbitrarily small their common radius.

A metric space (X, ρ) is called *totally bounded* if the set X is totally bounded in (X, ρ).

Remarks 2.58.
- A set A is totally bounded in a metric space (X, ρ) *iff* totally bounded is its closure $\overline{A}$.
- For a totally bounded set A in a metric space (X, ρ) and any $\varepsilon > 0$, a finite ε-net for A can be chosen to consist entirely of points of A.
- The total boundedness of a nonempty set A in a metric space (X, ρ) is equivalent to the total boundedness of (A, ρ) as a subspace of (X, ρ).

Exercise 2.92. Verify.

Hint. Consider a finite $\varepsilon/2$-net $\{x_1, \ldots, x_N\} \subseteq X$ ($N \in \mathbb{N}$) for A, and construct an ε-net $\{y_1, \ldots, y_N\} \subseteq A$ for A.

Examples 2.36.
1. *Finite sets*, including the empty set $\emptyset$, are *totally bounded* in an arbitrary metric space (X, ρ), and only finite sets are totally bounded in a *discrete* metric space (X, ρ_d).
2. A *bounded* set A in the n-space $\mathbb{R}^n$ ($n \in \mathbb{N}$) with the *Euclidean metric*, i. e., in $l_2^{(n)}(\mathbb{R})$, is *totally bounded*.

   Indeed, being *bounded*, the set A is contained in a *hypercube*

   $$J_m = [-m, m]^n$$

   with some $m \in \mathbb{N}$. Then, for any $\varepsilon > 0$, $N_\varepsilon' := N_\varepsilon \cap J_m$, where N_ε is the ε-net for $l_2^{(n)}(\mathbb{R})$ from Examples 2.35, is a *finite* ε-net for A.

   Remark 2.59. The same is true for any (real or complex) $l_p^{(n)}$ ($1 \le p \le \infty$) (see Remark 2.56).

   In particular, $(0, 1]$ and $\{1/n\}_{n \in \mathbb{N}}$ are *totally bounded* sets in $\mathbb{R}$.
3. The set $E := \{e_n := (\delta_{nk})_{k \in \mathbb{N}}\}_{n \in \mathbb{N}}$, where δ_{nk} is the *Kronecker delta*, is *bounded*, but *not totally bounded* in (c_{00}, ρ_∞) (also in (c_0, ρ_∞), (c, ρ_∞), and l_∞) since there is no finite $1/2$-net for E.

   The same example works in l_p ($1 \le p < \infty$), in which case there is no finite $2^{1/p-1}$-net for E.

Exercise 2.93. Verify.

2.16.1.2 Properties and Characterizations

Theorem 2.43 (Properties of Totally Bounded Sets). *The totally bounded sets in a metric space (X, ρ) have the following properties:*

(1) *a totally bounded set is necessarily bounded, but not vice versa;*
(2) *a subset of a totally bounded set is totally bounded;*
(3) *an arbitrary intersection of totally bounded sets is totally bounded;*
(4) *a finite union of totally bounded sets is totally bounded.*

Exercise 2.94.
(a) Prove.
(b) Give an example showing that an infinite union of totally bounded sets need not be totally bounded.

Exercise 2.95. Using the set E from Examples 2.36, show that a nontrivial sphere/ball in (c_{00}, ρ_∞) (also in (c_0, ρ_∞), (c, ρ_∞), and l_p $(1 \le p \le \infty)$) is *not totally bounded*.

From the prior proposition and Examples 2.36, we obtain

Corollary 2.6 (Characterization of Total Boundedness in the n-Space). *A set A is totally bounded in the (real or complex) space $l_p^{(n)}$ $(n \in \mathbb{N}, 1 \le p \le \infty)$ iff it is bounded.*

It is remarkable that *total boundedness* in a metric space can be characterized in terms of *fundamentality* as follows:

Theorem 2.44 (Characterization of Total Boundedness). *A nonempty set A is totally bounded in a metric space (X, ρ) iff every sequence $(x_n)_{n \in \mathbb{N}}$ in A contains a fundamental subsequence $(x_{n(k)})_{k \in \mathbb{N}}$.*

Proof. "*Only if*" part. Suppose a set A is totally bounded in a metric space (X, ρ) and let $(x_n)_{n \in \mathbb{N}}$ be an arbitrary sequence in A.

If the set A is *finite*, i. e., $A = \{y_1, \dots, y_N\}$ with some $N \in \mathbb{N}$, then $(x_n)_{n \in \mathbb{N}}$ necessarily assumes the same value y_i with some $i = 1, \dots, N$ for infinitely many indices $n \in \mathbb{N}$ (i. e., "*frequently*").

Exercise 2.96. Explain.

Hence, $(x_n)_{n \in \mathbb{N}}$ contains a *constant* subsequence, which is *fundamental*.
Now, assume that the set A is *infinite* and let $(x_n)_{n \in \mathbb{N}}$ be an arbitrary sequence in A. If $(x_n)_{n \in \mathbb{N}}$ assumes only a finite number of distinct values, we arrive at the prior case.
Suppose that $(x_n)_{n \in \mathbb{N}}$ assumes infinite many distinct values. Then, without loss of generality, we can regard all values of the sequence as distinct, i. e.,

$$x_m \ne x_n, \ m, n \in \mathbb{N}, m \ne n.$$

Exercise 2.97. Explain.

By the total boundedness of A, since it is coverable by a finite number of 1-balls, there must exist a 1-ball $B(y_1, 1)$ with some $y_1 \in X$, which contains *infinitely many* terms of $(x_n)_{n \in \mathbb{N}}$, and hence, a subsequence $(x_{1,n})_{n \in \mathbb{N}}$ of $(x_n)_{n \in \mathbb{N}}$.

Similarly, there must exist a 1/2-ball $B(y_2, 1/2)$ with some $y_2 \in X$, which contains *infinitely many* terms of $(x_{1,n})_{n\in\mathbb{N}}$, and hence, a subsequence $(x_{2,n})_{n\in\mathbb{N}}$ of $(x_{1,n})_{n\in\mathbb{N}}$. Continuing inductively, we obtain a countable collection of sequences

$$\{(x_{m,n})_{n\in\mathbb{N}} \mid m \in \mathbb{Z}_+\}$$

such that

$$(x_{0,n})_{n\in\mathbb{N}} := (x_n)_{n\in\mathbb{N}}$$

for each $m \in \mathbb{N}$, $(x_{m,n})_{n\in\mathbb{N}}$ is a subsequence of $(x_{(m-1),n})_{n\in\mathbb{N}}$, and

$$(x_{m,n})_{n\in\mathbb{N}} \subseteq B(y_m, 1/m)$$

with some $y_m \in X$.

Then the *"diagonal subsequence"* $(x_{n,n})_{n\in\mathbb{N}}$ is a *fundamental subsequence* of $(x_n)_{n\in\mathbb{N}}$ since, by the *triangle inequality*,

$$\forall\, n \in \mathbb{N}, \forall\, m \geq n : \rho(x_{m,m}, x_{n,n}) \leq \rho(x_{m,m}, y_n) + \rho(y_n, x_{n,n}) < 1/n + 1/n = 2/n.$$

"If" part. Let us prove this part *by contrapositive*, assuming that a set A is *not* totally bounded in a metric space (X, ρ). Then there is an $\varepsilon > 0$ such that there does not exist a finite ε-net for A.

In particular, for an arbitrary $x_1 \in A$, the ε-ball $B(x_1, \varepsilon)$ does not cover A, i. e.,

$$A \nsubseteq B(x_1, \varepsilon),$$

and hence,

$$\exists\, x_2 \in A : x_2 \notin B(x_1, \varepsilon).$$

Similarly, the ε-balls $B(x_1, \varepsilon)$ and $B(x_2, \varepsilon)$ do not cover A, i. e.,

$$A \nsubseteq B(x_1, \varepsilon) \cup B(x_2, \varepsilon),$$

and hence,

$$\exists\, x_3 \in A : x_3 \notin B(x_1, \varepsilon) \cup B(x_2, \varepsilon).$$

Continuing inductively, we obtain a sequence $(x_n)_{n\in\mathbb{N}}$ of points of A such that

$$x_n \notin \bigcup_{k=1}^{n-1} B(x_k, \varepsilon), \; n \geq 2,$$

i. e., whose distinct terms are at least at distance ε from each other:

$$\rho(x_m, x_n), \; m, n \in \mathbb{N}, m \neq n.$$

Clearly, this sequence has no fundamental subsequence, which completes the proof by contrapositive. $\qquad\square$

Based on the prior characterization and *Fundamentality and Uniform Continuity Proposition* (Proposition 2.2), one can easily prove the following statement:

Proposition 2.6 (Total Boundedness and Uniform Continuity). *Let (X,ρ) and (Y,σ) be metric spaces and a function $f : X \to Y$ be uniformly continuous on X. If A is a totally bounded set in (X,ρ), then $f(A)$ is a totally bounded set in (Y,σ), i.e., uniformly continuous functions map totally bounded sets to totally bounded sets.*

Exercise 2.98.

(a) Prove.

(b) Give an example showing that a continuous function need not preserve total boundedness.

Proposition 2.7 (Total Boundedness and Separability). *A totally bounded metric space is separable.*

Proof. Let (X,ρ) be a totally bounded metric space. Then, for any $n \in \mathbb{N}$, there is a finite $1/n$-net N_n for X.

The union $\bigcup_{n=1}^{\infty} N_n$ is a *countable dense set* in (X,ρ).

Exercise 2.99. Explain. □

Remarks 2.60.

– Thus, a nonseparable metric space, e. g., l_∞, is not totally bounded, which also follows from the fact that it is unbounded.

– As the example of the space $\mathbb{R}$ with the usual metric shows, the converse statement is not true, i. e., a separable metric space need not be totally bounded.

– A totally bounded metric space need not be complete. For instance, the space $X := \{1/n\}_{n\in\mathbb{N}}$ with the usual metric ρ is totally bounded, but incomplete. The same is true for the space

$$X := \left\{ \sum_{k=0}^{n} \frac{x^k}{k!} \,\middle|\, n \in \mathbb{N}, x \in [a, b] \right\}$$

of the partial sums of the Maclaurin series of e^x as a subspace of $(C[a,b],\rho_\infty)$ $(-\infty < a < b < \infty)$.

– Similarly to boundedness and completeness, total boundedness is *not a topological property* of the space, i. e., it is *not homeomorphically invariant*.

Indeed, as is discussed in Remark 2.39, for $X := \{1/n\}_{n\in\mathbb{N}}$, the spaces (X,ρ) and (X,ρ_d), where ρ is the regular metric and ρ_d is the discrete metric, are *homeomorphic* relative the identity mapping $Ix := x$. However, the former space is totally bounded (but incomplete), whereas the latter is not totally bounded (but complete).

Exercise 2.100. Explain.

2.16.2 Compactness, Precompactness

2.16.2.1 Definitions and Examples

Definition 2.38 (Cover, Subcover, Open Cover). A collection $\mathscr{C} := \{C_i\}_{i \in I}$ of subsets of a nonempty set X is said to be a *cover* of (for) a set $A \subseteq X$ (or *to cover A*) if

$$A \subseteq \bigcup_{i \in I} C_i. \tag{2.22}$$

A subcollection $\mathscr{C}'$ of a cover $\mathscr{C}$ of A, which is also a cover of A, is called a *subcover* of $\mathscr{C}$.

If (X, ρ) is a metric space, a cover of a set $A \subseteq X$, consisting of open sets is called an *open cover* of A.

Remark 2.61. In particular, when $A = X$, (2.22) acquires the form

$$X = \bigcup_{i \in I} C_i.$$

Examples 2.37.
1. An arbitrary set $A \subseteq X$ has a cover. E. g., $\mathscr{C} := \{A\}$ is a finite *cover* for A or, provided $A \neq \emptyset$, $\mathscr{C} := \{\{x\}\}_{x \in A}$ is also a *cover* for A.
2. The collection $\{[n, n + 1)\}_{n \in \mathbb{Z}}$ is a *cover* for $\mathbb{R}$.
3. The collection $\{(n, n + 1)\}_{n \in \mathbb{Z}}$ is *not* a *cover* for $\mathbb{Z}$.
4. The collection of all concentric open balls in a metric space (X, ρ) centered at a fixed point $x \in X$:

$$\{B(x, r) \mid r > 0\},$$

   is an *open cover* of X, the subcollection

$$\{B(x, n) \mid n \in \mathbb{N}\}$$

   being its countable *subcover*.
5. Let A be a *dense set* in a metric space (X, ρ). For any $\varepsilon > 0$, the collection of ε-balls:

$$\{B(x, \varepsilon) \mid x \in A\},$$

   is an open cover of X, i. e., A is an ε-net for X (see Definition 2.36 and Examples 2.35).
6. Let $\{r_n\}_{n \in \mathbb{N}}$ be a countably infinite subset of $\mathbb{R}$. The collection of intervals

$$\{[r_n - 1/2^{n+1}, r_n + 1/2^{n+1}] \mid n \in \mathbb{N}\}$$

   does not cover $\mathbb{R}$. This is true even when the set is *dense* in $\mathbb{R}$, as is the case, e. g., for $\mathbb{Q}$.

Exercise 2.101. Verify.

Definition 2.39 (Compactness). A set A is said to be *compact* in a metric space (X, ρ) if each open cover $\mathcal{O}$ of A contains a finite subcover $\mathcal{O}'$.

A metric space (X, ρ) is called *compact* if the set X is compact in (X, ρ).

Remarks 2.62.
- Compactness, in the sense of the prior definition, is also called compactness in the *Heine*[15]*–Borel*[16] *sense*.
- Being formulated entirely in terms of open sets and union, *compactness* is a *topological property*, i. e., is *homeomorphically invariant*.
- The compactness of a nonempty set A in a metric space (X, ρ) is equivalent to the compactness of (A, ρ) as a subspace of (X, ρ).

Examples 2.38.
1. *Finite sets*, including the empty set $\emptyset$, are *compact* in an arbitrary metric space (X, ρ), and only finite sets are compact in a *discrete* metric space (X, ρ_d).
2. The sets $[0, \infty)$, $(0, 1]$, and $\{1/n\}_{n \in \mathbb{N}}$ are *not compact* in $\mathbb{R}$, and the set $\{0\} \cup \{1/n\}_{n \in \mathbb{N}}$ is.
3. The set $E := \{e_n := (\delta_{nk})_{k \in \mathbb{N}}\}_{n \in \mathbb{N}}$, where δ_{nk} is the *Kronecker delta*, is *closed* and *bounded*, but *not compact* in (c_{00}, ρ_∞) (also in (c_0, ρ_∞), (c, ρ_∞), and l_∞) since its open cover by $1/2$-balls:

$$\{B(e_n, 1/2)\}_{n \in \mathbb{N}},$$

has no finite subcover.

Observe that E is also *not totally bounded* (see Examples 2.36).

The same example works in l_p $(1 \le p < \infty)$, in which case the open cover of E by $2^{1/p-1}$-balls:

$$\{B(e_n, 2^{1/p-1})\}_{n \in \mathbb{N}},$$

has no finite subcover.

Exercise 2.102. Verify.

Definition 2.40 (Precompactness). A set A is said to be *precompact* (also *relatively compact*) in a metric space (X, ρ) if its closure $\overline{A}$ is compact.

Remarks 2.63.
- For a closed set in a metric space (X, ρ), in particular for X, precompactness is equivalent to compactness.

---

15 Heinrich Heine (1821–1881).
16 Émile Borel (1871–1956).

- Being formulated entirely in terms of closure and compactness, *precompactness* is a *topological property*, i. e., is *homeomorphically invariant*.

Examples 2.39.
1. The set $(0, \infty)$ is *not precompact* in $\mathbb{R}$.
2. The set $\{1/n\}_{n \in \mathbb{N}}$ is *precompact*, but not compact in $\mathbb{R}$, and the same is true for the set $\{\sum_{k=0}^{n} \frac{x^k}{k!} \mid n \in \mathbb{N}, x \in [a, b]\}$ of the partial sums of the Maclaurin series of e^x in $(C[a, b], \rho_\infty)$ $(-\infty < a < b < \infty)$.

Exercise 2.103. Verify.

2.16.2.2 Properties

Theorem 2.45 (Properties of Compact Sets). *The compact sets in a metric space (X, ρ) have the following properties:*
(1) *a compact set is necessarily totally bounded, but not vice versa;*
(2) *a compact set is necessarily closed, but not vice versa;*
(3) *a closed subset of a compact set is compact. In particular, a closed set in a compact metric space (X, ρ) is compact;*
(4) *an arbitrary intersection of compact sets is compact;*
(5) *a finite union of compact sets is compact.*

Proof. Here, we prove properties (2) and (3) only, proving properties (1), (4), and (5) is left to the reader as an exercise.

(2) Let A be an arbitrary compact set in a metric space (X, ρ).

If $A = \emptyset$ or $A = X$, then A is, obviously, closed.

Suppose that A is a nontrivial proper subset of X and let $x \in A^c$ be arbitrary. Then, for any $y \in A$, by the *Separation Property* (Proposition 2.1), there are *disjoint* open balls $B(x, r(y))$ and $B(y, r(y))$ (e. g., $r(y) := \rho(x, y)/2$).

By the *compactness* of A, its open cover:

$$\{B(y, r(y))\}_{y \in A},$$

contains a finite subcover $\{B(y_1, r(y_1)), \ldots, B(y_N, r(y_N))\}$ with some $N \in \mathbb{N}$.

Then the open set

$$V := \bigcup_{k=1}^{N} B(y_k, r(y_k))$$

contains A and is *disjoint* from the open ball centered at x:

$$B(x, r) = \bigcap_{k=1}^{N} B(x, r(y_k)), \text{ where } r := \min_{1 \le k \le n} r(y_k).$$

Hence,

$$B(x,r) \cap V = \emptyset, \tag{2.23}$$

and the more so,

$$B(x,r) \cap A = \emptyset.$$

That means, x is an *exterior* point of A, which proves that the complement A^c is *open* in (X,ρ), implying that by the *Characterizations of Closed Sets* (Theorem 2.19), the set A is *closed*.

A closed set in a metric space need not be compact. For instance, the set $[0,\infty)$ is closed, but not compact in $\mathbb{R}$ (see Examples 2.38).

(3) Let B be a closed subset of a compact set A in (X,ρ) and $\mathcal{O} := \{O_i\}_{i \in I}$ be an arbitrary *open cover* of B.

Then, by the *Characterizations of Closed Sets* (Theorem 2.19), B^c is an *open set*, and $\mathcal{O} \cup \{B^c\}$ is an open cover of A, which, by the compactness of A, has a *finite subcover*. Thus, there exists a *finite subcollection* $\mathcal{O}'$ of $\mathcal{O}$ such that $\mathcal{O}' \cup \{B^c\}$ is an *open cover of A*, which, in view of

$$B \cap B^c = \emptyset,$$

implies that $\mathcal{O}'$ is a finite subcover of B, proving its compactness. $\qquad\square$

Exercise 2.104.
(a) Prove (1), (4), and (5).
(b) Give an example showing that an infinite union of compact sets need not be compact.

Remark 2.64. From property (1) and Exercise 2.95, we infer that a nontrivial sphere/ball in (c_{00}, ρ_∞) (also in (c_0, ρ_∞), (c, ρ_∞), and l_p ($1 \le p \le \infty$)) is *not compact*.

From property (1), in view of Remarks 2.58, we infer

Proposition 2.8 (Precompactness and Total Boundedness). *A precompact set A in metric space (X,ρ) is totally bounded, but not vice versa.*

Example 2.40. The set $\{1/n\}_{n \in \mathbb{N}}$ is *precompact* in the complete metric space $\mathbb{R}$ (see Example 2.39). However, the same set, while *totally bounded* (see Example 2.36), is *not precompact* in the incomplete subspace $\mathbb{R} \setminus \{0\}$ of $\mathbb{R}$.

Exercise 2.105. Verify.

Remark 2.65. The reason why the *completeness* of the underlying space is important becomes apparent from the *Hausdorff Criterion* (Theorem 2.46) (see also Remarks 2.67).

From property (2), we also arrive at the following:

Proposition 2.9 (Characterization of Compactness). *A set A in a metric space (X, ρ) is compact iff it is closed and precompact.*

From property (1), and the *Total Boundedness and Separability Proposition* (Proposition 2.7), we obtain the following statement:

Proposition 2.10 (Compactness and Separability). *A compact metric space is separable.*

Furthermore,

Proposition 2.11 (Compactness and Completeness). *A compact metric space is complete.*

Proof. Let us prove the statement *by contradiction*, assuming that there exists a *compact* metric space (X, ρ) that is *incomplete*. Then there exists a *fundamental* sequence $(x_n)_{n \in \mathbb{N}}$ with no limit in (X, ρ). Then

$$\forall x \in X \; \exists \varepsilon = \varepsilon(x) > 0 \; \forall N \in \mathbb{N} \; \exists n \geq N : \; \rho(x, x_n) \geq \varepsilon. \tag{2.24}$$

That means, the sequence $(x_n)_{n \in \mathbb{N}}$ is *frequently* not in the ball $B(x, \varepsilon(x))$.

Exercise 2.106. Explain.

On the other hand, by the fundamentality of $(x_n)_{n \in \mathbb{N}}$,

$$\exists M = M(\varepsilon(x)) \in \mathbb{N} \; \forall m, n \geq M : \; \rho(x_m, x_n) < \varepsilon/2. \tag{2.25}$$

Fixing an $n \geq M$, for which (2.24) holds, by (2.24), (2.25), and *triangle inequality*, we have

$$\forall m \geq M : \; \varepsilon \leq \rho(x, x_n) \leq \rho(x, x_m) + \rho(x_m, x_n) < \rho(x, x_m) + \varepsilon/2,$$

and hence,

$$\forall m \geq M : \; \rho(x, x_m) \geq \varepsilon/2.$$

Thus, for each $x \in X$, there exists a ball $B(x, \varepsilon(x)/2)$ such that the sequence $(x_n)_{n \in \mathbb{N}}$ is *eventually* not in it.

By the *compactness* of (X, ρ), the open cover

$$\{B(x, \varepsilon(x)/2)\}_{x \in X}$$

of X contains a finite subcover, which implies that

$$\exists K \in \mathbb{N} \; \forall k \geq K : \; x_k \notin X.$$

Exercise 2.107. Explain.

The latter is a *contradiction* proving the statement. □

Remarks 2.66.

– Thus, a metric space, which is *nonseparable*, as a *discrete metric space* (X, ρ_d) with an *uncountable X*, or is *incomplete*, as $(0, 1)$ with the regular distance, is *not compact*.

– As the example of the space $\mathbb{R}$ with the regular metric, which is both separable and complete (see Examples 2.17 and 2.22), shows that the converses to Proposition 2.10 and Proposition 2.11 are not true, i. e., a separable and complete metric space need not be compact.

Observe that, when proving property (2) (see (2.23)), we, in fact, prove the following *separation property*:

Proposition 2.12 (Separation Property for a Compact Set and a Point). *A compact set and a point disjoint from it in a metric space (X, ρ) can be separated by disjoint open sets.*

The latter allows the following generalization:

Proposition 2.13 (Separation Property for Compact Sets). *Disjoint compact sets in a metric space (X, ρ) can be separated by disjoint open sets.*

Proof. Let A and B be arbitrary disjoint compact sets in a metric space (X, ρ).
The statement is trivially true if at least one of the sets is *empty*.

Exercise 2.108. Explain.

Suppose that $A, B \neq \emptyset$. Then, by the *Separation Property for a Compact Set and a Point* (Proposition 2.12), for any $x \in A$, there exist disjoint open sets U_x and V_x such that $x \in U_x$ and $B \subseteq V_x$.

By the compactness of A, its open cover $\{U_x\}_{x \in A}$ contains a finite subcover $\{U_{x_1}, \ldots, U_{x_N}\}$ with some $N \in \mathbb{N}$.
Then the open set

$$U := \bigcup_{k=1}^{N} U_{x_k}$$

containing A is *disjoint* from the open set

$$V := \bigcap_{k=1}^{N} V_{x_k},$$

containing B.

Exercise 2.109. Explain. □

2.16.3 Hausdorff Criterion

The following essential statement establishes the equivalence of total boundedness and precompactness in a complete metric space, and allows us to describe compactness in certain complete metric spaces. The celebrated *Heine–Borel Theorem* (Theorem 2.47), which characterizes precompactness in the *n*-space, is one of its immediate implications.

Theorem 2.46 (Hausdorff Criterion). *In a complete metric space* (X, ρ),[17]
(1) *a set A is precompact iff it is totally bounded;*
(2) *a set A is compact iff it is closed and totally bounded.*

Proof. (1) *"Only if"* part follows from the *Precompactness and Total Boundedness Proposition* (Proposition 2.8), which is a more general statement not requiring the completeness of the space.

"If" part. Let us prove this part *by contradiction*, assuming that there exists a totally bounded set A in a *complete* metric space (X, ρ) that is *not* precompact, implying that the closure $\overline{A}$ is *not* compact. Then there exists an *open cover* $\mathcal{O} = \{O_i\}_{i \in I}$ of the closure $\overline{A}$ with no finite subcover.

Due to the *total boundedness* of A, and hence, of $\overline{A}$ (see Remarks 2.58), for each $n \in \mathbb{N}$, there exists a *finite* $1/2^{n-1}$-net N_n for $\overline{A}$. Then, for $n = 1$, at least one of the sets

$$\{B(x, 1) \cap \overline{A}\}_{x \in N_1}$$

cannot be covered by a finite number of sets from $\mathcal{O}$.

Exercise 2.110. Explain.

Suppose this is a set

$$A_1 := B(x_1, 1) \cap \overline{A}$$

with some $x_1 \in N_1$.
Similarly, for $n = 2$, at least one of the sets

$$\{B(x, 1/2) \cap A_1\}_{x \in N_2}$$

cannot be covered by a finite number of sets from $\mathcal{O}$.

Exercise 2.111. Explain.

Suppose this is a set

$$A_2 := B(x_2, 1/2) \cap A_1$$

with some $x_2 \in N_2$.

---

17 Felix Hausdorff (1868–1942).

Continuing inductively, we obtain a set sequence $(A_n)_{n\in\mathbb{N}}$ such that

(a) $A_{n+1} \subseteq A_n \subseteq \overline{A}$, $n \in \mathbb{N}$;

(b) $A_n \subseteq B(x_n, 1/2^{n-1})$, $n \in \mathbb{N}$, with some $x_n \in N_n$;

(c) for each $n \in \mathbb{N}$, the set A_n cannot be covered by a finite number of sets from $\mathcal{O}$, which, in particular, implies that A_n is *infinite*.

Consider a sequence $(y_n)_{n\in\mathbb{N}}$ of elements of $\overline{A}$ chosen as follows:

$$y_1 \in A_1, \ y_n \in A_n \setminus \{y_1, \ldots, y_{n-1}\}, \ n = 2, 3, \ldots.$$

The sequence $(y_n)_{n\in\mathbb{N}}$ is *fundamental* in (X, ρ) since, for any $m, n \in \mathbb{N}$ with $n \geq m$,

$$\rho(y_m, y_n) \qquad\qquad \text{by the } \textit{triangle inequality;}$$
$$\leq \rho(y_m, x_m) + \rho(x_m, y_n) \qquad \text{by the inclusions} \quad A_n \subseteq A_m \subseteq B(x_m, 1/2^{m-1});$$
$$< \frac{1}{2^{m-1}} + \frac{1}{2^{m-1}} = \frac{1}{2^{m-2}}.$$

Hence, by the *completeness* of (X, ρ), there exists an element $y \in X$ such that

$$y_n \to y, \ n \to \infty, \ \text{in } (X, \rho).$$

In view of the *closedness* of $\overline{A}$, by the *Sequential Characterization of Closed Sets* (Theorem 2.19), we infer that

$$y \in \overline{A}.$$

Since $\mathcal{O} = \{O_i\}_{i\in I}$ is an open cover of $\overline{A}$.

$$\exists\, j \in I : \ y \in O_j,$$

and hence, by the *openness* of O_j:

$$\exists\, \delta > 0 : \ B(y, \delta) \subseteq O_j.$$

Choosing an $n \in \mathbb{N}$ sufficiently large so that

$$\frac{1}{2^{n-1}} < \frac{\delta}{3} \quad \text{and} \quad \rho(y_n, y) < \frac{\delta}{3},$$

for any $x \in B(x_n, 1/2^{n-1})$, we have:

$$\rho(x, y) \qquad\qquad \text{by the } \textit{generalized triangle inequality} \text{ (Theorem 2.6);}$$
$$\leq \rho(x, x_n) + \rho(x_n, y_n) + \rho(y_n, y) \qquad \text{since } y_n \in A_n \subseteq B(x_n, 1/2^{n-1});$$
$$< \frac{1}{2^{n-1}} + \frac{1}{2^{n-1}} + \rho(y_n, y) < \frac{\delta}{3} + \frac{\delta}{3} + \frac{\delta}{3} = \delta.$$

Hence, we also have the inclusions

$$A_n \subseteq B(x_n, 1/2^{n-1}) \subseteq B(y, \delta) \subseteq O_j,$$

which imply that the set A_n is covered by *one* set $O_j \in \mathcal{O}$.

The obtained *contradiction* proves that $\bar{A}$ is *compact*, and hence, A is *precompact*.

(2) Part (2) follows from part (1) by the *Characterization of Compactness* (Proposition 2.9). $\square$

Remarks 2.67.

- Thus, in a complete metric space, *"totally bounded"* is synonymous to *"precompact"*.
- As Example 2.40 shows, in an incomplete metric space, a totally bounded set need not be precompact. Hence, the requirement of the completeness on the space in the *Hausdorff Criterion* is essential and cannot be dropped.

The following is an immediate corollary of the *Compactness and Completeness Proposition* (Proposition 2.11), the *Properties of Compact Sets* (Theorem 2.45), and the *Hausdorff Criterion* (Theorem 2.46):

Corollary 2.7 (Characterization of Compactness of a Metric Space). *A metric space is compact iff it is complete and totally bounded.*

2.16.4 Compactness in Certain Complete Metric Spaces

The following descriptions of precompactness in certain complete metric spaces are direct implications of the *Hausdorff Criterion* (Theorem 2.46).

Theorem 2.47 (Heine–Borel Theorem (Compactness in the n-Space)). *Let $n \in \mathbb{N}$ and $1 \leq p \leq \infty$.*

(1) *A set A is precompact in the (real or complex) space $l_p^{(n)}$ iff it is bounded.*

(2) *A set A is compact in the (real or complex) space $l_p^{(n)}$ iff it is closed and bounded.*

Proof. The statement follows, by the *Hausdorff Criterion*, considering the *completeness* of $l_p^{(n)}$ (see Examples 2.22) and the equivalence of boundedness and total boundedness in it (Corollary 2.6). $\square$

Remark 2.68. As follows from the *Heine–Borel Theorem*, any closed and bounded interval $[a, b]$ ($-\infty < a < b < \infty$) is a *compact set* in $\mathbb{R}$. More generally, any closed and bounded box

$$[a_1, b_1] \times \cdots \times [a_n, b_n]$$

$(-\infty < a_i < b_i < \infty, i = 1, \ldots, n)$ is a *compact set* in $l_p^{(n)}$ ($n \in \mathbb{N}$, $1 \le p \le \infty$) (see *Existence and Uniqueness Theorem for Fredholm Integral Equations of the Second Kind* (Theorem 2.37) and *Picard's Existence and Uniqueness Theorem* (Theorem 2.38)).

Theorem 2.48 (Compactness in l_p ($1 \le p < \infty$)). *A set A is precompact (a closed set A is compact) in l_p ($1 \le p < \infty$) if*
1. *A is bounded;*
2. *$\forall \varepsilon > 0 \; \exists N \in \mathbb{N} \; \forall x := (x_n)_{n \in \mathbb{N}} \in A : \sum_{n=N+1}^{\infty} |x_n|^p < \varepsilon.$*

Proof. In view of the *completeness* of l_p ($1 \le p < \infty$) (Theorem 2.24), by the *Hausdorff Criterion* (Theorem 2.46), we show that the two conditions imply *total boundedness* in l_p for a set A satisfying them.

Indeed, by condition 2, for any

$$\forall \varepsilon > 0 \; \exists N \in \mathbb{N} \; \forall x := (x_n)_{n \in \mathbb{N}} \in A : \left\| (0, \ldots, 0, x_{N+1}, \ldots) \right\|_p^p$$
$$= \sum_{n=N+1}^{\infty} |x_n|^p < \varepsilon^p / 2 \tag{2.26}$$

(see Definition 2.4).

By condition 1, the set

$$A_N := \{ (x_1, \ldots, x_N, 0, \ldots) \mid x := (x_n)_{n \in \mathbb{N}} \in A \}$$

is *bounded* in l_p.

Exercise 2.112. Explain.

Being *isometric* to the bounded set

$$\tilde{A}_N := \{ (x_1, \ldots, x_N) \mid x := (x_n)_{n \in \mathbb{N}} \in A \},$$

in $l_p^{(N)}$, where boundedness is equivalent to total boundedness (Corollary 2.6), the set A_N is *totally bounded* in l_p.

Therefore, the set A_N has a finite $\varepsilon/2^{1/p}$-net $\{y_1, \ldots, y_m\} \subseteq A_N$ (see Remarks 2.58) with some $m \in \mathbb{N}$, where

$$y_i := (y_1^{(i)}, \ldots, y_N^{(i)}, 0, \ldots), \; i = 1, \ldots, m,$$

which is an ε-net for A since, for any $x := (x_n)_{n \in \mathbb{N}} \in A$, there exists an $i = 1, \ldots, m$ such that

$$\rho_p((x_1, \ldots, x_N, 0, \ldots), y_i) < \varepsilon/2^{1/p}.$$

Hence, in view of (2.26),

$$\rho_p^p(x, y_i) = \sum_{n=1}^{N} |x_n - y_n^{(i)}|^p + \sum_{n=N+1}^{\infty} |x_n|^p = \rho_p^p((x_1, \ldots, x_N, 0, \ldots), y_i)$$

$$+ \|(0, \ldots, 0, x_{N+1}, \ldots)\|_p^p < \varepsilon^p/2 + \varepsilon^p/2 = \varepsilon^p.$$

Therefore, A is *totally bounded* in l_p. ☐

The proof of the following statement can be obtained by mimicking the proof of the prior proposition.

Theorem 2.49 (Compactness in (c_0, ρ_∞)). *A set A is precompact (a closed set A is compact) in (c_0, ρ_∞) if*

1. *A is bounded;*
2. *$\forall \varepsilon > 0 \, \exists N \in \mathbb{N} \, \forall x := (x_n)_{n \in \mathbb{N}} \in A : \sup_{n \geq N+1} |x_n| < \varepsilon.$*

Exercise 2.113. Prove.

Remark 2.69. In fact, Theorems 2.48 and 2.49 provide not only *sufficient*, but also *necessary* conditions for precompactness in the corresponding spaces, being particular cases of the precompactness characterization in a *Banach space* with a *Schauder*[18] *basis* (see, e. g., [43], also Section 3.2.5).

2.16.5 Other Forms of Compactness

Here, we introduce two other forms of compactness for, and to prove that, in a metric space setting, they are equivalent to that in the Heine–Borel sense.

2.16.5.1 Sequential Compactness

Definition 2.41 (Sequential Compactness). A *nonempty set A in a metric space (X, ρ) is said to be *sequentially compact* in (X, ρ) if every sequence $(x_n)_{n \in \mathbb{N}}$ in A has a subsequence $(x_{n(k)})_{k \in \mathbb{N}}$ convergent to a limit in A.
A metric space (X, ρ) is called *sequentially compact* if the set X is sequentially compact in (X, ρ).

Remarks 2.70.
– Sequential compactness is also called compactness in the *Bolzano*[19]*–Weierstrass*[20] *sense*.
– Being formulated entirely in terms of convergence, *sequential compactness* is a topological property, i. e., it is *homeomorphically invariant*.

---

[18] Juliusz Schauder (1899–1943).
[19] Bernard Bolzano (1781–1848).
[20] Karl Weierstrass (1815–1897).

– The sequential compactness of a nonempty set A in a metric space (X,ρ) is equivalent to the sequential compactness of (A,ρ) as a subspace of (X,ρ).

Examples 2.41.
1. *Finite sets* are *sequentially compact* in an arbitrary metric space (X,ρ), and only finite sets are sequentially compact in a *discrete* metric space (X,ρ_d).
2. The sets $[0,\infty)$, $(0,1]$, and $\{1/n\}_{n\in\mathbb{N}}$ are *not sequentially compact* in $\mathbb{R}$, and the set $\{0\}\cup\{1/n\}_{n\in\mathbb{N}}$ is.
3. The set $E := \{e_n := (\delta_{nk})_{k\in\mathbb{N}}\}_{n\in\mathbb{N}}$ in (c_{00},ρ_∞) is *not sequentially* compact in (c_{00},ρ_∞) (also in (c_0,ρ_∞), (c,ρ_∞), and l_∞).
   The same example works in l_p $(1\le p<\infty)$.

Exercise 2.114. Verify (see Examples 2.38).

2.16.5.2 Limit-Point Compactness

Definition 2.42 (Limit-Point Compactness). An *infinite set A* in a metric space (X,ρ) is said to be *limit-point compact* if every infinite subset of A has a limit point in A.

A metric space (X,ρ) is called *limit-point compact* if the set X is limit-point compact in (X,ρ).

Remarks 2.71.
– Being characterized in terms of sequential convergence (see Proposition 2.17), *limit-point compactness* is a topological property, i. e., it is *homeomorphically invariant*.
– The limit-point compactness of an infinite set A in a metric space (X,ρ) is equivalent to the limit-point compactness of (A,ρ) as a subspace of (X,ρ).

Examples 2.42.
1. No set is limit-point compact in a *discrete metric space* (X,ρ_d).
2. The sets $[0,\infty)$, $(0,1]$, and $\{1/n\}_{n\in\mathbb{N}}$ are *not limit-point compact* in $\mathbb{R}$, and the set $\{0\}\cup\{1/n\}_{n\in\mathbb{N}}$ is.
3. The set $E := \{e_n := (\delta_{nk})_{k\in\mathbb{N}}\}_{n\in\mathbb{N}}$ in (c_{00},ρ_∞) is *not limit-point compact* (also in (c_0,ρ_∞), (c,ρ_∞), and l_∞).
   The same example works in l_p $(1\le p<\infty)$.

Exercise 2.115. Verify (see Examples 2.38 and 2.41).

2.16.6 Equivalence of Different Forms of Compactness

Theorem 2.50 (Equivalence of Different Forms of Compactness). *For an infinite set A in a metric space (X,ρ), the following statements are equivalent.*

1. *A is compact.*
2. *A is sequentially compact.*
3. *A is limit-point compact.*

Proof. To prove the statement, let us show that the following closed chain of implications:

$$1 \Rightarrow 2 \Rightarrow 3 \Rightarrow 1,$$

holds true.

$1 \Rightarrow 2$. Suppose that an infinite set A is *compact* and let $(x_n)_{n \in \mathbb{N}}$ be an arbitrary sequence in A.

By the *Characterization of Compactness of a Metric Space* (Corollary 2.7), the subspace (A, ρ) is *complete* and *totally bounded*.

By the *Characterization of Total Boundedness* (Theorem 2.44), the sequence $(x_n)_{n \in \mathbb{N}}$ in (A, ρ) contains a *fundamental* subsequence $(x_{n(k)})_{k \in \mathbb{N}}$, which, by the completeness of (A, ρ), implies that the latter converges to a limit $x \in A$.

Hence, A is *sequentially compact*.

$2 \Rightarrow 3$. Suppose that an infinite set A is *sequentially compact* and let B be an arbitrary infinite subset of A. Then, by the *Properties of Countable Sets* (Theorem 1.3), we can choose a sequence $(x_n)_{n \in \mathbb{N}}$ in B, whose values are distinct, i. e.,

$$x_m \neq x_n, \ m, n \in \mathbb{N}, m \neq n. \tag{2.27}$$

By the sequential compactness of A, $(x_n)_{n \in \mathbb{N}}$ contains a subsequence $(x_{n(k)})_{k \in \mathbb{N}}$, which converges to a limit $x \in A$.

In view of (2.27), each open ball centered at x contains infinitely many elements of B distinct from x.

Exercise 2.116. Explain.

Hence, x is a limit point of B, which proves that A is *limit-point compact*.

$3 \Rightarrow 1$. Suppose that an infinite set A is *limit-point compact* and let $(x_n)_{n \in \mathbb{N}}$ be an arbitrary sequence in A, with B being the set of its values.

If B is *infinite*, then, by the limit-point compactness of A, B has a limit point $x \in A$, and hence, by the *Sequential Characterization of Limit Points* (Theorem 2.17), there is a subsequence $(x_{n(k)})_{k \in \mathbb{N}}$ convergent to x.

Exercise 2.117. Explain.

If B is *finite*, then there is a *constant* subsequence $(x_{n(k)})_{k \in \mathbb{N}}$, which is also convergent to a limit in A.

Thus, an arbitrary sequence in A contains a *convergent* to a limit in A, and hence, *fundamental*, subsequence, which, by the *Characterization of Total Boundedness* (Theorem 2.44), implies that the subspace (A, ρ) is *totally bounded* and, by the *Fundamen-*

tal Sequence with Convergent Subsequence Proposition (Proposition 2.22, Section 2.18, Problem 25), also implies that the subspace (A, ρ) is *complete*.

Exercise 2.118. Explain.

Hence, by the *Characterization of Compactness of a Metric Space* (Corollary 2.7), the subspace (A, ρ) is *compact*, i.e., the set A is *compact* in (X, ρ). □

As an immediate implication we obtain the following:

Corollary 2.8 (Compact Set Consisting of Isolated Points). *Any compact set A in a metric space (X, ρ), consisting of isolated points, is necessarily finite.*

Exercise 2.119. Prove.

Remark 2.72. Equivalently, any infinite set A in a metric space (X, ρ), consisting of isolated points, is *not compact*.

Hence, another reason for the noncompactness of the set $\mathbb{Z}$ of the integers in $\mathbb{R}$, besides being *unbounded*, is its being an infinite set consisting of isolated points (see Examples 2.38).

Similarly, the infinite set $E := \{e_n := (\delta_{nk})_{k \in \mathbb{N}}\}_{n \in \mathbb{N}}$, where δ_{nk} is the *Kronecker delta*, consisting of isolated points, although *closed* and *bounded*, is *not compact* in (c_{00}, ρ_∞) (also in (c_0, ρ_∞), (c, ρ_∞), and l_p ($1 \leq p \leq \infty$)) (see Examples 2.38).

Other direct implications of the *Equivalence of Different Forms of Compactness Theorem* (Theorem 2.50) are the following results:

Theorem 2.51 (Sequential Characterization of Precompactness). *A nonempty set A is precompact in a metric space (X, ρ) iff every sequence $(x_n)_{n \in \mathbb{N}}$ in A contains a subsequence $(x_{n(k)})_{k \in \mathbb{N}}$ convergent to a limit in (X, ρ).*

Exercise 2.120. Prove.

Theorem 2.52 (Bolzano–Weierstrass Theorem). *Each bounded sequence of real or complex numbers contains a convergent subsequence.*

Exercise 2.121. Prove.

2.16.7 Compactness and Continuity

In this section, we consider the profound and beautiful interplay between *compactness* and *continuity*.

Theorem 2.53 (Continuous Image of a Compact Set). *Let (X, ρ) and (Y, σ) be metric spaces and a function $f : X \to Y$ be continuous on X. If A is a compact set in (X, ρ), then $f(A)$ is a compact set in (Y, σ), i.e., continuous functions map compact sets to compact sets.*

Proof. Let A be an arbitrary compact set in (X,ρ).

For an arbitrary open cover:

$$\mathcal{O} = \{O_i\}_{i \in I},$$

of $f(A)$ in (Y,σ), by the *Properties of Inverse Image* (Theorem 1.4),

$$A \subseteq f^{-1}(f(A)) \subseteq f^{-1}\left(\bigcup_{i \in I} O_i\right) = \bigcup_{i \in I} f^{-1}(O_i).$$

Hence, by the *Characterization of Continuity* (Theorem 2.61) (see Section 2.18, Problem 15), the collection

$$f^{-1}(\mathcal{O}) := \{f^{-1}(O_i)\}_{i \in I}$$

is an *open cover* of A in (X,ρ).

By the compactness of A in (X,ρ), there is a finite subcover $\{f^{-1}(O_{i_1}), \ldots, f^{-1}(O_{i_N})\}$ of A with some $N \in \mathbb{N}$, i. e.,

$$A \subseteq \bigcup_{k=1}^{N} f^{-1}(O_{i_k}).$$

Then, since image preserves unions (see Exercise 1.4),

$$f(A) \subseteq f\left(\bigcup_{k=1}^{N} f^{-1}(O_{i_k})\right) = \bigcup_{k=1}^{N} f(f^{-1}(O_{i_k})) = \bigcup_{k=1}^{N} O_{i_k},$$

and hence, $\{O_{i_1}, \ldots, O_{i_N}\}$ is a finite subcover of $f(A)$, which proves the compactness of $f(A)$, and completes the proof. $\qquad\square$

Exercise 2.122.
(a) Give an example showing that the requirement of *compactness* in the prior theorem is essential and cannot be dropped.
(b) Prove the prior theorem via the *sequential approach*.

   Hint. Use the *Sequential Characterization of Local Continuity* (Theorem 2.11).

Remark 2.73. Thus, continuous functions preserve compactness, uniformly continuous functions preserve total boundedness (see the *Total Boundedness and Uniform Continuity Proposition* (Proposition 2.6)), and Lipschitz continuous functions preserve boundedness (see the *Boundedness and Lipschitz Continuity Proposition* (Proposition 2.18, Section 2.18, Problem 13)).

Theorem 2.54 (Homeomorphism Theorem). *Let* (X,ρ), (Y,σ) *be metric spaces and* $f : X \to Y$ *be a bijective continuous function. If the space* (X,ρ) *is compact, then* f *is a homeomorphism.*

Exercise 2.123. Prove.

Hint. To prove the continuity of the *inverse* f^{-1}, apply the *Properties of Compact Sets* (Theorem 2.45), the prior theorem, and *Characterization of Continuity* (Theorem 2.61) and show that the images of *closed sets* in (X, ρ) are closed in (Y, σ).

Remark 2.74. The requirement of the compactness of the domain space (X, ρ) in the *Homeomorphism Theorem* is essential and cannot be dropped.

Indeed, let (X, ρ) be the set $\mathbb{Z}_+$ as a subspace of $\mathbb{R}$, which is *not compact*, and (Y, σ) be the set $\{0\} \cup \{1/n\}_{n \in \mathbb{N}}$ as a subspace of $\mathbb{R}$. The function $f : X \to Y$:

$$f(0) := 0, \ f(n) = 1/n, \ n \in \mathbb{N},$$

is bijective and continuous (see Exercise 2.39), but is *not* a homeomorphism since its inverse $f : Y \to X$:

$$f^{-1}(0) := 0, \ f^{-1}(1/n) = n, \ n \in \mathbb{N},$$

is *not* continuous at 0.

Exercise 2.124. Verify.

From the *Continuous Image of a Compact Set Theorem* (Theorem 2.53) and the *Properties of Compact Sets* (Theorem 2.45), we obtain

Corollary 2.9 (Boundedness of Continuous Functions on Compact Sets). *A continuous function from one metric space to another is bounded on compact sets.*

The following generalized version of a well-known result from calculus is referred to in *Existence and Uniqueness Theorem for Fredholm Integral Equations of the Second Kind* (Theorem 2.37) and *Picard's Existence and Uniqueness Theorem* (Theorem 2.38) (see also Remark 2.68).

Theorem 2.55 (Weierstrass Extreme Value Theorem). *Let (X, ρ) be a compact metric space. A continuous real-valued function $f : X \to \mathbb{R}$ attains its absolute minimum and maximum values on X, i. e., there exist $x_*, x^* \in X$ such that*

$$f(x_*) = \inf_{x \in X} f(x) \quad and \quad f(x^*) = \sup_{x \in X} f(x).$$

Proof. As follows from the *Continuous Image of a Compact Set Theorem* (Theorem 2.53), the *image set* $f(X)$ is *compact* in $\mathbb{R}$, and hence, by the *Heine–Borel Theorem* (Theorem 2.47), being *closed* and *bounded* in $\mathbb{R}$, contains

$$\inf_{x \in X} f(x) \quad and \quad \sup_{x \in X} f(x).$$

This implies that there exist $x_*, x^* \in X$ such that

$$f(x_*) = \inf_{x\in X} f(x) \quad \text{and} \quad f(x^*) = \sup_{x\in X} f(x),$$

which completes the proof. ☐

From the prior theorem and the *Continuity of Composition Theorem* (Theorem 2.13), we obtain

Corollary 2.10 (Extreme Value Theorem for Modulus). *Let (X,ρ) be a compact metric space. For a continuous complex-valued function $f : X \to \mathbb{C}$, the modulus function $|f|$ attains its absolute minimum and maximum values on X.*

We also have the following

Corollary 2.11 (Nearest and Farthest Points Property Relative to Compact Sets). *Let A be a nonempty compact set in a metric space (X,ρ). Then, for any $x \in X$, in A, there exist a nearest point to x, i. e.,*

$$\exists y \in A : \rho(x,y) = \inf_{u\in A} \rho(x,u) =: \rho(x,A),$$

and a farthest point from x, implying,

$$\exists z \in A : \rho(x,z) = \sup_{u\in A} \rho(x,u).$$

Exercise 2.125. Prove.

Hint. Use the *Lipschitz continuity* of the distance function

$$f(y) := \rho(x,y), \ y \in A,$$

on the compact metric space (A,ρ) (see Section 2.18, Problem 12).

Remark 2.75. If $x \in A$, then the *unique* nearest to x point in A is, obviously, x itself. In general, the statement says nothing about the *uniqueness* of the nearest and farthest points.

Exercise 2.126.
(a) Give an example showing that nearest and farthest points in a compact set need not be unique.
(b) If A is a *nonempty compact set* in a *discrete* metric space (X,ρ_d), which points in A are the *nearest* to and the *farthest* from a point $x \in X$?

The following celebrated result shows that compactness jointly with continuity yield uniform continuity.

Theorem 2.56 (Heine–Cantor Uniform Continuity Theorem). *Let (X,ρ) and (Y,σ) be metric spaces and $f \in C(X,Y)$. If the space (X,ρ) is compact, f is uniformly continuous on X.*

Proof. Let us prove the statement *by contradiction.*

Assume that f is *not* uniformly continuous on X (see Definition 2.8). Hence,

$$\exists \varepsilon > 0 \ \forall n \in \mathbb{N} \ \exists x_n', x_n'' \in X \text{ with } \rho(x_n', x_n'') < 1/n : \ \sigma(f(x_n'), f(x_n'')) \ge \varepsilon. \qquad (2.28)$$

Since the space (X,ρ) is *compact*, by the *Equivalence of Different Forms of Compactness Theorem* (Theorem 2.50), it is *sequentially compact*, and hence, the sequence $(x_n')_{n\in\mathbb{N}}$ contains a subsequence $(x_{n(k)}')_{k\in\mathbb{N}}$ convergent to an element $x \in X$ in (X,ρ), i. e.,

$$\rho(x, x_{n(k)}') \to 0, \ k \to \infty.$$

In view of (2.28), by the *triangle inequality*, we have:

$$0 \le \rho(x, x_{n(k)}'') \le \rho(x, x_{n(k)}') + \rho(x_{n(k)}', x_{n(k)}'') < \rho(x, x_{n(k)}') + 1/n(k) \to 0, \ k \to \infty,$$

which, by the *Squeeze Theorem*, implies that

$$x_{n(k)}'' \to x, \ k \to \infty, \text{ in } (X,\rho).$$

By the *continuity* of f (see the *Sequential Characterization of Local Continuity* (Theorem 2.11) and Remark 2.38),

$$f(x_{n(k)}') \to f(x) \text{ and } f(x_{n(k)}'') \to f(x), \ k \to \infty, \text{ in } (Y,\sigma),$$

which *contradicts* (2.28), showing that the assumption is false and completing the proof. □

Exercise 2.127. Give an example showing that the requirement of *compactness* is essential and cannot be dropped.

2.17 Arzelà–Ascoli Theorem

In this section, we prove the *Arzelà[21]–Ascoli[22] Theorem*, which characterizes precompactness in the space $(C(X), \rho_\infty)$, where the domain space (X,ρ) is compact.

---

21 Cesare Arzelà (1847–1912).
22 Giulio Ascoli (1843–1896).

2.17.1 Space $(C(X, Y), \rho_\infty)$

Here, we introduce and take a closer look at certain abstract sets of functions that are naturally metrizable.

Theorem 2.57 (Space $(M(X, Y), \rho_\infty)$). *Let X be a nonempty set and (Y, σ) be a metric space. The set $M(X, Y)$ of all bounded functions $f : X \to Y$ is a metric space relative to the supremum metric (or uniform metric):*

$$M(X, Y) \ni f, g \mapsto \rho_\infty(f, g) := \sup_{x \in X} \sigma(f(x), g(x)), \tag{2.29}$$

which is complete, provided (Y, σ) is complete.

Exercise 2.128. Prove the statement.
(a) Show that $(M(X, Y), \rho_\infty)$ is a metric space, including the fact that ρ_∞ is well defined by (2.29).
(b) Suppose that the space (Y, σ) is *complete* and, by mimicking the proof of the *Completeness of $(M(T), \rho_\infty)$ Theorem* (Theorem 2.25), prove that the space $(M(X, Y), \rho_\infty)$ is *complete*.

Remarks 2.76.
– In particular, for $Y = \mathbb{R}$ or $Y = \mathbb{C}$, we have the (real/complex) space $(M(X), \rho_\infty)$ (see Examples 2.4) with

$$M(X) \ni f, g \mapsto \rho_\infty(f, g) = \sup_{x \in X} |f(x) - g(x)|,$$

which turns into l_∞ for $X = \mathbb{N}$. Thus, as immediate corollaries, we obtain the *Completeness of $(M(T), \rho_\infty)$ Theorem* (Theorem 2.25) and the *Completeness of l_∞ Corollary* (Corollary 2.1) (see Examples 2.22).
– Convergence in $(M(X, Y), \rho_\infty)$ is the *uniform convergence* on X, i. e.,

$$f_n \to f, \ n \to \infty, \ \text{in} \ (M(X, Y), \rho_\infty)$$

iff

$$\forall \varepsilon > 0 \ \exists N \in \mathbb{N} \ \forall n \geq N \ \forall x \in X : \ \sigma(f_n(x), f(x)) < \varepsilon$$

(see Examples 2.5).
– If (X, ρ) is a *compact* metric space, by the *Boundedness of Continuous Functions on Compact Sets Corollary* (Corollary 2.9), we have the inclusion:

$$C(X, Y) \subseteq M(X, Y),$$

and, by the *Weierstrass Extreme Value Theorem* (Corollary 2.55) applied to the continuous real-valued function

$$X \ni x \mapsto \sigma(f(x), g(x)) \in \mathbb{R}$$

for any $f, g \in C(X, Y)$, in (2.29), we can use max instead of sup, i. e.,

$$C(X, Y) \ni f, g \mapsto \rho_\infty(f, g) = \max_{x \in X} \sigma(f(x), g(x)),$$

which makes $(C(X, Y), \rho_\infty)$ a *subspace* of $(M(X, Y), \rho_\infty)$. Hence, convergence in $(C(X, Y), \rho_\infty)$ is the *uniform convergence*.

In particular, for $Y = \mathbb{R}$ or $Y = \mathbb{C}$, we have the (real/complex) space $(C(X), \rho_\infty)$ with

$$C(X) \ni f, g \mapsto \rho_\infty(f, g) = \max_{x \in X} |f(x) - g(x)|,$$

which turns into $(C[a, b], \rho_\infty)$ for $X = [a, b]$ $(-\infty < a < b < \infty)$.

Theorem 2.58 (Completeness of $(C(X, Y), \rho_\infty)$). *Let (X, ρ) and (Y, σ) be metric spaces, the former being compact and the latter being complete. Then the metric space $(C(X, Y), \rho_\infty)$ is complete.*

In particular, for $Y = \mathbb{R}$ or $Y = \mathbb{C}$, complete is the (real/complex) space $(C(X), \rho_\infty)$.

Proof. By the *Characterization of Completeness* (Theorem 2.27), to prove the completeness of the space $(C(X, Y), \rho_\infty)$, it suffices to prove the *closedness* of the set $C(X, Y)$ in the metric space $(M(X, Y), \rho_\infty)$, which, by the prior theorem, is *complete* due to the completeness of (Y, σ).

To show that $C(X, Y)$ is *closed* in $(M(X, Y), \rho_\infty)$ (regardless whether (Y, σ) is complete or not), consider an arbitrary sequence $(f_n)_{n \in \mathbb{N}}$ in $C(X, Y)$ convergent to a function f in $(M(X, Y), \rho_\infty)$. Then

$$\forall \varepsilon > 0 \, \exists N \in \mathbb{N} \, \forall n \geq N : \rho_\infty(f, f_n) := \sup_{x \in X} \sigma(f(x), f_n(x)) < \varepsilon/3. \qquad (2.30)$$

Furthermore, in view of the compactness of the domain space (X, ρ), by the *Heine–Cantor Uniform Continuity Theorem* (Theorem 2.56), the continuous function f_N is *uniformly continuous* on X:

$$\exists \delta > 0 \, \forall x', x'' \in X \text{ with } \rho(x', x'') < \delta : \sigma(f_N(x'), f_N(x'')) < \varepsilon/3. \qquad (2.31)$$

By (2.30), (2.31), and the *generalized triangle inequality* (Theorem 2.6), we infer that

$$\forall x', x'' \in X \text{ with } \rho(x', x'') < \delta : \sigma(f(x'), f(x'')) \leq \sigma(f(x'), f_N(x'))$$
$$+ \sigma(f_N(x'), f_N(x'')) + \sigma(f_N(x''), f(x'')) < \varepsilon/3 + \varepsilon/3 + \varepsilon/3 = \varepsilon,$$

which implies that f is *uniformly continuous* on X, and hence, the more so (see Remark 2.17), $f \in C(X, Y)$.

Thus, by the *Sequential Characterization of Closed Sets* (Theorem 2.19), $C(X, Y)$ is *closed* in $(M(X, Y), \rho_\infty)$, which completes the proof. $\qquad\square$

Remark 2.77. As a particular case, we obtain the *Completeness of $(C[a, b], \rho_\infty)$ Theorem* (Theorem 2.26) (see Exercise 2.59).

2.17.2 Uniform Boundedness and Equicontinuity

To proceed, we need to introduce the important notions of *uniform boundedness* and *equicontinuity*.

Definition 2.43 (Uniform Boundedness). Let X be a nonempty set. A nonempty set $F \subseteq M(X)$ is called *uniformly bounded* on X if

$$\exists C > 0 \ \forall f \in F, \ \forall x \in X : |f(x)| \leq C$$

or, equivalently,

$$\exists C > 0 \ \forall f \in F : \sup_{x \in X} |f(x)| \leq C.$$

Remark 2.78. The uniform boundedness of $F \subseteq M(X)$ on X is, obviously, equivalent to the boundedness of F in the space $(M(X), \rho_\infty)$.

Example 2.43.
1. Any nonempty finite set $F \subseteq M(X)$ is *uniformly bounded* on X.
2. The set of all constant functions on X is *not* uniformly bounded on X.
3. The set $\{x^n\}_{n \in \mathbb{N}}$ is *uniformly bounded* on $[0,1]$, but not on $[0,2]$.

Exercise 2.129. Verify.

Definition 2.44 (Equicontinuity). Let (X, ρ) be a metric space. A nonempty set $F \subseteq C(X)$ is called *equicontinuous* on X if

$$\forall \varepsilon > 0 \ \exists \delta > 0 \ \forall f \in F, \ \forall x', x'' \in X \text{ with } \rho(x', x'') < \delta : |f(x') - f(x'')| < \varepsilon.$$

Remark 2.79. Thus, the equicontinuity of F on X is the *uniform continuity* of all functions from F on X in, so to speak, equal extent.

Examples 2.44.
1. If X is *finite*, any nonempty set $F \subseteq C(X)$ is *equicontinuous* on X.
2. The set of all constant functions on X is *equicontinuous*.
3. More generally, the set of all Lipschitz continuous functions on X (see Definition 2.9) with the same Lipschitz constant L is *equicontinuous* on X.
4. The set $\{x^n\}_{n \in \mathbb{N}}$ is *not* equicontinuous on $[0,1]$.

Exercise 2.130. Verify.

Remark 2.80. If (X, ρ) is a *compact* metric space, a set $F \subseteq C(X)$ shares the properties of *uniform boundedness* and *equicontinuity*, respectively, with its closure $\overline{F}$ in $(C(X), \rho_\infty)$.

Exercise 2.131. Verify (see Section 2.18, Problem 49).

2.17.3 Arzelà–Ascoli Theorem

Theorem 2.59 (Arzelà–Ascoli Theorem). *Let (X, ρ) be a compact metric space. A set F is precompact (a closed set F is compact) in the space $(C(X), \rho_\infty)$ iff F is uniformly bounded and equicontinuous on X.*

Proof. "*Only if*" part. Let a set F be *precompact* in the space $(C(X), \rho_\infty)$. In view of the *completeness* of the latter (see the *Completeness of* $(C(X, Y), \rho_\infty)$ *Theorem* (Theorem 2.58)), by the *Hausdorff Criterion* (Theorem 2.46), this is equivalent to the fact that F is *totally bounded* in $(C(X), \rho_\infty)$. Hence, by the *Properties of Totally Bounded Sets* (Proposition 2.43), the set F is *bounded* in $(C(X), \rho_\infty)$, i. e., *uniformly bounded* on X.

By the total boundedness of F, for an arbitrary $\varepsilon > 0$, we can choose a finite $\varepsilon/3$-net $\{f_1, \ldots, f_N\} \subseteq F$ with some $N \in \mathbb{N}$ for F (see Remarks 2.58).

Since, by the *Heine–Cantor Uniform Continuity Theorem* (Theorem 2.56), the functions $f_1, \ldots, f_N$ are *uniformly continuous* on X:

$$\forall\, i = 1, \ldots, N \;\; \exists\, \delta(i) > 0 \;\; \forall\, x', x'' \in X \text{ with } \rho(x', x'') < \delta(i): \; \left| f_i(x') - f_i(x'') \right| < \varepsilon/3.$$

Thus, setting $\delta := \min_{1 \le i \le N} \delta(i) > 0$, we have

$$\forall\, i = 1, \ldots, N, \;\; \forall\, x', x'' \in X \text{ with } \rho(x', x'') < \delta: \; \left| f_i(x') - f_i(x'') \right| < \varepsilon/3, \tag{2.32}$$

i. e., the set $\{f_1, \ldots, f_N\}$ is *equicontinuous* on X.

Since, for each $f \in F$,

$$\exists\, k = 1, \ldots, N: \; \rho_\infty(f, f_k) := \max_{x \in X} \left| f(x) - f_k(x) \right| < \varepsilon/3,$$

in view of (2.32), we infer that

$$\forall\, x', x'' \in X \text{ with } \rho(x', x'') < \delta: \; \left| f(x') - f(x'') \right| \le \left| f(x') - f_k(x') \right| + \left| f_k(x') - f_k(x'') \right|$$
$$+ \left| f_k(x'') - f(x'') \right| < \varepsilon/3 + \varepsilon/3 + \varepsilon/3 = \varepsilon.$$

Hence, the set F is *equicontinuous* on X.

"*If*" part. Suppose that a set F in $(C(X), \rho_\infty)$ is *uniformly bounded* and *equicontinuous* on X. In view of the completeness of $(C(X), \rho_\infty)$ (see the *Completeness of* $(C(X, Y), \rho_\infty)$ *Theorem* (Theorem 2.58)), by the *Hausdorff Criterion* (Theorem 2.46), we show the *total boundedness* for F.

By the *Compactness and Separability Proposition* (Proposition 2.10), there is a *countable dense set* $\{x_i\}_{i \in I}$, where $I \subseteq \mathbb{N}$, in (X, ρ).

If the set $\{x_i\}_{i \in I}$ is *finite*, i. e., $I = \{1, \ldots, n\}$ with some $n \in \mathbb{N}$, then

$$X = \overline{\{x_1, \ldots, x_n\}} = \{x_1, \ldots, x_n\}$$

is finite itself. Hence $(C(X),\rho_\infty) = (M(X),\rho_\infty)$, which is *isometric* to $l_\infty^{(n)}$, where, by the *Heine–Borel Theorem* (Theorem 2.47), boundedness alone implies precompactness. Hence, set F, being *bounded*, is precompact in $(C(X),\rho_\infty)$.

Suppose now that the set $\{x_i\}_{i \in I}$ is *infinite*, i.e., we can acknowledge that $I = \mathbb{N}$, and consider an arbitrary sequence $(f_n)_{n \in \mathbb{N}} \subseteq F$. Since, by the *uniform boundedness* of F, the *numeric sequence* $(f_n(x_1))_{n \in \mathbb{N}}$ is *bounded*, by the *Bolzano–Weierstrass Theorem* (Theorem 2.52), the sequence $(f_n)_{n \in \mathbb{N}}$ contains a subsequence $(f_{1,n})_{n \in \mathbb{N}}$ such that

$$f_{1,n}(x_1) \to f(x_1) \in \mathbb{C}, \ n \to \infty.$$

Similarly, since the *numeric sequence* $(f_{1,n}(x_2))_{n \in \mathbb{N}}$ is *bounded*, the subsequence $(f_{1,n})_{n \in \mathbb{N}}$ contains a subsequence $(f_{2,n})_{n \in \mathbb{N}}$ such that

$$f_{2,n}(x_2) \to f(x_2) \in \mathbb{C}, \ n \to \infty.$$

Continuing inductively, we obtain a countable collection of sequences:

$$\{(f_{m,n})_{n \in \mathbb{N}} \mid m \in \mathbb{Z}_+\},$$

such that

$$(f_{0,n})_{n \in \mathbb{N}} := (f_n)_{n \in \mathbb{N}},$$

for each $m \in \mathbb{N}$, $(f_{m,n})_{n \in \mathbb{N}}$ is a subsequence of $(f_{(m-1),n})_{n \in \mathbb{N}}$, and

$$\forall m \in \mathbb{N}, \ \forall i = 1,\dots,m : f_{m,n}(x_i) \to f(x_i) \in \mathbb{C}, \ n \to \infty.$$

Hence, for the *"diagonal subsequence"* $(f_{n,n})_{n \in \mathbb{N}}$ of the sequence $(f_n)_{n \in \mathbb{N}}$, we have

$$\forall i \in \mathbb{N} : f_{n,n}(x_i) \to f(x_i), \ n \to \infty. \tag{2.33}$$

By the *equicontinuity* of F on X,

$$\forall \varepsilon > 0 \ \exists \delta > 0 \ \forall n \in \mathbb{N}, \ \forall x', x'' \in X \text{ with } \rho(x',x'') < \delta :$$
$$|f_{n,n}(x') - f_{n,n}(x'')| < \varepsilon. \tag{2.34}$$

Since, due to its denseness in (X,ρ), the set $\{x_i\}_{i \in \mathbb{N}}$ is a δ-net for X (see Examples 2.35), by the *compactness* of X, there is a *finite* subnet $\{x_{i_1},\dots,x_{i_N}\} \subseteq \{x_i\}_{i \in \mathbb{N}}$ with some $N \in \mathbb{N}$.

Hence,

$$\forall x \in X \ \exists k = 1,\dots,N : \rho(x, x_{i_k}) < \delta.$$

In view of this and (2.34), for any $m,n \in \mathbb{N}$ and an arbitrary $x \in X$, we have

$$|f_{m,m}(x) - f_{n,n}(x)| \le |f_{m,m}(x) - f_{m,m}(x_{i_k})| + |f_{m,m}(x_{i_k}) - f_{n,n}(x_{i_k})|$$
$$+ |f_{n,n}(x_{i_k}) - f_{n,n}(x)| < \varepsilon + \sum_{j=1}^{N} |f_{m,m}(x_{i_j}) - f_{n,n}(x_{i_j})| + \varepsilon.$$

Hence,

$$\forall\, m, n \in \mathbb{N} : \rho_\infty(f_{m,m}, f_{n,n}) := \max_{x \in X} |f_{m,m}(x) - f_{n,n}(x)|$$

$$< \varepsilon + \sum_{j=1}^{N} |f_{m,m}(x_{i_j}) - f_{n,n}(x_{i_j})| + \varepsilon.$$

The middle term in the right-hand side *vanishing* as $m, n \to \infty$ by (2.33), we conclude that the subsequence $(f_{n,n})_{n \in \mathbb{N}}$ is *fundamental* in $(C(X), \rho_\infty)$. Therefore, by the *Characterization of Total Boundedness* (Theorem 2.44), the set F is *totally bounded* in $(C(X), \rho_\infty)$, which completes the proof for the precompactness.

The compactness part of the theorem follows from the above by the *Hausdorff Criterion* (Theorem 2.46). $\qquad\qquad\qquad\qquad\qquad\qquad\qquad\qquad\qquad\qquad\qquad\qquad\qquad\qquad$ □

Remarks 2.81.
- In view of the *completeness* of the space $(C(X), \rho_\infty)$ (Theorem 2.58), by the *Hausdorff Criterion* (Theorem 2.46), in the *Arzelà–Ascoli Theorem*, "precompact" can be replaced with "totally bounded".
- In particular, if X is finite, as follows from Examples 2.44, and the corresponding segment of the proof of the "if" part, the *Arzelà–Ascoli Theorem* is consistent with the *Heine–Borel Theorem* (Theorem 2.47).

Examples 2.45.
1. The set $\{(\frac{x-a}{b-a})^n\}_{n \in \mathbb{N}}$ is uniformly bounded, but not equicontinuous on $[a, b]$ ($-\infty < a < b < \infty$) (see Examples 2.44), and hence, by the *Arzelà–Ascoli Theorem*, is *not precompact* (*not totally bounded*) in $(C[a, b], \rho_\infty)$.
2. The set of all Lipschitz continuous functions on $[a, b]$ ($-\infty < a < b < \infty$) with the same Lipschitz constant L is equicontinuous, but, containing all constants, not uniformly bounded on $[a, b]$, and hence, by the *Arzelà–Ascoli Theorem*, is not precompact (not totally bounded) in $(C[a, b], \rho_\infty)$.

2.17.4 Application: Peano's Existence Theorem

As a somewhat unexpected application of the *Arzelà–Ascoli Theorem* (Theorem 2.59), we obtain of the following profound classical result:

Theorem 2.60 (Peano's Existence Theorem). *If a real-valued function $f(\cdot, \cdot)$ is continuous on a closed rectangle*[23]

$$R := [x_0 - a, x_0 + a] \times [y_0 - b, y_0 + b],$$

---

23 Giuseppe Peano (1858–1932).

with some $(x_0, y_0) \in \mathbb{R}^2$ and $a, b > 0$, then, there exists an $h \in (0, a]$ such that the initial-value problem

$$\frac{dy}{dx} = f(x, y), \ y(x_0) = y_0, \tag{2.35}$$

has a solution on the interval $[x_0 - h, x_0 + h]$.

Proof. Since, by the *Heine–Borel Theorem* (Theorem 2.47), the closed rectangle R is *compact* in $l_\infty^{(2)}$ (see Remark 2.68), by the *Weierstrass Extreme Value Theorem* (Theorem 2.55),

$$\exists M > 0 : \ \max_{(x,y)\in R} |f(x, y)| \le M.$$

Let us fix such an M, and set

$$h := \min\left[a, \frac{b}{M}\right] > 0.$$

By the *Heine–Cantor Uniform Continuity Theorem* (Theorem 2.56), the function f is *uniformly continuous* on the *compact* set R, and hence,

$$\forall n \in \mathbb{N} \ \exists \delta(n) \in (0, b] \ \forall (x', y'), (x'', y'') \in R \text{ with } |x' - x''| \le \delta(n) \text{ and}$$

$$|y' - y''| \le \delta(n) : \ |f(x', y') - f(x'', y'')| < 1/n. \tag{2.36}$$

For each $n \in \mathbb{N}$, let $x_0 = x_0^{(n)} < \cdots < x_{k(n)}^{(n)} = x_0 + h \ (k(n) \in \mathbb{N})$ be a partition of $[x_0, x_0 + h]$ such that

$$\max_{1 \le i \le k(n)} [x_i^{(n)} - x_{i-1}^{(n)}] \le \min\left[\delta(n), \frac{\delta(n)}{M}\right]. \tag{2.37}$$

Let us define the *polygonal approximation* $y_n(\cdot)$ for the desired solution $y(\cdot)$ on $[x_0, x_0 + h]$ as follows:

$$y_n(x) := y_0 + f(x_0, y_0)(x - x_0), \ x_0 \le x \le x_1^{(n)}, \ n \in \mathbb{N},$$

and, for $n \in \mathbb{N}, i = 2, \ldots, k(n)$:

$$y_n(x) := y_n(x_{i-1}^{(n)}) + f(x_{i-1}^{(n)}, y_n(x_{i-1}^{(n)}))(x - x_{i-1}^{(n)}), \ x_{i-1}^{(n)} \le x \le x_i^{(n)}.$$

Exercise 2.132. Verify that $(x_i^{(n)}, y_n(x_i^{(n)})) \in R$, for all $n \in \mathbb{N}, i = 0, \ldots, k(n)$.

Assigning arbitrary real values to $y_n'(x_i^{(n)})$, $n \in \mathbb{N}, i = 0, \ldots, k(n)$, we have the following integral representation:

$$y_n(x) = y_0 + \int_{x_0}^{x} y_n'(t) \, dt, \ n \in \mathbb{N}, x \in [x_0, x_0 + h]. \tag{2.38}$$

Whence, we conclude that

$$\left| y_n(x) - y_0 \right| = \left| \int_{x_0}^{x} y_n'(t)\, dt \right| \le M|x - x_0| \le M\frac{b}{M} = b, \ n \in \mathbb{N}, x \in [x_0, x_0 + h].$$

The latter implies that, for all $n \in \mathbb{N}$,

$$(x, y_n(x)) \in R, \ x \in [x_0, x_0 + h],$$

and that the set of functions $F := \{y_n\}_{n \in \mathbb{N}}$ is *uniformly bounded* on $[x_0, x_0 + h]$.

Since, for each $n \in \mathbb{N}$, the absolute value of the *slope* of y_n, except, possibly, at the partition points $x_i^{(n)}$, $i = 0, \ldots, k(n)$, does not exceed M, as follows from the *Mean Value Theorem*:

$$\forall n \in \mathbb{N}, \ \forall x', x'' \in [x_0, x_0 + h] : \ |y_n(x') - y_n(x'')| \le M|x' - x''|,$$

which implies that the functions of the set F are Lipschitz continuous on $[x_0, x_0 + h]$, with the same Lipschitz constant M. Hence, the set F is *equicontinuous* on $[x_0, x_0 + h]$ (see Examples 2.44).

By the *Arzelà–Ascoli Theorem* (Theorem 2.59), the set $\{y_n\}_{n \in \mathbb{N}}$ is *precompact* in the space $(C[x_0, x_0 + h], \rho_\infty)$, i. e., its *closure* $\overline{F}$ is compact. Hence, by the *Equivalence of Different Forms of Compactness Theorem* (Theorem 2.50), the sequence $(y_n)_{n \in \mathbb{N}}$ contains a subsequence $(y_{n(i)})_{n \in \mathbb{N}}$ *uniformly convergent* on $[x_0, x_0 + h]$ to a function $y \in \overline{F} \subseteq C[x_0, x_0 + h]$.

Fixing an arbitrary $n \in \mathbb{N}$, for any $x \in [x_0, x_0 + h]$ *distinct* from the partition points $x_i^{(n)}$, $n \in \mathbb{N}$, $i = 0, \ldots, k(n)$, and choosing a $j = 1, \ldots, k(n)$ such that $x_{j-1}^{(n)} < x < x_j^{(n)}$, by the *Mean Value Theorem* and in view of (2.37), we have

$$\left| y_n(x) - y_n(x_{j-1}^{(n)}) \right| \le M|x - x_{j-1}^{(n)}| \le M\frac{\delta(n)}{M} = \delta(n),$$

which, in view of (2.36), implies that

$$\left| f(x, y_n(x)) - f(x_{j-1}^{(n)}, y_n(x_{j-1}^{(n)})) \right| < 1/n.$$

Since, on the interval $(x_{j-1}^{(n)}, x_j^{(n)})$, $y_n'(x) = f(x_{j-1}^{(n)}, y_n(x_{j-1}^{(n)}))$, for all $n \in \mathbb{N}$ and $i = 0, \ldots, k(n)$, we have

$$\left| f(x, y_n(x)) - y_n'(x) \right| < 1/n, \ x \in [x_0, x_0 + h], \ x \ne x_i^{(n)}. \tag{2.39}$$

By integral representation (2.38), for all $n \in \mathbb{N}$ and $x \in [x_0, x_0 + h]$ we have

$$y_n(x) = y_0 + \int_{x_0}^{x} y_n'(t)\, dt = y_0 + \int_{x_0}^{x} f(t, y_n(t))\, dt + \int_{x_0}^{x} [y_n'(t) - f(t, y_n(t))]\, dt. \tag{2.40}$$

Since, by the *uniform continuity* of f on R, $(f(t, y_{n(i)}(t)))_{i \in \mathbb{N}}$ *uniformly converges to* $f(t, y(t))$ on $[x_0, x_0 + h]$ and, due to (2.39),

$$\int_{x_0}^{x} |y_n'(t) - f(t, y_n(t))| \, dt \le h/n,$$

passing to the limit in (2.40) with $n = n(i)$ as $i \to \infty$, we obtain the following integral representation for y:

$$y(x) = y_0 + \int_{x_0}^{x} f(t, y(t)) \, dt, \quad x \in [x_0, x_0 + h].$$

The latter, by the *Fundamental Theorem of Calculus*, implies that y is a solution of initial value problem (2.35) on $[x_0, x_0 + h]$.

Using the same construct on $[x_0 - h, x_0]$, we extend the obtained solution to the interval $[x_0 - h, x_0 + h]$. □

Corollary 2.12 (Existence of a Local Solution). *If $f(\cdot, \cdot)$ is a real-valued function continuous on an open subset $D \subseteq \mathbb{R}^2$. Then, for each $(x_0, y_0) \in D$, the initial-value problem (2.35) has a local solution, i. e., a solution on an interval $[x_0 - h, x_0 + h]$ with some $0 < h < \infty$.*

Remark 2.82. As the following example demonstrates, *Peano's Existence Theorem*, void of the additional condition of Lipschitz continuity in y, guarantees the *local existence* of the initial-value problem solutions only, without securing their *uniqueness* (see *Picard's Existence and Uniqueness Theorem* (Theorem 2.38), see Remarks 2.54).

Example 2.46. The initial value problem

$$y' = |y|^{1/2}, \quad y(0) = 0,$$

whose right-hand side is *continuous* in x and y, but *not Lipschitz continuous* in y on $\mathbb{R}^2$, has *infinitely many* solutions on $(-\infty, \infty)$: $y = 0$ and

$$y = \begin{cases} \frac{(x-c)^2}{4} & \text{for } x \ge c, \\ 0 & \text{for } x < c \end{cases} \quad \text{or} \quad y = \begin{cases} -\frac{(x+c)^2}{4} & \text{for } x \le -c, \\ 0 & \text{for } x > -c, \end{cases} \quad c \ge 0.$$

Exercise 2.133. Verify.

2.18 Problems

1. Determine all values of $p > 0$ such that the mapping

$$\mathbb{R} \ni x, y \mapsto |x - y|^p \in \mathbb{R}$$

is a *metric* on $\mathbb{R}$.

2. Is the mapping

$$x := (x_1, \ldots, x_n), y := (y_1, \ldots, y_n) \mapsto \rho_p(x, y) := \left[\sum_{i=1}^{n} |x_i - y_i|^p \right]^{1/p}$$

with $0 < p < 1$ a metric on the n-space ($n \in \mathbb{N}, n \geq 2$)?

3. Let (X, ρ) be a metric space. Determine which of the following mappings are *metrics* on X and which are not:

   (a) $X \ni x, y \mapsto d(x, y) := c\rho(x, y)$ ($c > 0$),

   (b) $X \ni x, y \mapsto d(x, y) := \sqrt{\rho(x, y)}$,

   (c) $X \ni x, y \mapsto d(x, y) := \rho^2(x, y)$,

   (d) $X \ni x, y \mapsto d(x, y) := \min(\rho(x, y), 1)$,

   (e) $X \ni x, y \mapsto d(x, y) := \frac{\rho(x,y)}{\rho(x,y)+1}$.

4. Let $1 \leq p \leq \infty$. Show that $C[a, b]$ ($-\infty < a < b < \infty$) is a *metric space* relative to p-metric

$$C[a, b] \ni f, g \mapsto \rho_p(f, g) := \begin{cases} [\int_a^b |f(t) - g(t)|^p \, dt]^{1/p} & \text{if } 1 \leq p < \infty, \\ \max_{a \leq t \leq b} |f(t) - g(t)| & \text{if } p = \infty. \end{cases}$$

Hint. The cases of $p = 1$ and $p = \infty$ are considered in Examples 2.4. To consider the case of $1 < p < \infty$, first prove *Hölder's inequality* for the following functions:

$$\int_a^b |f(t)g(t)| \, dt \leq \|f\|_p \|g\|_q, \quad f, g \in C[a, b],$$

where $1 \leq p, q \leq \infty$ are the *conjugate indices*, and $\|f\|_p$ is the *p-norm* of f, i. e., the distance of f from the *zero function*:

$$C[a, b] \ni f \mapsto \|f\|_p := \rho_p(f, 0) = \begin{cases} [\int_a^b |f(t)|^p \, dt]^{1/p} & \text{if } 1 \leq p < \infty, \\ \max_{a \leq t \leq b} |f(t)| & \text{if } p = \infty, \end{cases}$$

which is clear for the symmetric pairs $p = 1$, $q = \infty$ and $p = \infty$, $q = 1$, and follows from *Young's Inequality* (Theorem 2.1) for $1 < p, q < \infty$ in the same manner as *Minkowski's Inequality for n-Tuples* (Theorem 2.2). Then deduce *Minkowski's inequality for the following functions*:

$$\|f + g\|_p \leq \|f\|_p + \|g\|_p, \quad f, g \in C[a, b],$$

with $1 < p < \infty$ similarly to how it is done for n-tuples.

5. (Cartesian Product of Metric Spaces)

Let (X_1, ρ_1), (X_2, ρ_2) be metric spaces. The *Cartesian product* $X = X_1 \times X_2$ is a *metric space*, relative to the *product metric*

$$X \ni (x_1, x_2), (y_1, y_2) \mapsto \rho((x_1, x_2), (y_1, y_2)) := \sqrt{\rho_1^2(x_1, y_1) + \rho_2^2(x_2, y_2)}.$$

The product space (X, ρ) is naturally called the *Cartesian product* of the spaces (X_1, ρ_1) and (X_2, ρ_2).

Verify that (X, ρ) is a *metric space*.

6. Prove

Proposition 2.14 (Characterization of Convergence in Product Space). *Let* (X, ρ) *be the Cartesian product of metric spaces* (X_1, ρ_1) *and* (X_2, ρ_2) *(see Problem 5). Then*

$$(x_1^{(n)}, x_2^{(n)}) \to (x_1, x_2), \ n \to \infty, \ in \ (X, \rho)$$

iff

$$x_i^{(n)} \to x_i, \ n \to \infty, \ in \ (X_i, \rho_i), \ i = 1, 2.$$

I. e., the convergence of a sequence in the Cartesian product of metric spaces of (X_1, ρ_1) *and* (X_2, ρ_2) *is equivalent to the componentwise convergence in the corresponding spaces.*

Remark 2.83. The statement can be naturally extended to arbitrary finite products.

7. Prove

Proposition 2.15 (Joint Continuity of Metric). *If* $x_n \to x$ *and* $y_n \to y$, $n \to \infty$, *in a metric space* (X, ρ), *then*

$$\rho(x_n, y_n) \to \rho(x, y), \ n \to \infty.$$

Remark 2.84. Joint continuity of metric implies its continuity in each argument:

$$\rho(x_n, y) \to \rho(x, y) \ and \ \rho(x, y_n) \to \rho(x, y), \ n \to \infty.$$

8. Show that the power sequence $\{t^n\}_{n=0}^{\infty}$ converges in $(C[0, 1/2], \rho_\infty)$, but does not converge in $(C[0, 1], \rho_\infty)$.

9. * Prove

Proposition 2.16 (Characterization of Convergence in (c_0, ρ_∞)). *In* (c_0, ρ_∞), $(x_k^{(n)})_{k \in \mathbb{N}} =: x^{(n)} \to x := (x_k)_{k \in \mathbb{N}}, \ n \to \infty, \ iff$
(a) $\forall k \in \mathbb{N} : x_k^{(n)} \to x_k, \ n \to \infty$, *and*
(b) $\forall \varepsilon > 0 \ \exists N \in \mathbb{N} \ \forall n \in \mathbb{N} : \sup_{k \geq N+1} |x_k^{(n)}| < \varepsilon.$

10. * Prove

Proposition 2.17 (Characterization of Convergence in l_p ($1 \leq p < \infty$)). *In* l_p ($1 \leq p < \infty$), $(x_k^{(n)})_{k \in \mathbb{N}} =: x^{(n)} \to x := (x_k)_{k \in \mathbb{N}}, \ n \to \infty, \ iff$
(a) $\forall k \in \mathbb{N} : x_k^{(n)} \to x_k, \ n \to \infty$, *and*
(b) $\forall \varepsilon > 0 \ \exists N \in \mathbb{N} \ \forall n \in \mathbb{N} : \sum_{k=N+1}^{\infty} |x_k^{(n)}|^p < \varepsilon.$

11. Let (X,ρ) be a metric space and $f : X \to l_p^{(n)}$ $(n \in \mathbb{N}, 1 \le p \le \infty)$:

$$X \ni x \mapsto f(x) := (f_1(x),\ldots,f_n(x)) \in l_p^{(n)}.$$

Prove that $f(\cdot)$ is *continuous* at a point $x_0 \in X$ *iff* each scalar component function $f_i(\cdot)$, $i = 1,\ldots,n$, is continuous at x_0.

12. **Definition 2.45** (Distance to a Set). Let A be a nonempty set in a metric space (X,ρ). For any $x \in X$, the nonnegative number

$$\rho(x,A) := \inf_{y \in A} \rho(x,y)$$

is called the *distance from the point x to the set A*.

Prove that the *distance-to-a-set function* $f(x) := \rho(x,A)$, $x \in X$, is *Lipschitz continuous* on X.

13. (a) Prove

> **Proposition 2.18** (Boundedness and Lipschitz Continuity). *Let (X,ρ) and (Y,σ) be metric spaces and a function $f : X \to Y$ be Lipschitz continuous on X. If A is a bounded set in (X,ρ), then $f(A)$ is a bounded set in (Y,σ), i. e., Lipschitz continuous functions map bounded sets to bounded sets.*

(b) Give an example showing that the image of a bounded set under a uniformly continuous function need not be bounded.

Hint. Consider the *identity mapping*

$$(\mathbb{R},d) \ni x \mapsto Ix := x \in (\mathbb{R},\rho),$$

where ρ is the regular metric, and

$$\mathbb{R} \ni x,y \mapsto d(x,y) := \min(\rho(x,y),1)$$

is the *standard bounded metric* generated by ρ (see Problem 3, Examples 2.18, and Remarks 2.32).

14. Prove that, for a set A in a metric space (X,ρ), the *interior* of A is the largest open set contained in A, i. e.,

$$\mathrm{int}(A) = \bigcup_{O \in \mathscr{G},\, O \subseteq A} O,$$

where $\mathscr{G}$ is the metric topology of (X,ρ)

Hint. Show first that $\mathrm{int}(A)$ is an open set.

15. Prove

Theorem 2.61 (Characterization of Continuity). *Let (X,ρ) and (Y,σ) be metric spaces.*
A function $f : X \to Y$ is continuous iff, for each open set O (closed set C) in (Y,σ), the inverse image $f^{-1}(O)$ ($f^{-1}(C)$) is an open (respectively, closed) set in (X,ρ).

16. Prove that, for a set A in a metric space (X,ρ), the *closure* of A is the smallest closed set containing A, i. e.,

$$\overline{A} = \bigcap_{C \in \mathscr{C}, A \subseteq C} C,$$

where $\mathscr{C}$ is the collection of all closed sets in (X,ρ).

Hint. Show first that $\overline{A}$ is a closed set.

17. Prove that, for a nonempty set A in a metric space (X,ρ),

$$\overline{A} = \{x \in X \mid \rho(x,A) = 0\}$$

(see Definition 2.45).
18. Prove that, for a set A in a metric space (X,ρ),

$$\overline{A} = \text{ext}(A)^c, \text{ i. e., } \overline{A} = \text{int}(A) \cup \partial A.$$

19. Prove

Proposition 2.19 (Subspace of Separable Metric Space). *Each subspace of a separable metric space (X,ρ) is separable.*

20. Prove

Proposition 2.20 (Characterization of the Separability of Product Space). *The Cartesian product (X,ρ) of metric spaces (X_1,ρ_1) and (X_2,ρ_2) (see Problem 5) is separable iff each metric space (X_i,ρ_i), $i = 1, 2$, is separable.*

21. Prove

Proposition 2.21 (Continuous Image of Separable Metric Space). *Let (X,ρ) and (Y,σ) be metric spaces, the former one being separable, and $f \in C(X,Y)$. Then the subspace $(f(X),\sigma)$ of (Y,σ) is separable, i. e., the image of a separable space under a continuous mapping is separable.*

22. Prove that the space $(M(X),\rho_\infty)$, where X is an infinite set, is *not separable*.
23. Determine whether on $C[a,b]$ $(-\infty < a < b < \infty)$, the metrics

$$\rho_1(f,g) := \int_a^b |f(t) - g(t)| \, dt, \ f,g \in C[a,b],$$

and

$$\rho_\infty(f,g) := \max_{a \le t \le b} |f(t) - g(t)|, \ f,g \in C[a,b],$$

are equivalent.

24. Show in *two different ways* that the real line $\mathbb{R}$ with the regular distance is not isometric to the plane $\mathbb{R}^2$ with the Euclidean distance.
25. Prove

Proposition 2.22 (Fundamental Sequence with Convergent Subsequence). *If a fundamental sequence $(x_n)_{n \in \mathbb{N}}$ in a metric space (X,ρ) contains a subsequence $(x_{n(k)})_{k \in \mathbb{N}}$ such that*

$$\exists x \in X: \ x_{n(k)} \to x, \ k \to \infty, \ in \ (X,\rho),$$

then

$$x_n \to x, \ n \to \infty, \ in \ (X,\rho).$$

26. Prove

Proposition 2.23 (Characterization of Completeness of Product Space). *The Cartesian product (X,ρ) of metric spaces (X_1,ρ_1) and (X_2,ρ_2) (see Problem 5) is complete iff each metric space (X_i,ρ_i), $i = 1, 2$, is complete.*

27. Let s be the set of all real/complex *sequences*.
    (a) Prove that s is a metric space, relative to the mapping

$$s \ni x := (x_k)_{k \in \mathbb{N}}, y := (y_k)_{k \in \mathbb{N}} \mapsto \rho(x,y) := \sum_{k=1}^\infty \frac{1}{2^k} \frac{|x_k - y_k|}{|x_k - y_k| + 1}.$$

    (b) Describe *convergence* in (s,ρ).
    (c) Prove that the space (s,ρ) is *complete*.
28. Show that

$$\mathbb{R} \ni x, y \mapsto d(x,y) := |\arctan x - \arctan y|$$

is metric on $\mathbb{R}$, and that the metric space $(\mathbb{R}, d)$ is *incomplete*.
29. Let (X,ρ) and (Y,σ) be *isometric* metric spaces. Prove that
    (a) (X,ρ) is *separable* iff (Y,σ) is *separable*;
    (b) (X,ρ) is *complete* iff (Y,σ) is *complete*.
30. (a) Prove

Proposition 2.24 (Finite Intersections of Open Dense Sets). *In a metric space (X,ρ), any finite intersection of open dense sets is dense.*

    (b) Give an example showing that the condition of *openness* is essential and cannot be dropped.

(c) Give an example showing that an infinite intersection of open dense sets need not be dense.

31. Show that the set P of all *polynomials* with real/complex coefficients is of the *first category* in $(C[a,b], \rho_\infty)$ ($-\infty < a < b < \infty$).

32. (a) Prove

> **Proposition 2.25** (Complement of a First-Category Set). *In a complete metric space* (X, ρ), *for any first-category set A, the complement A^c is a second-category set.*

(b) Give an example showing that the converse statement is not true, i. e., in a complete metric space, a set, whose complement is a second-category set, need not be of first category.

33. Prove

> **Proposition 2.26** (Unique Fixed Point). *Let X be a nonempty set and $T : X \to X$ be a mapping. If, for an $n \in \mathbb{N}$, $n \geq 2$, there exists a unique fixed point $x \in X$ for T^n, then x is also a unique fixed point for T.*

34. From the *Banach Fixed-Point Theorem* (Theorem 2.33) and the proposition of Problem 33, we derive the following:

> **Corollary 2.13** (Unique Fixed Point). *If (X, ρ) is a complete metric space, and a mapping $T : X \to X$ is such that, for some $n \in \mathbb{N}$, $n \geq 2$, T^n is a contraction on X, then T has a unique fixed point in X.*

35. Show that the *Banach Fixed-Point Theorem* (Theorem 2.33) fails to hold if the *contractiveness* condition for T is relaxed to *weak contractiveness* (see Exercises 2.79 (b) and 2.80).

36. Apply the *Linear Systems Theorem* (Theorem 2.36) to show that the linear system

$$\begin{bmatrix} -3/4 & -1/4 & 1/4 \\ -1/4 & -3/4 & 1/4 \\ 1/4 & -1/4 & -3/4 \end{bmatrix} \begin{bmatrix} x \\ y \\ z \end{bmatrix} = \begin{bmatrix} b_1 \\ b_2 \\ b_3 \end{bmatrix},$$

has a *unique solution* for any $(b_1, b_2, b_3) \in \mathbb{C}^3$.

37. Apply *Picard's Global Existence and Uniqueness Theorem* (Theorem 2.39) to find the solution the initial value problem

$$y' = y, \ y(0) = 1,$$

on $\mathbb{R}$.

38. Find the solution of the initial value problem

$$y'(t) = \begin{bmatrix} -1 & 0 \\ 0 & 1 \end{bmatrix} y(t), \ y(0) = (1, 1),$$

on $\mathbb{R}$.

39. **Definition 2.46** (Centered Collections of Sets). A collection $\mathscr{C}$ of subsets of a set X is said to have the *finite intersection property*, or to be *centered*, if the intersection of any finite subcollection of $\mathscr{C}$ is nonempty.

Examples 2.47.
(a) The collection $\mathscr{C} = \{[n, \infty)\}_{n \in \mathbb{N}}$ is *centered*.
(b) The collection $\mathscr{C} = \{[n, n+1)\}_{n \in \mathbb{N}}$ is *not centered*.
(c) The collection of all open/closed balls in a metric space (X, ρ), centered at a fixed point $x \in X$, is *centered*.

Remark 2.85. A centered collection of sets cannot contain $\emptyset$.

(a) Verify the examples.
(b) Prove

> **Theorem 2.62** (Centered Collection Characterization of Compactness).
> *A metric space (X, ρ) is compact iff every centered collection $\mathscr{C}$ of its closed subsets has a nonempty intersection.*

> **Corollary 2.14** (Nested Sequences of Closed Sets in Compact Spaces).
> *If $(C_n)_{n \in \mathbb{N}}$ is a sequence of nonempty closed sets in a compact metric space (X, ρ) such that*
> $$C_n \supseteq C_{n+1}, \ n \in \mathbb{N},$$
> *then*
> $$\bigcap_{n=1}^{\infty} C_n \neq \emptyset.$$

40. Prove

> **Theorem 2.63** (Centered Collections of Compact Sets). *In a metric space (X, ρ), every centered collection of compact sets has a nonempty intersection*

Hint. For an arbitrary centered collection $\mathscr{C} = \{C_i\}_{i \in I}$ of compact sets in (X, ρ), fix a $j \in I$ and, in the *compact subspace* (C_j, ρ), consider the centered collection of closed subsets $\{C_i \cap C_j\}_{i \in I}$.

41. Use the prior theorem to prove

> **Theorem 2.64** (Cantor's Intersection Theorem). *If $(C_n)_{n \in \mathbb{N}}$ is a sequence of nonempty compact sets in a metric space (X, ρ) such that*
> $$C_n \supseteq C_{n+1}, \ n \in \mathbb{N},$$
> *then*
> $$\bigcap_{n=1}^{\infty} C_n \neq \emptyset.$$

42. Determine which of the following sets are compact in $l_p^{(2)}$ $(1 \le p \le \infty)$.
    (a) $\{(x, y) \mid x = 0\}$,
    (b) $\{(x, y) \mid 0 < x^2 + y^2 \le 1\}$,
    (c) $\{(x, y) \mid |y| \le 1, x^2 + y^2 \le 4\}$,
    (d) $\{(x, y) \mid y = \sin\frac{1}{x}, 0 < x \le 1\} \cup \{(x, y) \mid x = 0, -1 \le y \le 1\}$.

43. Show that the set

$$\{(x_n)_{n \in \mathbb{N}} \mid |x_n| \le 1/n\}_{n \in \mathbb{N}}$$

is *compact* in l_p $(1 < p < \infty)$ and (c_0, ρ_∞).

44. Prove

Proposition 2.27 (Characterization of Compactness of Product Space).
The Cartesian product (X, ρ) of metric spaces (X_1, ρ_1) and (X_2, ρ_2) (see Problem 5) is compact iff each metric space (X_i, ρ_i), $i = 1, 2$, is compact.

Hint. Use the *sequential approach*.

45. Let (X_1, ρ_1) and (X_2, ρ_2) be metric spaces and $T : X_1 \to X_2$ be a *continuous transformation*. Prove that, if (X_1, ρ_1) is *compact*, then the graph of T

$$G_T := \{(x, Tx) \in X_1 \times X_2 \mid x \in X_1\}$$

is a *compact set* in the product space (X, ρ) of (X_1, ρ_1) and (X_2, ρ_2) (see Problem 5).

46. **Definition 2.47** (Distance Between Sets). For nonempty sets A and B in a metric space (X, ρ), the nonnegative number

$$\rho(A, B) := \inf_{x \in A, y \in B} \rho(x, y)$$

is called the *distance between the sets A and B*.

Prove

Proposition 2.28 (Distance Between Compact Sets). *Prove that, for nonempty compact sets A and B in a metric space (X, ρ), the distance between A and B is attained, i. e.,*

$$\exists x_0 \in A, y_0 \in B : \rho(x_0, y_0) = \rho(A, B),$$

which, provided $A \cap B = \emptyset$, implies that $\rho(A, B) > 0$.

Hint. Use the results of Problem 7 and 44.

47. (a) Prove

Proposition 2.29 (Nearest Point Property in the n-Space). *Let A be a nonempty closed set in the (real or complex) space $l_p^{(n)}$ $(n \in \mathbb{N}, 1 \le p \le \infty)$. Then, for any $x \in X$, in A, there is a nearest point to x, i. e.,*

$$\exists y \in A : \rho(x, y) = \inf_{p \in A} \rho(x, p) =: \rho(x, A).$$

(b) Give an example showing that a *farthest point* from x in A need not exist.

48. * Prove

Theorem 2.65 (Weak Contraction Principle on a Compact). *A weak contraction T on a compact metric space (X,ρ) has a unique fixed point.*

Hint. Consider the function $f(x) := \rho(x, Tx)$, $x \in X$.

Corollary 2.15 (Weak Contraction). *A weak contraction T on a metric space (X,ρ) such that $T(X)$ is compact in (X,ρ) has a unique fixed point.*

49. Let (X,ρ) be a metric space and $(f_n)_{n\in\mathbb{N}}$ be a sequence in $C(X)$ equicontinuous on X, i.e.,

$$\forall \varepsilon > 0 \, \exists \delta > 0 \, \forall n \in \mathbb{N}, \ \forall x', x'' \in X \text{ with } \rho(x', x'') < \delta : \ |f_n(x') - f_n(x'')| < \varepsilon.$$

Show that, if $(f_n)_{n\in\mathbb{N}}$ *pointwise converges* on X to a function f, i.e.,

$$\forall x \in X : f_n(x) \to f(x), \ n \to \infty,$$

then f is *uniformly continuous* on X.

50. Show in *two different ways* that a nontrivial sphere/closed ball is *not* a compact set in $(C[a,b], \rho_\infty)$ $(-\infty < a < b < \infty)$.

Hint. Consider the unit sphere $S(0,1)$ in $(C[0,1], \rho_\infty)$.

3 Vector Spaces, Normed Vector Spaces, and Banach Spaces

In this chapter, we introduce *normed vector spaces* and study their properties emerging from the remarkable interplay between their linear and topological structures.

3.1 Vector Spaces

First, we introduce and study *vector spaces* endowed with linear structure alone.

3.1.1 Definition, Examples, Properties

Definition 3.1 (Vector Space). A *vector space* (also a *linear space*) over the scalar field $\mathbb{F}$ of real or complex numbers (i. e., $\mathbb{F} = \mathbb{R}$ or $\mathbb{F} = \mathbb{C}$) is a set X of elements, also called *vectors*, equipped with the two *linear operations* of

- *vector addition:* $X \ni x, y \mapsto x + y \in X$, and
- *scalar multiplication:* $\mathbb{F} \ni \lambda, X \ni x \mapsto \lambda x \in X$,

subject to the following *vector space axioms*:

1.	$x + y = y + x,\, x, y \in X.$	*Commutativity*
2.	$(x + y) + z = x + (y + z),\, x, y, z \in X.$	*Associativity*
3.	$\exists\, 0 \in X : x + 0 = x,\, x \in X.$	Existence of *additive identity* (*zero vector*)
4.	$\forall x \in X \,\exists\, -x \in X : x + (-x) = 0.$	Existence of *additive inverses* or *opposite vectors*
5.	$\lambda(\mu x) = (\lambda\mu)x,\, \lambda, \mu \in \mathbb{F},\, x \in X.$	*Associativity of Scalar Multiplication*
6.	$1x = x,\, x \in X.$	*Neutrality of Scalar Identity*
7.	$\lambda(x + y) = \lambda x + \lambda y,\, \lambda \in \mathbb{F},\, x, y \in X.$	*Right Distributivity*
8.	$(\lambda + \mu)x = \lambda x + \mu x,\, \lambda, \mu \in \mathbb{F},\, x \in X.$	*Left Distributivity*

Remarks 3.1.

- Henceforth, without further specifying, we understand that $\mathbb{F}$ stands for $\mathbb{R}$ or $\mathbb{C}$, the underlying vector space being called *real* in the former case and *complex* in the latter.
- For any complex vector space X, the *associated real space* $X_{\mathbb{R}}$ is obtained by restricting the scalars to $\mathbb{R}$.

https://doi.org/10.1515/9783110600988-003

Examples 3.1.

1. The sets of all *bound vectors* (directed segments with a common initial point) on a line, in a plane, or in the 3-space with the ordinary addition (by the *parallelogram law*) and the usual scalar multiplication are *real vector spaces*.

2. The sets of all *free vectors* (directed segments) on a line, in a plane, or in the 3-space (any two vectors of the same direction and length considered identical) with the ordinary addition (by the *triangle* or *parallelogram law*) and the usual scalar multiplication are *real vector spaces*.

3. The set $\mathbb{R}$ is a *real vector space*, and the set $\mathbb{C}$ is a *complex vector space*.

4. The *n-space* $\mathbb{F}^n$ ($n \in \mathbb{N}$) of all ordered *n-tuples* of numbers with the *componentwise linear operations*

$$(x_1, \ldots, x_n) + (y_1, \ldots, y_n) := (x_1 + y_1, \ldots, x_n + y_n),$$
$$\lambda(x_1, \ldots, x_n) := (\lambda x_1, \ldots, \lambda x_n),$$

is a vector space over $\mathbb{F}$.

5. The set $M_{m \times n}$ ($m, n \in \mathbb{N}$) of all $m \times n$ *matrices* with entries from $\mathbb{F}$ and the *entrywise linear operations*

$$[a_{ij}] + [b_{ij}] := [a_{ij} + b_{ij}],$$
$$\lambda[a_{ij}] := [\lambda a_{ij}]$$

is a vector space over $\mathbb{F}$.

6. The set s of all $\mathbb{F}$-termed sequences with the *termwise linear operations*

$$(x_n)_{n \in \mathbb{N}} + (x_n)_{n \in \mathbb{N}} := (x_n + y_n)_{n \in \mathbb{N}},$$
$$\lambda(x_n)_{n \in \mathbb{N}} := (\lambda x_n)_{n \in \mathbb{N}}$$

is a vector space over $\mathbb{F}$.

7. Due to *Minkowski's Inequality for Sequences* (Theorem 2.4), the set $l_p(\mathbb{F})$ ($1 \leq p \leq \infty$) with the *termwise linear operations* is a vector space over $\mathbb{F}$.

8. The set $F(T)$ of all $\mathbb{F}$-valued functions on a nonempty set T with the *pointwise linear operations*

$$(f + g)(t) := f(t) + g(t), \ t \in T,$$
$$(\lambda f)(t) := \lambda f(t), \ t \in T,$$

is a vector space over $\mathbb{F}$.

   However, the set $F_+(T)$ of all *nonnegative* functions on a nonempty set T with the *pointwise linear operations* is *not* a real vector space.

9. The set $M(T)$ of all $\mathbb{F}$-valued functions *bounded* on a nonempty set T, with the pointwise linear operations, is a vector space over $\mathbb{F}$.

   However, the set $U(T)$ of all $\mathbb{F}$-valued functions *unbounded* on an *infinite* set T, with the pointwise linear operations, is *not* a vector space over $\mathbb{F}$.

10. The set $R[a, b]$ ($-\infty < a < b < \infty$) of all $\mathbb{F}$-valued functions *Riemann integrable* on $[a, b]$ with the pointwise linear operations is a vector space over $\mathbb{F}$.
11. Let (X, ρ) be a *metric space*. The set $C(X)$ of all $\mathbb{F}$-valued functions *continuous* on X with the pointwise linear operations is a vector space over $\mathbb{F}$.

   In particular, for $X = [a, b]$ ($-\infty < a < b < \infty$), $C[a, b]$ is a vector space over $\mathbb{F}$.
12. The set P of all *polynomials* with coefficients from $\mathbb{F}$, and pointwise linear operations, is a vector space over $\mathbb{F}$.
13. The set P_n of all polynomials of degree *at most* n ($n \in \mathbb{Z}_+$) with coefficients from $\mathbb{F}$, and the pointwise linear operations, is also a vector space over $\mathbb{F}$.

   However, the set $\hat{P}_n$ of all polynomials of degree n ($n \in \mathbb{N}$) with coefficients from $\mathbb{F}$, and the pointwise linear operations, is *not* a vector space over $\mathbb{F}$.

Exercise 3.1. Verify.

Theorem 3.1 (Properties of Vector Spaces). *In a vector space X over $\mathbb{F}$, the following hold:*

(1) *the zero vector 0 is unique;*
(2) *for each vector $x \in X$, the opposite vector $-x$ is unique;*
(3) *$\forall x \in X: -(-x) = x$;*
(4) *for $\lambda \in \mathbb{F}$ and $x \in X$, $\lambda x = 0$ iff $\lambda = 0$ or $x = 0$;*   *(Zero Product Rule)*
(5) *$(-1)x = -x$.*

Remark 3.2. Observe that the same notation 0 is used to designate both the *scalar zero* and the *zero vector*, such an economy of symbols being a rather common practice.

Proof.

(1) Assume that a vector $0'$ is also an additive identity. Then, by the *vector space axioms*,

$$0' = 0' + 0 = 0.$$

(2) Assume that y and z are *additive inverses* of x, i.e., $x + y = x + z = 0$. Then, by the *vector space axioms*,

$$y = y + 0 = y + (x + z) = (y + x) + z = 0 + z = z.$$

(3) Immediately follows from (2).
(4) "*If*" part. If $\lambda = 0$, for any $x \in X$, by the *vector space axioms*, the following hold:

$$Ox = (0 + 0)x = 0x + 0x,$$
$$0x + (-0x) = [0x + 0x] + (-0x),$$
$$0 = 0x + [0x + (-0x)],$$
$$0 = 0x + 0,$$
$$0 = 0x.$$

The case of $x = 0$ is considered similarly.

Exercise 3.2. Consider.

"Only if" part. Let us prove this part *by contradiction* assuming that, for some $\lambda \neq 0$ and $x \neq 0$,

$$\lambda x = 0.$$

Multiplying through by $1/\lambda$, by the *vector space axioms* and the *"if"* part, we have the following:

$$(1/\lambda)(\lambda x) = (1/\lambda)0,$$
$$((1/\lambda)\lambda)x = 0,$$
$$1x = 0,$$
$$x = 0,$$

which is a *contradiction*, proving the statement.

(5) By the *vector space axioms* and (4),

$$x + (-1)x = 1x + (-1)x = \big[1 + (-1)\big]x = 0x = 0.$$

Whence, by (2), $(-1)x = -x$. $\qquad\square$

3.1.2 Homomorphisms and Isomorphisms

Important in the theory of vector spaces are the following notions of *homomorphism* and *isomorphism*:

Definition 3.2 (Homomorphism of Vector Spaces). Let X and Y be vector spaces over $\mathbb{F}$. A *homomorphism of X to Y* is a mapping $T : X \rightarrow Y$ preserving linear operations:

$$T(\lambda x + \mu y) = \lambda Tx + \mu Ty, \ \lambda, \mu \in \mathbb{F}, x, y \in X.$$

When $Y = X$, a homomorphism $T : X \rightarrow X$ is called an *endomorphism of X*.

Examples 3.2.
1. Multiplication by an arbitrary number $\lambda \in \mathbb{F}$ in a vector space X over $\mathbb{F}$,

$$X \ni x \rightarrow Ax := \lambda x \in X,$$

is an *endomorphism* of X.
2. Multiplication by an $m \times n$ $(m, n \in \mathbb{N})$ matrix $[a_{ij}]$ with entries from $\mathbb{F}$,

$$\mathbb{F}^n \ni x \mapsto Ax := [a_{ij}]x \in \mathbb{F}^m,$$

is a *homomorphism* of $\mathbb{F}^n$ to $\mathbb{F}^m$ and, provided $m = n$, is an *endomorphism* of $\mathbb{F}^n$.

Exercise 3.3. Verify.

Definition 3.3 (Isomorphism of Vector Spaces). Let X and Y be vector spaces over $\mathbb{F}$.

An *isomorphism of X to Y* is a *one-to-one homomorphism: $T : X \to Y$.* It is said to *isomorphically embed X in Y.*

If an isomorphism $T : X \to Y$ is *onto* (i. e., *surjective*), it is called an *isomorphism between X and Y*, and the spaces are called *isomorphic*.

An isomorphism between X and itself is called an *automorphism of X.*

Examples 3.3.
1. Multiplication by a *nonzero* number $\lambda \in \mathbb{F} \setminus \{0\}$ in a vector space X is an *automorphism* of X.
2. Multiplication by a *nonsingular* $n \times n$ ($n \in \mathbb{N}$) matrix $[a_{ij}]$ with entries from $\mathbb{F}$ is an *automorphism* of $\mathbb{F}^n$.
3. For any $n \in \mathbb{Z}_+$, the mapping

$$\mathbb{F}^{n+1} \ni (a_0, a_1, \ldots, a_n) \mapsto \sum_{k=0}^{n} a_k t^k \in P_n$$

is an *isomorphism* between $\mathbb{F}^{n+1}$ and the space P_n of all polynomials of degree at most n with coefficients from $\mathbb{F}$.

Exercise 3.4. Verify.

Remarks 3.3.
– Being isomorphic is an *equivalence relation* (*reflexive, symmetric,* and *transitive*) on the set of all vector spaces.

   Exercise 3.5. Verify.

– Isomorphic vector spaces are *linearly indistinguishable*, i. e., identical as vector spaces.

3.1.3 Subspaces

As well as for metric spaces (see Section 2.3), we can consider *subspaces* of vector spaces.

Definition 3.4 (Subspace of a Vector Space). A subset Y of a vector space X over $\mathbb{F}$, which is itself a vector space over $\mathbb{F}$ relative to the induced linear operations is called a *subspace* of X.

Remarks 3.4.

- Thus, $Y \subseteq X$ is a *subspace* of a vector space X over $\mathbb{F}$ *iff* Y is *closed under the linear operations*, i. e.,
  (a) $Y + Y \subseteq Y$ and
  (b) $\lambda Y \subseteq Y, \lambda \in \mathbb{F}$,
  or equivalently,

$$\forall x, y \in Y, \ \forall \lambda, \mu \in \mathbb{F} : \lambda x + \mu y \in Y.$$

- Each nonzero vector space $X \neq \{0\}$ has at least two subspaces: the *zero subspace* $\{0\}$ and the whole X, called *trivial*.
- Each subspace of a vector space always contains at least the *zero vector* 0, and hence, cannot be empty.
- A subspace Y of X such that $Y \neq X$ is called a *proper subspace* of X.

Examples 3.4.

1. The *isomorphic embedding*

$$\{(x, 0) \mid x \in \mathbb{R}\}$$

   of $\mathbb{R}$ into the real space $\mathbb{C}_{\mathbb{R}}$ associated with $\mathbb{C}$ (see Remarks 3.1) is a *subspace* of $\mathbb{C}_{\mathbb{R}}$, the latter and the space $\mathbb{R}^2$ being isomorphic.

2. Due to the set-theoretic inclusions

$$c_{00} \subset l_p \subset l_q \subset c_0 \subset c \subset l_\infty,$$

   where $1 \le p < q < \infty$, (see Examples 2.3), and the closedness under the termwise linear operations, each of the above sequence spaces is a *proper subspace* of the next one.

3. Due to the set-theoretic inclusion

$$M(T) \subseteq F(T),$$

   and the closedness under the pointwise linear operations, $M(T)$ is a subspace of $F(T)$.

   Exercise 3.6. When is it a *proper subspace*?

4. Due to the set-theoretic inclusions

$$P_n \subset P \subset C[a, b] \subset M[a, b] \subset F[a, b],$$

   ($n \in \mathbb{Z}_+$, $-\infty < a < b < \infty$), and the closedness under the pointwise linear operations, each of the above function spaces is a *proper subspace* of the next one.

5. The set
$$Y := \{(x, y) \in \mathbb{R}^2 \mid xy \geq 0\}$$
   is *not* a subspace of $\mathbb{R}^2$.

Exercise 3.7.
(a) Verify.
(b) Describe all subspaces in $\mathbb{R}$, $\mathbb{R}^2$, and $\mathbb{R}^3$.

Exercise 3.8. Let Y be a *subspace* in a vector space X over $\mathbb{F}$. Show that
(a) $Y + Y = Y$,
(b) $\lambda Y = Y$, $\lambda \in \mathbb{F} \setminus \{0\}$.

Proposition 3.1 (Sum and Intersection of Subspaces). *For a vector space X, the following hold:*
(1) *the sum of an arbitrary finite collection* $\{Y_1, \ldots, Y_n\}$ $(n \in \mathbb{N})$ *of subspaces of X*
$$\sum_{i=1}^{n} Y_i := \left\{ \sum_{i=1}^{n} y_i \,\middle|\, y_i \in Y_i, \ i = 1, \ldots, n \right\}$$
   is a subspace of X;
(2) *the intersection of an arbitrary collection* $\{Y_i\}_{i \in I}$ *of subspaces of X*
$$\bigcap_{i \in I} Y_i$$
   is a subspace of X.

Exercise 3.9.
(a) Prove.
(b) Give an example showing that the *union* of subspaces need not be a subspace.

The following statement gives conditions *necessary and sufficient* for the union of two subspaces of a vector space to be a subspace.

Proposition 3.2 (Union of Subspaces). *In a vector space X, the union $Y \cup Z$ of subspaces Y and Z is a subspace iff $Y \subseteq Z$, or $Z \subseteq Y$.*

Exercise 3.10. Prove.

Hint. Prove the *"only if"* part *by contrapositive* or *by contradiction*.

3.1.4 Spans and Linear Combinations

By the *Sum and Intersection of Subspaces Proposition* (Proposition 3.1), the following notion is well defined:

Definition 3.5 (Linear Span). Let S be a subset of a vector space X. Then

$$\text{span}(S) := \bigcap_{Y \text{ is a subspace of } X,\, S \subseteq Y} Y$$

is the *smallest subspace of X containing S*, called the *span*. It is also called *linear span* or *linear hull* of S, or the *subspace generated by S*.

Exercise 3.11. Show that, in a vector space X,
(a) $\text{span}(\emptyset) = \{0\}$,
(b) $\text{span}(X) = X$, and
(c) for an arbitrary finite collection $\{Y_1, \ldots, Y_n\}$ ($n \in \mathbb{N}$) of subspaces of X,

$$\text{span}\left(\bigcup_{k=1}^{n} Y_k\right) = \sum_{k=1}^{n} Y_k.$$

Definition 3.6 (Linear Combination). Let X be a vector space over $\mathbb{F}$. A *linear combination* of vectors $x_1, \ldots, x_n \in X$ ($n \in \mathbb{N}$) with coefficients $\lambda_1, \ldots, \lambda_n \in \mathbb{F}$ is the sum

$$\sum_{k=1}^{n} \lambda_k x_k.$$

Remark 3.5. The linear combination with $\lambda_1 = \cdots = \lambda_n = 0$ is called *trivial*. Obviously, any *trivial* linear combination is the *zero vector*. The *converse* is not true.

Exercise 3.12. Give a corresponding example.

Proposition 3.3 (Span's Structure). *For a nonempty subset S of a vector space X over $\mathbb{F}$, span(S) is the set of all possible linear combinations of the elements of S. I. e.,*

$$\text{span}(S) = \left\{ \sum_{k=1}^{n} \lambda_k x_k \mid x_1, \ldots, x_n \in S,\, \lambda_1, \ldots, \lambda_n \in \mathbb{F},\, n \in \mathbb{N} \right\}.$$

Exercise 3.13. Prove.

Remark 3.6. In particular, for a singleton $\{x\} \subseteq X$,

$$\text{span}(\{x\}) = \{\lambda x \mid \lambda \in \mathbb{F}\}.$$

Exercise 3.14.
(a) In c_0, describe $\text{span}(\{e_n := (\delta_{nk})_{k \in \mathbb{N}}\}_{n \in \mathbb{N}})$, where δ_{nk} is the *Kronecker delta*.
(b) In $C[a, b]$ ($-\infty < a < b < \infty$), describe $\text{span}(\{t^n\}_{n \in \mathbb{Z}_+})$.

3.1.5 Linear Independence, Hamel Bases, Dimension

Fundamental for vector spaces is the concept of *linear (in)dependence*.

3.1.5.1 Linear Independence/Dependence

Definition 3.7 (Linearly Independent/Dependent Set). A nonempty subset S of a vector space X is called *linearly independent* if none of its vectors is a linear combination of some other vectors of S, i. e.,

$$\forall x \in S : x \notin \operatorname{span}(S \setminus \{x\}),$$

and is said to be *linearly dependent* otherwise.

We also say that the vectors of S are *linearly independent/dependent*.

Remarks 3.7.

-   A singleton $\{x\}$ is a linearly dependent set in a vector space X iff $x = 0$.
    Thus, the notion of linear independence is well defined only for a *nonzero* vector space $X \neq \{0\}$.
-   A nonempty subset S of a vector space X is linearly independent *iff* any finite subset of S is linearly independent and linearly dependent *iff* there exists a finite subset of S that is linearly dependent.
-   Furthermore, a nonempty subset S of a vector space X is linearly independent *iff* any nonempty subset of S is linearly independent and linearly dependent *iff* there exists a nonempty subset of S that is linearly dependent.
-   Since a linearly independent set S in a vector space X cannot contain a linearly dependent subset, in particular, it cannot contain the *zero vector* 0.

Examples 3.5.

1. A two-vector set $\{x, y\}$ is a linearly dependent in a nonzero vector space X *iff* one of the two vectors is a constant multiple of the other, i. e.,

$$\exists \lambda \in \mathbb{F} : y = \lambda x, \text{ or } x = \lambda y.$$

2. The set $\{(1, 0), (1, 1)\}$ is linearly independent, and the set $\{(1, 0), (2, 0)\}$ is linearly dependent in $\mathbb{R}^2$.
3. The same sets $\{(1, 0), (1, 1)\}$ and $\{(1, 0), (2, 0)\}$ are linearly dependent in $\mathbb{C}^2$.

Exercise 3.15. Verify the prior remarks and examples.

Proposition 3.4 (Characterization of Linear Independence). *A nonempty subset S of a nonzero vector space X is linearly independent iff only the trivial linear combinations of its vectors are equal to 0, i. e.,*

$$\forall \{x_1, \ldots, x_n\} \subseteq S \ (n \in \mathbb{N}) : \sum_{k=1}^{n} \lambda_k x_k = 0 \iff \lambda_1 = \cdots = \lambda_n = 0,$$

and is linearly dependent iff there exists a nontrivial linear combination of its vectors equal to 0, i.e.,

$$\exists \{x_1, \ldots, x_n\} \subseteq S \ (n \in \mathbb{N}), \ \exists c_1, \ldots, c_n \in \mathbb{F} \ not \ all \ zero : \sum_{k=1}^{n} \lambda_k x_k = 0.$$

Exercise 3.16. Prove.

Hint. Reason *by contrapositive*.

Remark 3.8. The prior characterization is often used as a definition of linear independence, especially for finite sets of vectors.

3.1.5.2 Hamel Bases

Definition 3.8 (Basis of a Vector Space). A *basis* B (also *Hamel*[1] *basis* or *algebraic basis*) for a nonzero vector space X is a *maximal linearly independent* subset of X, or equivalently, a linearly independent set of X spanning the entire X.

Remark 3.9. The *"maximality"* is understood relative to the set-theoretic inclusion $\subseteq$ (see Section A.2).

Exercise 3.17. Prove the equivalence of the two definitions.

Examples 3.6.
1. The singleton $\{1\}$ is a basis for both $\mathbb{R}$ and $\mathbb{C}$, but not for the real space $\mathbb{C}_{\mathbb{R}}$ (see Remarks 3.1), whose bases coincide with those of $\mathbb{R}^2$.
2. The sets $\{(1,0), (0,1)\}$ and $\{(1,0), (1,1)\}$ are both bases for $\mathbb{R}^2$.
3. The set of n $(n \in \mathbb{N})$ ordered n-tuples

$$\{e_1 := (1,0,\ldots,0), e_2 := (0,1,\ldots,0),\ldots, e_n := (0,0,\ldots,1)\}$$

   is a basis for the n-space $\mathbb{F}^n$ called the *standard basis*.
4. The set $E := \{e_n := (\delta_{nk})_{k \in \mathbb{N}}\}_{n \in \mathbb{N}}$, where δ_{nk} is the *Kronecker delta*, is a basis for c_{00}, but not for c_0.
5. The set $\{1, t, \ldots, t^n\}$ $(n \in \mathbb{Z}_+)$ is a basis for P_n, but not for P.
6. The set $\{t^n\}_{n \in \mathbb{Z}_+}$ is a basis for P, but not for $C[a,b]$ $(-\infty < a < b < \infty)$.

Exercise 3.18. Verify.

The following fundamental statement, whose proof is based on *Zorn's Lemma* (see Section A.3), establishes the existence of a basis for any nonzero vector space.

---

[1] Georg Karl Wilhelm Hamel (1877–1954).

Theorem 3.2 (Basis Theorem). *Each linearly independent set S in a nonzero vector space X ≠ {0} can be extended to a basis B for X.*

Proof. Let $\mathscr{L}$ be the collection of all linearly independent sets in X partially ordered by the set-theoretic inclusion $\subseteq$ (see Section A.2) and $S \in \mathscr{L}$ be arbitrary.

For an arbitrary *chain* $\mathscr{C}$ in $(\mathscr{L}, \subseteq)$, the set

$$L := \bigcup_{C \in \mathscr{C}} C \subseteq X,$$

is *linearly independent*, and is an *upper bound* of $\mathscr{C}$ in $(\mathscr{L}, \subseteq)$.

Exercise 3.19. Verify.

Hence, by *Zorn's Lemma (Precise Version)* (Theorem A.6), there is a *maximal element* B in $(\mathscr{L}, \subseteq)$, i. e., a *maximal linearly independent* subset in X such that $S \subseteq B$, which completes the proof. □

Remark 3.10. A basis $B := \{x_i\}_{i \in I}$ for a nonzero vector space X is never unique. One can always produce a new basis $B' := \{x_i'\}_{i \in I}$ via multiplying each basis vector by a scalar $\lambda \neq 0, 1$:

$$x_i' := \lambda x_i, \ i \in I.$$

Exercise 3.20. Explain.

Theorem 3.3 (Basis Representation Theorem). *A nonempty subset $B := \{x_i\}_{i \in I}$ of a nonzero vector space $X \neq \{0\}$ over $\mathbb{F}$ is a basis for X iff any vector $x \in X$ can be uniquely represented as a sum*

$$\sum_{i \in I} c_i x_i,$$

in which only a finite number of the coefficients $c_i \in \mathbb{F}, i \in I$, are nonzero.

Exercise 3.21. Prove.

Remark 3.11. The prior representation is called the *representation of x relative to basis B* and the numbers $c_i, i \in I$, are called the *coordinates of x relative to basis B*.

Corollary 3.1 (Representation of Nonzero Vectors). *Each nonzero vector x of a nonzero vector space X ≠ {0} with a basis B allows a unique representation as a linear combination of vectors of B with nonzero coefficients.*

3.1.5.3 Dimension

Theorem 3.4 (Dimension Theorem). *All bases of a nonzero vector space have equally many elements.*

Proof. Let B and B' be two arbitrary bases for X.

We show that their *cardinalities* $|B|$ and $|B'|$ are equal (see Section 1.1.2).

The case when B or B' is a *finite set* is considered in *linear algebra* (see, e. g., [54]) by reducing it to the question of the existence of nontrivial solutions of a homogeneous linear system, which has more unknowns than equations.

Exercise 3.22. Fill in the details.

Suppose that both B or B' are *infinite*. By the *Representation of Nonzero Vectors Corollary* (Corollary 3.1), each $y \in B'$ is a linear combination with nonzero coefficients of some elements $x_1, \ldots, x_n \in B$ $(n \in \mathbb{N})$, and at most n elements of B' can be associated with the set $\{x_1, \ldots, x_n\}$, or its subset in such a way.

Exercise 3.23. Explain.

Since B is infinite, by the *Cardinality of the Collection of Finite Subsets* (Proposition 1.2), the cardinality of the collection of all its finite subsets is $|B|$. Hence, considering that $\aleph_0 \leq |B|$, where $\aleph_0$ is the cardinality of $\mathbb{N}$ (see Examples 1.1), by the arithmetic of cardinals (see, e. g., [29, 33, 41, 52]),

$$|B'| \leq \aleph_0 |B| = |B|$$

Similarly, $|B| \leq |B'|$. Whence, by the *Schröder–Bernstein Theorem* (Theorem 1.1) (see Remark 1.4), we infer that $|B| = |B'|$, which completes the proof. $\square$

By the *Dimension Theorem*, the following fundamental notion is well defined.

Definition 3.9 (Dimension of a Vector Space). The *dimension* of a nonzero vector space $X \neq \{0\}$ is the *common cardinality* of all bases of X.

Notation. $\dim X$.

The dimension of a trivial space is naturally defined to be 0.

We call a vector space *finite-dimensional* if its dimension is a finite number and *infinite-dimensional* otherwise.

If $\dim X = n$ with some $n \in \mathbb{Z}_+$, the vector space is called *n-dimensional*.

Examples 3.7.
1. $\dim X = 0$ *iff* $X = \{0\}$.
2. $\dim \mathbb{C} = \begin{cases} 1 & \text{over } \mathbb{C}, \\ 2 & \text{over } \mathbb{R}. \end{cases}$
3. $\dim F^n = n$ over $\mathbb{F}$ $(n \in \mathbb{N})$.
4. $\dim P_n = n + 1$ $(n \in \mathbb{Z}_+)$.
5. $\dim P = \dim c_{00} = \aleph_0$.
6. $\dim C[a, b] = \mathfrak{c}$ $(-\infty < a < b < \infty)$, where $\mathfrak{c}$ is the *cardinality of the continuum*, i. e., $\mathfrak{c} = |\mathbb{R}|$ (see Examples 1.1 and Remark 3.29).

Exercise 3.24. Explain 1–5. For 6, show that $\dim C[a, b] \geq \mathfrak{c}$.

As the following theorem shows, all vector spaces of the same dimension are *linearly indistinguishable*.

Theorem 3.5 (Isomorphism Theorem). *Two nonzero vector spaces X and Y over $\mathbb{F}$ are isomorphic iff $\dim X = \dim Y$.*

Proof. "Only if" part. Suppose that X and Y are isomorphic and let $T : X \to Y$ be an *isomorphism* between X and Y.

Then the set $B \subseteq X$ is a *basis* for X iff $T(B)$ is a *basis* for Y.

Exercise 3.25. Verify.

This, considering that T is a *bijection*, implies that $\dim X = \dim Y$.

"*If*" part. Suppose $\dim X = \dim Y$.

Then we can choose bases $B_X := \{x_i\}_{i \in I}$ for X and $B_Y := \{y_i\}_{i \in I}$ for Y sharing the *indexing set* I whose *cardinality* $|I| = \dim X = \dim Y$ and establish an *isomorphism* T between X and Y by matching the vectors with the *identical basis representations* relative to B_X and B_Y, respectively (see the *Basis Representation Theorem* (Theorem 3.3)):

$$X \ni x = \sum_{i \in I} c_i x_i \mapsto Tx := \sum_{i \in I} c_i y_i \in Y;$$

in particular, $Tx_i = y_i$, $i \in I$.

Exercise 3.26. Verify that T is an *isomorphism* between X and Y. □

Corollary 3.2 (*n*-Dimensional Vector Spaces). *Each n-dimensional vector space ($n \in \mathbb{N}$) over $\mathbb{F}$ is isomorphic to $\mathbb{F}^n$.*

3.1.6 New Spaces from Old

Here, we discuss several ways of generating new vector spaces.

3.1.6.1 Direct Products and Sums

Definition 3.10 (Direct Product). Let $\{X_i\}_{i \in I}$ be a nonempty collection of nonempty sets.

The *direct product* (also *Cartesian product*) of the sets X_i, $i \in I$, is the set of all choice functions on the indexing set I:

$$\prod_{i \in I} X_i := \left\{ x : I \to \bigcup_{i \in I} X_i \,\middle|\, x(i) = x_i \in X_i, \; i \in I \right\}.$$

For each $i \in I$, the value $x(i) = x_i$ of a choice function $x \in \prod_{i \in I} X_i$ at i is also called the *ith coordinate* of x and denoted x_i, the function x is also called an *I-tuple* and denoted $(x_i)_{i \in I}$, the set X_i is called the *ith factor space*, and, for each $j \in I$, the mapping

$$\pi_j : \prod_{i \in I} X_i \to X_j$$

assigning to each *I*-tuple its *j*th coordinate:

$$\pi_j((x_i)_{i \in I}) := x_j, \quad (x_i)_{i \in I} \in \prod_{i \in I} X_i,$$

is called the *projection mapping* of $\prod_{i \in I} X_i$ onto X_j, or simply, the *jth projection mapping*.

Each set X_i, $i \in I$, being a *vector space* over $\mathbb{F}$, the product space $\prod_{i \in I} X_i$ is also a vector space over $\mathbb{F}$ relative to the *coordinatewise linear operations*

$$X \ni (x_i)_{i \in I}, (y_i)_{i \in I} \mapsto (x_i)_{i \in I} + (y_i)_{i \in I} := (x_i + y_i)_{i \in I},$$
$$\mathbb{F} \ni \lambda, X \ni (x_i)_{i \in I}, (y_i)_{i \in I} \mapsto \lambda(x_i)_{i \in I} := (\lambda x_i)_{i \in I}.$$

Remarks 3.12.
- The *nonemptiness* of $\prod_{i \in I} X_i$ is guaranteed by the *Axiom of Choice* (see Section A.1).
- In particular, if $X_i = X$, $i \in I$, then $\prod_{i \in I} X_i$ is the set of all *X*-valued functions on *I*:

$$x : I \to X$$

and we use the notation X^I.

Examples 3.8.
1. For $I = \{1, \ldots, n\}$ $(n \in \mathbb{N})$,

$$\prod_{i=1}^{n} X_i = X_1 \times \cdots \times X_n := \{(x_1, \ldots, x_n) \mid x_i \in X_i, \, i = 1, \ldots, n\}$$

   is the set of all *choice n-tuples*.
   In particular, if $X_i = X$, $i = 1, \ldots, n$, then we use the notation X^n, which includes the case of the *n*-space $\mathbb{F}^n$.
2. For $I = \mathbb{N}$,

$$\prod_{i=1}^{\infty} X_i = \{(x_i)_{i \in \mathbb{N}} \mid x_i \in X_i, \, i \in \mathbb{N}\}$$

   is the set of all *choice sequences*.
   In particular, if $X_i = X$, $i \in \mathbb{N}$, then we obtain $X^{\mathbb{N}}$, the set of all *X*-valued sequences, which includes the case of the space $s := \mathbb{F}^{\mathbb{N}}$ of all $\mathbb{F}$-valued sequences.

3. For $I = [0, 1]$ and $X_i = \mathbb{F}$, $i \in [0, 1]$,

$$\prod_{i \in [0,1]} X_i = \mathbb{F}^{[0,1]} = F[0, 1],$$

where $F[0, 1]$ is the set of all $\mathbb{F}$-valued functions on $[0, 1]$.

Definition 3.11 (Direct Sum). Let $\{X_i\}_{i \in I}$ be a nonempty collection of vector spaces. The *direct sum* of X_i, $i \in I$, is a *subspace* of the direct product $\prod_{i \in I} X_i$, defined as follows:

$$\bigoplus_{i \in I} X_i := \left\{ (x_i)_{i \in I} \in \prod_{i \in I} X_i \,\middle|\, x_i = 0 \text{ for all but a finite number of } i \in I \right\}.$$

Remark 3.13. As is easily seen,

$$\bigoplus_{i \in I} X_i = \prod_{i \in I} X_i \iff I \text{ is a } \textit{finite set.}$$

Examples 3.9.

1. For $I = \mathbb{N}$ and $X_i = \mathbb{F}$, $i \in \mathbb{N}$,

$$\prod_{i=1}^{\infty} X_i = s \quad \text{and} \quad \bigoplus_{i=1}^{\infty} X_i = c_{00}.$$

2. For $I = [0, 1]$ and $X_i = \mathbb{F}$, $i \in [0, 1]$, $\prod_{i=1}^{\infty} X_i = F[0, 1]$, whereas $\bigoplus_{i=1}^{\infty} X_i$ is the subset of all $\mathbb{F}$-valued functions on $[0, 1]$ equal to zero for all but a finite number of values $t \in [0, 1]$.

3.1.6.2 Quotient Spaces

Let Y be a subspace of a vector space X.

The binary relation on X defined as

$$x \sim y \iff y - x \in Y$$

is an *equivalence relation* (*reflexive*, *symmetric*, and *transitive*) called the *equivalence modulo Y*.

Exercise 3.27.

(a) Verify.

(b) Show that every equivalence class $[x]$ modulo Y represented by an element $x \in X$ is of the form

$$[x] = x + Y,$$

i. e., the equivalence classes modulo Y are the *translations* of Y.

Definition 3.12 (Cosets, Quotient Space). For a subspace Y of a vector space X over $\mathbb{F}$, the set X/Y of all equivalence classes modulo Y, called the *cosets modulo Y*, is a vector space over $\mathbb{F}$ relative to the linear operations

$$X/Y \ni [x], [y] \mapsto [x] + [y] := [x + y] = (x + y) + Y,$$
$$\mathbb{F} \ni \lambda, X/Y \ni [x] \mapsto \lambda[x] := [\lambda x] = \lambda x + Y,$$

called the *quotient space of X modulo Y*.

Exercise 3.28.
(a) Verify that X/Y is a vector space.
(b) What is the *zero element* of X/Y?
(c) Describe $X/\{0\}$ and X/X.
(d) Let $m \in \mathbb{R}$. Describe $\mathbb{R}^2/\{(x, y) \in \mathbb{R}^2 \mid y = mx\}$.

Definition 3.13 (Canonical Homomorphism). The mapping

$$X \ni x \mapsto Tx := [x] = x + Y \in X/Y$$

is a homomorphism of X onto X/Y, called the *canonical homomorphism of X onto X/Y*.

Exercise 3.29.
(a) Verify that the mapping $T : X \to X/Y$ is a *homomorphism* of X onto X/Y.
(b) Determine the *kernel* (the *null space*) of the *canonical homomorphism T*

$$\ker T := \{x \in X \mid Tx = 0\}.$$

(c) Let $m \in \mathbb{R}$. Describe the *canonical homomorphism* of $\mathbb{R}^2$ onto

$$\mathbb{R}^2/\{(x, y) \in \mathbb{R}^2 \mid y = mx\}.$$

3.1.7 Disjoint and Complementary Subspaces, Direct Sum Decompositions, Deficiency and Codimension

Definition 3.14 (Disjoint and Complementary Subspaces). Two subspaces Y and Z of a vector space X are called *disjoint* if

$$Y \cap Z = \{0\}.$$

Two disjoint subspaces Y and Z of a vector space X are called *complementary* if every $x \in X$ allows a *unique decomposition*

$$x = y + z$$

with some $y \in Y$ and $z \in Z$.

Remarks 3.14.

– For complementary subspaces, nontrivial is only the *existence* of the decomposition, the *uniqueness* immediately following from the disjointness.

Exercise 3.30. Explain.

– The definitions naturally extend to any finite number of subspaces.

Examples 3.10.

1. The subspaces

$$\{(x,y,z) \in \mathbb{R}^3 \mid z = 0\} \text{ and } \{(x,y,z) \in \mathbb{R}^3 \mid x = y = 0\}$$

    are *complementary* in $\mathbb{R}^3$.

2. The subspaces

$$\{(x,y,z) \in \mathbb{R}^3 \mid y = z = 0\} \text{ and } \{(x,y,z) \in \mathbb{R}^3 \mid x = z = 0\}$$

    are *disjoint* but *not complementary* in $\mathbb{R}^3$.

3. The subspaces

$$\{(x,y,z) \in \mathbb{R}^3 \mid z = 0\} \text{ and } \{(x,y,z) \in \mathbb{R}^3 \mid y = 0\}$$

    are *not disjoint* in $\mathbb{R}^3$.

4. The subspaces c_0 and $\mathrm{span}(\{(1,1,1,\dots)\}) = \{(\lambda,\lambda,\lambda,\dots) \mid \lambda \in \mathbb{F}\}$ are *complementary* in c.

Theorem 3.6 (Existence of a Complementary Subspace). *Every subspace of a vector space X has a complementary subspace.*

Proof. Let Y be a subspace of X. Consider the collection $\mathscr{S}$ of all subspaces in X disjoint from Y, partially ordered by the set-theoretic inclusion $\subseteq$.

Exercise 3.31. Why is $\mathscr{S}$ *nonempty*?

Let $\mathscr{C}$ be an arbitrary *chain* in $(\mathscr{S}, \subseteq)$. Then

$$L := \bigcup_{C \in \mathscr{C}} C$$

is also a *subspace* of X disjoint from Y.

Exercise 3.32. Verify.

Clearly, L is an *upper bound* of $\mathscr{C}$ in $(\mathscr{S}, \subseteq)$.

By *Zorn's Lemma* (Theorem A.5), there is a *maximal element* Z in $(\mathscr{S}, \subseteq)$, i.e., a *maximal* disjoint from Y subspace of X.

Let us show that Z is *complementary* to Y *by contradiction*, assuming that

$$\exists x \in X \text{ such that } x \notin Y + Z.$$

Then

$$Z' := \text{span}(Z \cup \{x\}) = Z + \text{span}(\{x\})$$

is also a *subspace* of X disjoint from Y.

Exercise 3.33. Verify.

Since $Z \subset Z'$, this *contradicts* the *maximality* of Z in $(\mathscr{S}, \subseteq)$, showing that Z is *complementary* to Y. $\qquad\square$

Remark 3.15. The complementary subspace need not be unique.

Exercise 3.34. Give a corresponding example.

However, as we see below, all complementary subspaces of a given subspace are *isomorphic*.

Proposition 3.5 (Direct Sum Decompositions). *Let Y be a subspace of a vector space X. Then, for any subspace Z of X complementary to Y, X is isomorphic to the direct sum $Y \oplus Z$.*

Exercise 3.35. Prove.

Remark 3.16. We say that X *is* the direct sum of Y and Z, and write

$$X = Y \oplus Z,$$

calling the latter a *direct sum decomposition* of X, and, by the *Isomorphism Theorem* (Theorem 3.5), immediately obtain Corollary 3.3.

Corollary 3.3 (Sum of Dimensions). *Let Y be a subspace of a vector space X. Then, for any subspace Z complementary to Y,*

$$\dim X = \dim Y + \dim Z.$$

Exercise 3.36. Explain why

$$\dim(Y \oplus Z) = \dim Y + \dim Z.$$

Examples 3.11.
1. $\mathbb{R}^2 = \{(x,y) \in \mathbb{R}^2 \mid y = 0\} \oplus \{(x,y) \in \mathbb{R}^2 \mid x = 0\}$.
2. $\mathbb{R}^3 = \{(x,y,z) \in \mathbb{R}^3 \mid z = 0\} \oplus \{(x,y,z) \in \mathbb{R}^3 \mid x = y = 0\}$.
3. $\mathbb{R}^3 = \{(x,y,z) \in \mathbb{R}^3 \mid y = z = 0\} \oplus \{(x,y,z) \in \mathbb{R}^3 \mid x = z = 0\} \oplus \{(x,y,z) \in \mathbb{R}^3 \mid x = y = 0\}$.
4. $c = c_0 \oplus \text{span}(\{(1,1,1,\dots)\})$.

Definition 3.15 (Deficiency). The *deficiency* of a subspace Y in a vector space X is the *dimension* $\dim(X/Y)$ of the *quotient space* X/Y.

Exercise 3.37. What is the *deficiency*
(a) of $\{(x,y) \in \mathbb{R}^2 \mid y = 0\}$ in $\mathbb{R}^2$,
(b) of $\{(x,y,z) \in \mathbb{R}^3 \mid x = y = 0\}$ in $\mathbb{R}^3$,
(c) of P_n $(n \in \mathbb{Z}_+)$ in P?

Definition 3.16 (Hyperplane). A subspace Y in a vector space X of deficiency 1 is called a *hyperplane*.

Examples 3.12.
1. If $B = \{x_i\}_{i \in I}$ is a basis for X, then, for any $j \in I$, $Y := \text{span}\{B \setminus \{x_j\}\}$ is a hyperplane in X.
2. c_0 is a hyperplane in c.

Exercise 3.38.
(a) Verify.
(b) Describe the quotient space c/c_0.

Proposition 3.6 (Dimension of Complementary Subspaces). *Let Y be a subspace of a vector space X. Then, each complementary subspace Z of Y is isomorphic to the quotient space X/Y, and hence,*

$$\dim Z = \dim(X/Y).$$

Exercise 3.39. Prove.

Hint. Show that if B is a basis for Z, then $\{[z] = z + Y \mid z \in Z\}$ is a basis for X/Y.

Thus, by the *Isomorphism Theorem* (Theorem 3.5) and the *Sum of Dimensions Corollary* (Corollary 3.3), we obtain Corollary 3.4.

Corollary 3.4 (Codimension of a Subspace). *All complementary subspaces of a subspace Y of a vector space X are isomorphic sharing the same dimension, which is called the* codimension *of Y and coincides with its* deficiency:

$$\text{codim } Y = \dim(X/Y).$$

Furthermore,

$$\dim X = \dim Y + \text{codim } Y = \dim Y + \dim(X/Y).$$

Remark 3.17. Thus, a hyperplane can be equivalently defined as a subspace of codimension 1, and hence, can be described as in Examples 3.12.

3.2 Normed Vector and Banach Spaces

Here, we introduce the notion of *norm* on a vector space, combining its linear structure with topology, and study the surprising and beautiful results of the profound interplay between the two.

3.2.1 Definitions and Examples

Recall that $\mathbb{F}$ stands for the scalar field of real or complex numbers (i. e., $\mathbb{F} = \mathbb{R}$, or $\mathbb{F} = \mathbb{C}$).

Definition 3.17 (Normed Vector Space). A *normed vector space* over $\mathbb{F}$ is a *vector space* X over $\mathbb{F}$ equipped with a *norm*, i. e., a mapping

$$\| \cdot \| : X \to \mathbb{R}$$

subject to the following *norm axioms*:

1. $\|x\| \geq 0$, $x \in X$.   *Nonnegativity*
2. $\|x\| = 0$ iff $x = 0$.   *Separation*
3. $\|\lambda x\| = |\lambda| \|x\|$, $\lambda \in \mathbb{F}$, $x \in X$.   *Absolute Homogeneity/Scalability*
4. $\|x + y\| \leq \|x\| + \|y\|$, $x, y \in X$.   *Subadditivity/Triangle Inequality*

The space is said to be *real* if $\mathbb{F} = \mathbb{R}$, and *complex* if $\mathbb{F} = \mathbb{C}$.

Notation. $(X, \| \cdot \|)$.

Remarks 3.18.

– A function $\| \cdot \| : X \to \mathbb{R}$, satisfying the norm axioms of *absolute scalability* and *subadditivity* only, which immediately imply the following weaker version of the *separation axiom*:
    2w. $\|x\| = 0$ if $x = 0$,
    and hence, also the axiom of *nonnegativity*, is called a *seminorm* on X, and $(X, \| \cdot \|)$ is called a *seminormed vector space* (see the examples to follow).
– A norm $\| \cdot \|$ on a vector space X generates a metric on X, called the *norm metric*, as follows:

$$X \times X \ni (x, y) \mapsto \rho(x, y) := \|x - y\|, \tag{3.1}$$

which turns X into a *metric space*, endows it with the *norm metric topology*, and brings to life all the relevant concepts: *openness, closedness, denseness, category, boundedness, total boundedness*, and *compactness* for sets, various forms of *continuity* for functions, *fundamentality* and *convergence* for sequences, *separability* and *completeness* for spaces.

If $\| \cdot \|$ is a *seminorm*, (3.1) defines a *pseudometric* (see Remark 2.1).
- Due to the *axioms* of *subadditivity* and *absolute scalability*, the linear operations of *vector addition*

$$X \times X \ni (x, y) \mapsto x + y \in X$$

and *scalar multiplication*

$$\mathbb{F} \times X \ni (\lambda, x) \mapsto \lambda x \in X$$

are *jointly continuous*.
- The following implication of the *subadditivity*:

$$\big|\|x\| - \|y\|\big| \le \|x - y\|, \ x, y \in X, \tag{3.2}$$

showing that the norm is *Lipschitz continuous* on X, holds. Observe that this inequality holds for *seminorms* as well.

Exercise 3.40. Verify.

Definition 3.18 (Subspace of a Normed Vector Space). If $(X, \| \cdot \|)$ is a *normed vector space* and $Y \subseteq X$ is a *linear subspace* of X, then the restriction of the norm $\| \cdot \|$ to Y is a norm on Y, and the normed vector space $(Y, \| \cdot \|)$ is called a *subspace* of $(X, \| \cdot \|)$.

Definition 3.19 (Banach Space). A *Banach space* is a normed vector space $(X, \|\cdot\|)$ complete, relative to the norm metric, i. e., such that every *Cauchy sequence* (or *fundamental sequence*)

$$(x_n)_{n \in \mathbb{N}} \subseteq X : \rho(x_n, x_m) = \|x_n - x_m\| \to 0, \ n, m \to \infty,$$

converges to an element $x \in X$:

$$\lim_{n \to \infty} x_n = x \iff \rho(x_n, x) = \|x_n - x\| \to 0, \ n \to \infty.$$

Examples 3.13.
1. The vector space $\mathbb{F}$ (i. e., $\mathbb{R}$ or $\mathbb{C}$) is a *Banach space* relative to the *absolute-value norm*

$$\mathbb{F} \ni x \mapsto |x|,$$

which generates the regular metric (see Examples 2.22).
2. By the *Completeness of the n-Space Theorem* (Theorem 2.23), the vector space $l_p^{(n)}$ $(n \in \mathbb{N}, 1 \le p \le \infty)$ is a *Banach space* relative to *p-norm*

$$\mathbb{F}^n \ni x = (x_1, \dots, x_n) \mapsto \|x\|_p := \begin{cases} [\sum_{k=1}^n |x_k|^p]^{1/p} & \text{if } 1 \le p < \infty, \\ \max_{1 \le k \le n} |x_k| & \text{if } p = \infty, \end{cases}$$

which generates *p*-metric.

The *nonnegativity*, *separation*, and *absolute scalability axioms* are trivially verified. The *subadditivity axiom* is satisfied based on *Minkowski's Inequality for n-Tuples* (Theorem 2.3).

3. By the *completeness of l_p* $(1 \le p \le \infty)$ (Theorem 2.24, Corollary 2.1), the vector space l_p $(1 \le p \le \infty)$ is a *Banach space* relative to *p-norm*

$$l_p \ni x = (x_k)_{k \in \mathbb{N}} \mapsto \|x\|_p := \begin{cases} \left[\sum_{k=1}^{\infty} |x_k|^p\right]^{1/p} & \text{if } 1 \le p < \infty, \\ \sup_{k \ge 1} |x_k| & \text{if } p = \infty, \end{cases}$$

which generates *p-metric*.

Verifying the *nonnegativity*, *separation*, and *absolute scalability axioms* is trivial. The *subadditivity axiom* follows from *Minkowski's Inequality for Sequences* (Theorem 2.4).

4. By the *Characterization of Completeness* (Theorem 2.27), the normed vector spaces $(c_0, \| \cdot \|_\infty)$ and $(c, \| \cdot \|_\infty)$ are *Banach spaces*, being *closed proper subspaces* of l_∞ (see Exercise 2.63).

Remark 3.19. By the *Nowhere Denseness of Closed Proper Subspaces Proposition* (Proposition 3.16, Section 3.5, Problem 11), the subspace c_0 is *nowhere dense* in $(c, \| \cdot \|_\infty)$ and, in its turn, the subspace c is nowhere dense in l_∞. Both statements can also be verified directly.

Exercise 3.41. Verify directly.

5. By the *Characterization of Completeness* (Theorem 2.27), the normed vector space $(c_{00}, \| \cdot \|_\infty)$ is *incomplete* since it is a *subspace* of the Banach space $(c_0, \| \cdot \|_\infty)$ that is *not closed* (see Exercise 2.63).

6. Let T be a nonempty set. By the *completeness of $(M(T), \rho_\infty)$ Theorem* (Theorem 2.25), the vector space $M(T)$ (see Examples 3.1) is a *Banach space* relative to the *supremum norm*

$$M(T) \ni f \mapsto \|f\|_\infty := \sup_{t \in T} |f(t)|,$$

which generates the *supremum metric*, the norm axioms being readily verified.

Exercise 3.42. Verify.

7. Let (X, ρ) be a *compact metric space*. By the *completeness of $C(X, Y)$ Theorem* (Theorem 2.58), the vector space $C(X)$ (see Examples 3.1) is a *Banach space* relative to the *maximum norm*

$$C(X) \ni f \mapsto \|f\|_\infty := \max_{x \in X} |f(x)|,$$

which generates the *maximum metric*.

In particular, this includes the case of $(C[a, b], \| \cdot \|_\infty)$ $(-\infty < a < b < \infty)$.

8. By the *Characterization of Completeness* (Theorem 2.27), the vector space P of *all polynomials* is an *incomplete* normed vector space relative to the *maximum norm*

$$\|f\|_\infty := \max_{a \le t \le b} |f(t)|$$

$(-\infty < a < b < \infty)$ since it is a *subspace* of the Banach space $(C[a, b], \|\cdot\|_\infty)$ that is *not closed* (see Exercise 2.63).

9. For each $n \in \mathbb{Z}_+$, the *finite-dimensional subspace* P_n of all polynomial of degree at most n is a *Banach space* relative to *maximum norm*, which, as follows from the *Completeness of Finite-Dimensional Spaces Theorem* (Theorem 3.11) (see Section 3.3), is true in a more general context.

10. By the *Incompleteness of* $(C[a, b], \rho_p)$ $(1 \le p < \infty)$ *Proposition* (Proposition 2.3), the vector space $C[a, b]$ $(-\infty < a < b < \infty)$ is an *incomplete* normed vector space relative to the *p-norm*

$$C[a, b] \ni f \mapsto \|f\|_p := \left[\int_a^b |f(t)|^p \, dt \right]^{1/p} \quad (1 \le p < \infty).$$

11. The vector space $R[a, b]$ $(-\infty < a < b < \infty)$ of all $\mathbb{F}$-valued functions *Riemann integrable* on $[a, b]$ is a *seminormed vector space* relative to the *integral seminorm*

$$R[a, b] \ni f \mapsto \|f\|_1 := \int_a^b |f(t)| \, dt.$$

12. The set $BV[a, b]$ of all $\mathbb{F}$-valued functions of *bounded variation* on $[a, b]$ $(-\infty < a < b < \infty)$ (see Examples 2.4) with pointwise linear operations is *seminormed vector space* relative to the *total-variation seminorm*

$$BV[a, b] \ni f \mapsto \|f\| := V_a^b(f)$$

(see Exercise 2.14), and is a Banach space relative to the *total-variation norm*

$$BV[a, b] \ni f \mapsto \|f\| := |f(a)| + V_a^b(f)$$

(see the *Completeness of* $BV[a, b]$ *Proposition* (Proposition 3.14, Section 3.5, Problem 9)).

3.2.2 Series and Completeness Characterization

Unlike in a metric space, which is void of addition, in a normed vector space, one can consider convergence not only for sequences, but also for series and even characterize the completeness of the space in those terms.

Definition 3.20 (Convergence of Series in Normed Vector Space). Let $(X, \| \cdot \|)$ be a normed vector space and $(x_n)_{n \in \mathbb{N}}$ be a sequence of its elements.

The *series* $\sum_{k=1}^{\infty} x_k$ is said to *converge* (or to be *convergent*) in $(X, \| \cdot \|)$ if the sequence of its partial sums $(s_n := \sum_{k=1}^{n} x_k)_{n \in \mathbb{N}}$ converges to an element $x \in X$, i. e.,

$$\lim_{n \to \infty} \sum_{k=1}^{n} x_k = x.$$

We call x the *sum* of the series and write

$$\sum_{k=1}^{\infty} x_k = x.$$

The *series* $\sum_{k=1}^{\infty} x_k$ is said to *absolutely converge* (or to be *absolutely convergent*) in $(X, \| \cdot \|)$ if the numeric series $\sum_{k=1}^{\infty} \|x_k\|$ converges.

A series that does not converge in $(X, \| \cdot \|)$ is called *divergent*.

The following three statements generalize their well-known counterparts from classical analysis:

Proposition 3.7 (Necessary Condition of Convergence for Series). *If a series* $\sum_{k=1}^{\infty} x_k$ *converges in a normed vector space* $(X, \| \cdot \|)$, *then*

$$x_n \to 0, \; n \to \infty, \; in \; (X, \| \cdot \|).$$

Exercise 3.43. Prove.

The equivalent *contrapositive* statement is as follows:

Proposition 3.8 (Divergence Test for Series). *If, for a series* $\sum_{k=1}^{\infty} x_k$ *in a normed vector space* $(X, \| \cdot \|)$,

$$x_n \not\to 0, \; n \to \infty,$$

then the series diverges.

Theorem 3.7 (Cauchy's Convergence Test for Series). *For a series* $\sum_{k=1}^{\infty} x_k$ *in a normed vector space* $(X, \| \cdot \|)$ *to converge, it is necessary and, provided* $(X, \| \cdot \|)$ *is a Banach space, sufficient that*

$$\forall \varepsilon > 0 \; \exists N \in \mathbb{N} \; \forall n \geq N, \; \forall p \in \mathbb{N} : \left\| \sum_{k=n+1}^{n+p} x_k \right\| < \varepsilon.$$

Exercise 3.44. Prove.

It appears that one can characterize the completeness of the space in terms of series' convergence.

Theorem 3.8 (Series Characterization of Banach Spaces). *A normed vector space* $(X, \| \cdot \|)$ *is a Banach space iff every absolutely convergent series of its elements converges.*

Proof. "*Only if*" part. This part follows directly from the *Cauchy's Convergence Test for Series* (Theorem 3.7) by *subadditivity* of norm.

Exercise 3.45. Fill in the details.

"*If*" part. Let $(x_n)_{n \in \mathbb{N}}$ be an arbitrary *fundamental* sequence in $(X, \| \cdot \|)$. Then it contains a subsequence $(x_{n(k)})_{k \in \mathbb{N}}$ such that

$$\| x_{n(k+1)} - x_{n(k)} \| \le \frac{1}{2^k}, \; k \in \mathbb{N}.$$

Exercise 3.46. Explain.

This, by the *Comparison Test*, implies that the telescoping series

$$x_{n(1)} + [x_{n(2)} - x_{n(1)}] + [x_{n(3)} - x_{n(2)}] + \cdots$$

absolutely converges, and hence, by the premise, *converges* in $(X, \| \cdot \|)$, i. e.,

$$\exists x \in X : s_m := x_{n(1)} + [x_{n(2)} - x_{n(1)}] + [x_{n(3)} - x_{n(2)}] + \cdots + [x_{n(m)} - x_{n(m-1)}]$$
$$= x_{n(m)} \to x, \; n \to \infty, \text{ in } (X, \| \cdot \|).$$

Therefore, the subsequence $(x_{n(k)})_{k \in \mathbb{N}}$ converges to x in $(X, \| \cdot \|)$, which, by the *Fundamental Sequence with Convergent Subsequence Proposition* (Proposition 2.22, Section 2.18, Problem 25), implies that the fundamental sequence $(x_n)_{n \in \mathbb{N}}$ itself converges to x in $(X, \| \cdot \|)$, completing the proof. □

3.2.3 Comparing Norms, Equivalent Norms

Norms on a vector space can be naturally compared by their strength, which reflects the *strength* of the corresponding norm metric topology (see, e. g., [52, 56]).

Definition 3.21 (Comparing Norms). Let $\| \cdot \|_1$ and $\| \cdot \|_2$ be norms on a vector space X. Norm $\| \cdot \|_1$ is said to be *stronger* than norm $\| \cdot \|_2$, or norm $\| \cdot \|_2$ is said to be *weaker* than norm $\| \cdot \|_1$, if

$$\exists C > 0 \; \forall x \in X : \| x \|_2 \le C \| x \|_1.$$

Definition 3.22 (Equivalent Norms). Two norms $\| \cdot \|_1$ and $\| \cdot \|_2$ on a vector space X are called *equivalent* if

$$\exists c, C > 0 \; \forall x \in X : c \| x \|_1 \le \| x \|_2 \le C \| x \|_1,$$

i. e., each norm is both stronger and weaker than the other.

Remarks 3.20.

– As is easily verified, fundamentality/convergence of a sequence relative to the norm metric generated by a stronger norm is preserved by the norm metric generated by a weaker one.

Exercise 3.47. Verify.

However, as is shown in Examples 3.14 below, the converse is not true.

– Thus, relative to equivalent norms, same sequences are fundamental/convergent to the same limits, which implies that if $(X, \|\cdot\|)$ is a Banach space, then X is also a Banach space relative to any norm equivalent to $\|\cdot\|$.

– The equivalence of norms on a vector space X is an *equivalence relation* (*reflexive*, *symmetric*, and *transitive*) on the set of all norms on X.

– Equivalent norms generate bi-Lipschitz equivalent norm metrics (see Section 2.12.1), which define the same topology on X.

Exercise 3.48. Verify.

– Two norms $\|\cdot\|_1$ and $\|\cdot\|_2$ on a vector space X are equivalent *iff* the identity mapping $I : X \to X$,

$$(X, \|\cdot\|_1) \ni x \mapsto Ix := x \in (X, \|\cdot\|_2),$$

is a *bi-Lipschitzian isomorphism*, i. e., both the mapping I and its inverse

$$(X, \|\cdot\|_2) \ni x \mapsto I^{-1}x := x \in (X, \|\cdot\|_1)$$

are *Lipschitz continuous* (see Examples 2.18).

Examples 3.14.

1. On $C[a, b]$, the *maximum norm*

$$C[a, b] \ni f \mapsto \|f\|_\infty := \max_{a \le t \le b} |f(t)|$$

is *stronger* than the *integral norm*

$$C[a, b] \ni f \mapsto \|f\|_1 := \int_a^b |f(t)|\, dt,$$

but the two norms are *not equivalent*.

2. All p-norms ($1 \le p \le \infty$) on $\mathbb{F}^n$ ($n \in \mathbb{N}$) are equivalent (see Remarks 2.32).

Exercise 3.49. Verify.

3.2.4 Isometric Isomorphisms

Combining the notions of isometry and isomorphism, one can define *isometric isomorphism*.

Definition 3.23 (Isometric Isomorphism of Normed Vector Spaces). Let $(X, \| \cdot \|_X)$ and $(Y, \| \cdot \|_Y)$ be normed vector spaces over $\mathbb{F}$. A mapping $T : X \to Y$, which is *simultaneously* an *isometry* and an *isomorphism*, is called an *isometric isomorphism from X to Y*. It is said to *isometrically embed X in Y*.

If the isometric isomorphism T is *onto* (i. e., *surjective*), it is called an *isometric isomorphism between X and Y*, the spaces X and Y being called *isometrically isomorphic* or *linearly isometric*.

Remark 3.21. An isomorphism $T : (X, \| \cdot \|_X) \to (Y, \| \cdot \|_Y)$ is an isometry *iff* it is *norm preserving*, i. e.,

$$\forall x \in X : \|Tx\|_Y = \|x\|_X.$$

Exercise 3.50. Verify.

Examples 3.15.
1. For arbitrary $n \in \mathbb{N}, 1 \le p \le \infty$, the mapping

$$l_p^{(n)} \ni x = (x_1, \dots, x_n) \mapsto Tx := (x_1, \dots, x_n, 0, 0, \dots) \in l_p$$

is an *isometric embedding* of $l_p^{(n)}$ in l_p.
2. Let X be an *n-dimensional* vector space ($n \in \mathbb{N}$) with an ordered *basis* $B :=$ $\{x_1, \dots, x_n\}$ and $1 \le p \le \infty$. The *coordinate vector mapping*

$$X \ni x = \sum_{k=1}^{n} c_k x_k \mapsto Tx := [x]_B := (c_1, \dots, c_n) \in l_p^{(n)},$$

where

$$[x]_B := (c_1, \dots, c_n)$$

is the *coordinate vector* of x relative to basis B, is an *isomorphism* between the spaces X and $l_p^{(n)}$ (see Section 1.4), which also becomes an *isometry* when the former is equipped with the norm

$$X \ni x = \sum_{k=1}^{n} c_k x_k \mapsto \|x\|_X := \|Tx\|_p.$$

This follows from a more general construct (see Section 3.5, Problem 13).

Remarks 3.22.
- Being isometrically isomorphic is an *equivalence relation* (*reflexive, symmetric,* and *transitive*) on the set of all normed vector spaces.

  Exercise 3.51. Verify.

- Isometrically isomorphic metric spaces are both *metrically* and *linearly indistinguishable*, in particular, they have the *same dimension* and are *separable* or *complete* only simultaneously.

3.2.5 Topological and Schauder Bases

In a Banach space setting, the notion of *basis* acquires two other meanings discussed below.

Definition 3.24 (Topological Basis of a Normed Vector Space). A *topological basis* of a nonzero normed vector space $(X, \|\cdot\|)$ is a *linearly independent* subset B of X, whose *span* is *dense* in X:

$$\overline{\operatorname{span}(B)} = X.$$

Remarks 3.23.
- We require substantially less from a *topological basis* than from its *algebraic basis*:

$$\overline{\operatorname{span}(B)} = X \quad \text{as opposed to} \quad \operatorname{span}(B) = X.$$

- Clearly, every *algebraic basis* for a normed vector space is a *topological basis*, and hence, the *existence* of a topological basis for any nonzero normed vector space immediately follows from the *Basis Theorem* (Theorem 3.2), which also implies that a topological basis of a normed vector space is never unique (Remark 3.10).
- As the following examples show, a topological basis of a normed vector space need not be its algebraic basis.

Examples 3.16.
1. A subset B is a *topological basis* for a finite-dimensional normed vector space $(X, \|\cdot\|)$ *iff* it is an *algebraic basis* for X, which follows directly from the *Closedness of Finite-Dimensional Subspaces Theorem* (Theorem 3.12) (see Section 3.3.2), stating that each *finite-dimensional subspace* of a normed vector space is *closed*.
2. The set $E := \{e_n := (\delta_{nk})_{k\in\mathbb{N}}\}_{n\in\mathbb{N}}$, where δ_{nk} is the *Kronecker delta*, is the following:
   - an *algebraic basis* for $(c_{00}, \|\cdot\|_{\infty})$,
   - a *topological basis*, but not an *algebraic basis* for $(c_0, \|\cdot\|_{\infty})$ and l_p $(1 \le p < \infty)$, and
   - not even a *topological basis* for $(c, \|\cdot\|_{\infty})$ and l_{∞}.

3. The set $\{t^n\}_{n\in\mathbb{Z}_+}$ is the following:
   - an *algebraic basis* for P,
   - a *topological basis*, but not an *algebraic basis* for $(C[a,b], \|\cdot\|_\infty)$ $(-\infty < a < b < \infty)$, and
   - not even a *topological basis* for $(M[a,b], \|\cdot\|_\infty)$.

Exercise 3.52.
(a) Verify.

   Hint. Show that, in $(c_0, \|\cdot\|_\infty)$ and l_p, any sequence $x := (x_k)_{k\in\mathbb{N}}$ can be represented as the sum of the following series:

   $$x = \sum_{k=0}^\infty x_k e_k.$$

(b) Show that by appending the sequence $e_0 := (1,1,1,\dots)$ to the set $E := \{e_n\}_{n\in\mathbb{N}}$, we obtain the set $E' := \{e_n\}_{n\in\mathbb{Z}_+}$, which is a *topological basis* for $(c, \|\cdot\|_\infty)$, with each sequence $x := (x_k)_{k\in\mathbb{N}} \in c$ being represented as the sum of the following series:

   $$x = \left(\lim_{n\to\infty} x_n\right) e_0 + \sum_{k=1}^\infty \left(x_k - \lim_{n\to\infty} x_n\right) e_k.$$

Exercise 3.53. Does, in a normed vector space setting, the analogue of the *Dimension Theorem* (Theorem 3.4) hold for topological bases? In other words, is the notion of "topological dimension" well defined?

Hint. In $(C[a,b], \|\cdot\|_\infty)$, consider the topological basis $\{t^n\}_{n\in\mathbb{Z}_+} \cup \{e^{ct}\}_{c\in\mathbb{R}}$ and compare its *cardinality* to that of $\{t^n\}_{n\in\mathbb{Z}_+}$.

Definition 3.25 (Schauder Basis of a Banach Space). A *Schauder basis* (also a *countable basis*) of a nonzero Banach space $(X, \|\cdot\|)$ over $\mathbb{F}$ is a *countably infinite set* of elements $\{e_n\}_{n\in\mathbb{N}}$ in X such that

$$\forall x \in X \; \exists! (c_k)_{k\in\mathbb{N}} \subseteq \mathbb{F} : x = \sum_{k=1}^\infty c_k e_k.$$

The series is called the *Schauder expansion* of x, and the numbers $c_k \in \mathbb{F}$, $k \in \mathbb{N}$, uniquely determined by x, are called the *coordinates* of x relative to the Schauder basis $\{e_n\}_{n\in\mathbb{N}}$.

Remarks 3.24.
- Because of the uniqueness of the foregoing series representation, a *Schauder basis* of a Banach space, when existent, is automatically *linearly independent*. It is also *not unique*.

Exercise 3.54. Explain.

– A *Schauder basis* of a Banach space, when existent, is also its *topological basis*.

Exercise 3.55. Explain.

However, as the following examples show, a *topological basis* of a Banach need not be its *Schauder basis*.

– The *order* of vectors in a *Schauder basis* is *important*. A permutation of infinitely many elements may transform a *Schauder basis* into a set, which fails to be one (see, e. g., [43]).

Examples 3.17.
1. The set $E := \{e_n := (\delta_{nk})_{k \in \mathbb{N}}\}_{n \in \mathbb{N}}$ (see Examples 3.16) is
   – a (standard) *Schauder basis* for $(c_0, \|\cdot\|_\infty)$ and l_p $(1 \le p < \infty)$, but
   – not a *Schauder basis* for c and l_∞.
2. The set $\{t^n\}_{n \in \mathbb{Z}_+}$ is a *topological basis*, but not a *Schauder basis* for the space $(C[a,b], \|\cdot\|_\infty)$ $(-\infty < a < b < \infty)$, which does have Schauder bases of a more intricate structure (see, e. g., [43]).

Exercise 3.56.
(a) Verify.
(b) Show that the set $E' := \{e_n\}_{n \in \mathbb{Z}_+}$ (see Exercise 3.52) is a *Schauder basis* for $(c, \|\cdot\|_\infty)$.

Proposition 3.9 (Separability of Banach Space with Schauder Basis). *A Banach space X with a Schauder basis $\{e_n\}_{n \in \mathbb{N}}$ is separable.*

Exercise 3.57. Prove.

Hint. Consider the set

$$C := \left\{ \sum_{k=1}^{n} c_k e_k \,\middle|\, n \in \mathbb{N} \right\},$$

where c_k, $k = 1, \ldots, n$, are arbitrary rationals/complex rationals.

Remark 3.25. The converse statement is not true. The *basis problem* on whether every separable Banach space has a Schauder basis, posed by Stefan Banach, was negatively answered by Per Enflo[2] in [19].

3.3 Finite-Dimensional Spaces and Related Topics

In this section, we analyze certain features inherent to the important class of finite-dimensional normed vector spaces.

---

2 Per Enflo (1944–).

3.3.1 Norm Equivalence and Completeness

The following statement shows that the equivalence of all p-norms ($1 \le p \le \infty$) on $\mathbb{F}^n$ ($n \in \mathbb{N}$) (see Exercise 3.49) is not coincidental.

Theorem 3.9 (Norm Equivalence Theorem). *All norms on a finite-dimensional vector space are equivalent.*

Proof. Let X be a finite-dimensional vector space over $\mathbb{F}$ and $\|\cdot\|$ be an arbitrary norm on X. The case of $\dim X = 0$ being vacuous, suppose that with $\dim X = n$ with some $n \in \mathbb{N}$ and let $B := \{x_1, \ldots, x_n\}$ be an ordered *basis* of X. Then, by the *Basis Representation Theorem* (Theorem 3.3),

$$\forall x \in X \; \exists! \, (c_1, \ldots, c_n) \in \mathbb{F}^n : \; x = \sum_{k=1}^{n} c_k x_k,$$

and hence, the *coordinate vector mapping*

$$X \ni x = \sum_{k=1}^{n} c_k x_k \mapsto Tx := [x]_B := (c_1, \ldots, c_n) \in \mathbb{F}^n,$$

where

$$[x]_B := (c_1, \ldots, c_n)$$

is the *coordinate vector* of x relative to basis B, is an *isomorphism* between X and $\mathbb{F}^n$, and

$$X \ni x = \sum_{k=1}^{n} c_k x_k \mapsto \|x\|_2 := \|(c_1, \ldots, c_n)\|_2 = \left[\sum_{k=1}^{n} |c_k|^2 \right]^{1/2}$$

is a norm on X (see Examples 3.15 and Section 3.5, Problem 13), which is *stronger* than $\|\cdot\|$ since, for any $x \in X$,

$$\|x\| = \left\| \sum_{k=1}^{n} c_k x_k \right\| \qquad \text{by *subadditivity* and *absolute scalability* of norm;}$$

$$\le \sum_{k=1}^{n} |c_k| \|x_k\| \qquad \text{by the *Cauchy–Schwarz inequality* (see (2.2));}$$

$$\le \left[\sum_{k=1}^{n} \|x_k\|^2 \right]^{1/2} \left[\sum_{k=1}^{n} |c_k|^2 \right]^{1/2} = C\|x\|_2, \qquad (3.3)$$

where $C := [\sum_{k=1}^{n} \|x_k\|^2]^{1/2} > 0$.

Whence, in view of inequality (3.2),

$$\big| \|x\| - \|y\| \big| \le \|x - y\| \le C\|x - y\|_2, \quad x, y \in X,$$

which implies that the norm $\|\cdot\|$ is a *Lipschitz continuous* function on $(X, \|\cdot\|_2)$.

The spaces $(X, \| \cdot \|_2)$ and $l_2^{(n)} = (\mathbb{F}^n, \| \cdot \|_2)$ being *isometrically isomorphic*, by the *Heine–Borel Theorem* (Theorem 2.47), the *unit sphere* $S(0,1)$ in $(X, \| \cdot \|_2)$ is *compact*. Hence, by the *Weierstrass Extreme Value Theorem* (Theorem 2.55), the "old" norm $\| \cdot \|$ attains on $S(0,1)$ its *absolute minimum value* $c > 0$.

Exercise 3.58. Explain why $c > 0$.

For each $x \in X \setminus \{0\}$, $x/\|x\|_2 \in S(0,1)$ in $(X, \| \cdot \|_2)$, and hence, by *absolute scalability* of norm, we have

$$c \le \left\| \frac{1}{\|x\|_2} x \right\| = \frac{1}{\|x\|_2} \|x\|.$$

Combining the latter with (3.3), we have

$$c\|x\|_2 \le \|x\| \le C\|x\|_2, \quad x \in X,$$

which implies that the norms $\| \cdot \|$ and $\| \cdot \|_2$ are *equivalent*.

Since the equivalence of norms is an *equivalence relation* (see Remarks 3.20), we conclude that all norms on X are equivalent. □

Remark 3.26. As is seen from the proof of the prior theorem an n-dimensional ($n \in \mathbb{N}$) normed vector space $(X, \| \cdot \|)$ with an ordered basis $B := \{x_1, \dots, x_n\}$ can be *equivalently renormed* as follows:

$$X \ni x = \sum_{k=1}^n c_k x_k \mapsto \|x\|_2 := \|(c_1, \dots, c_n)\|_2 = \left[\sum_{k=1}^n |c_k|^2 \right]^{1/2},$$

which makes the spaces $(X, \| \cdot \|)$ and $l_2^{(n)}$ *isometrically isomorphic*.

Whence, we immediately obtain the following generalization of the *Heine–Borel Theorem* (Theorem 2.47).

Theorem 3.10 (Generalized Heine–Borel Theorem). *Let $(X, \| \cdot \|)$ be a (real or complex) finite-dimensional normed vector space.*
(1) *A set A is precompact in $(X, \| \cdot \|)$ iff it is bounded.*
(2) *A set A is compact in $(X, \| \cdot \|)$ iff it is closed and bounded.*

Exercise 3.59. Prove.

In the same manner, based on the *Norm Equivalence Theorem* (Theorem 3.9) and the renorming procedure described in Remark 3.26, one can prove the following important statement:

Theorem 3.11 (Completeness of Finite-Dimensional Spaces). *Every finite-dimensional normed vector space is a Banach space.*

Proof. Let $(X, \| \cdot \|)$ be a finite-dimensional normed vector space. The case of $\dim X = 0$ being vacuous, suppose that with $\dim X = n$ with some $n \in \mathbb{N}$.

By the *Norm Equivalence Theorem* (Theorem 3.9), equivalently renorming X as in Remark 3.26, we obtain a normed vector space $(X, \| \cdot \|_2)$, which is *isometrically isomorphic* to the Banach space $l_2^{(n)}$, and hence, is a Banach space.

The norms $\| \cdot \|$ and $\| \cdot \|_2$ being *equivalent*, the space $(X, \| \cdot \|)$ is also a Banach space (see Remarks 3.20). □

3.3.2 Finite-Dimensional Subspaces and Bases of Banach Spaces

3.3.2.1 Finite-Dimensional Subspaces

From the *Completeness of Finite-Dimensional Spaces Theorem* (Theorem 3.11) and *Characterization of Completeness* (Theorem 2.27), we immediately obtain the following corollary:

Theorem 3.12 (Closedness of Finite-Dimensional Subspaces). *Every finite-dimensional subspace of a normed vector space is closed.*

Exercise 3.60. Prove.

Theorem 3.13 (Nearest Point Property Relative to Finite-Dimensional Subspaces). *Let Y be a finite-dimensional subspace of a normed vector space $(X, \| \cdot \|)$. Then, for each $x \in X$, there exists a nearest point to x in Y, i. e.,*

$$\forall x \in X \, \exists y \in Y : \|x - y\| = \rho(x, Y) := \inf_{u \in Y} \|x - u\|.$$

Proof. If $x \in Y$, $\rho(x, Y) = 0$, and $y = x$.

Suppose that $x \in Y^c$. Then, since Y is *closed*, $\rho(x, Y) > 0$ (see Section 2.18, Problem 17).

There is a sequence $(y_n)_{n \in \mathbb{N}} \subset Y$ such that

$$\lim_{n \to \infty} \|x - y_n\| = \rho(x, Y), \tag{3.4}$$

which implies that $(y_n)_{n \in \mathbb{N}}$ is *bounded* in $(Y, \| \cdot \|)$.

Exercise 3.61. Explain.

Hence, in view of the *finite dimensionality* of Y, by the *Generalized Heine–Borel Theorem* (Theorem 3.10) and the *Equivalence of Different Forms of Compactness Theorem* (Theorem 2.50), there is a subsequence $(y_{n(k)})_{k \in \mathbb{N}}$ convergent to an element $y \in Y$. The latter, in respect to (3.4), by the *continuity* of norm, implies that

$$\|x - y\| = \lim_{k \to \infty} \|x - y_{n(k)}\| = \rho(x, Y),$$

completing the proof. □

Remark 3.27. The condition of the *finite dimensionality* of a subspace is essential and cannot be dropped. Indeed, as Examples 6.6 and 6.9 below demonstrate, in the space $(c_{00}, \|\cdot\|_2)$, there exists a *proper closed infinite-dimensional* subspace Y such that to no point $x \in Y^c$ is there a nearest point in Y.

Applying the prior statement to $(C[a,b], \|\cdot\|_\infty)$ $(-\infty < a < b < \infty)$, we obtain the following:

Corollary 3.5 (Best Approximation Polynomial in $(C[a,b], \|\cdot\|_\infty)$). *For each $f \in C[a,b]$ $(-\infty < a < b < \infty)$ and any $n \in \mathbb{Z}_+$, in $(C[a,b], \|\cdot\|_\infty)$, there is a best approximation polynomial $p_n \in P_n$:*

$$\max_{a \le t \le b} |f(t) - p_n(t)| = \|f - p_n\|_\infty = \rho(x, P_n) := \inf_{u \in P_n} \|f - u\|_\infty,$$

i. e., a polynomial nearest to f in the $(n+1)$-dimensional subspace P_n of all polynomials of degree at most n.

Remark 3.28. A nearest point relative to a finite-dimensional subspace, need not be unique.

Example 3.18. In $X = l_\infty^{(2)}(\mathbb{R})$, i. e., in $\mathbb{R}^2$ with the norm

$$\|(x_1, x_2)\|_\infty := \max[|x_1|, |x_2|], \quad (x_1, x_2) \in \mathbb{R}^2,$$

for the point $x := (0,1)$ and the *subspace* $Y := \mathrm{span}(\{(1,0)\}) = \{(\lambda, 0) \mid \lambda \in \mathbb{R}\}$,

$$\rho(x, Y) = \inf_{\lambda \in \mathbb{R}} \max[|0 - \lambda|, |1 - 0|] = \inf_{\lambda \in \mathbb{R}} \max[|\lambda|, 1] = \begin{cases} 1 & \text{for } |\lambda| \le 1, \\ > 1 & \text{for } |\lambda| > 1. \end{cases}$$

Hence, in Y, there are *infinitely many* nearest points to x of the form $y = (\lambda, 0)$ with $-1 \le \lambda \le 1$.

3.3.2.2 Bases of Banach Spaces

As is known (see Examples 3.16), a *topological basis* of a Banach space can be *countably infinite*. This, however, is not true for an *algebraic basis* of such a space. The *Closedness of Finite-Dimensional Subspaces Theorem* (Theorem 3.12) along with the *Nowhere Denseness of Closed Proper Subspaces Proposition* (Proposition 3.16, Section 3.5, Problem 11) allow us to prove the following interesting and profound fact:

Theorem 3.14 (Basis of a Banach Space). *A (Hamel) basis of a Banach space is either finite or uncountable.*

Proof. Let us prove the statement *by contradiction*, and assume that a Banach space $(X, \|\cdot\|)$ has a *countably infinite* Hamel basis $B := \{x_n\}_{n \in \mathbb{N}}$.

Then

$$X = \bigcup_{n=1}^{\infty} X_n, \tag{3.5}$$

where

$$X_n := \text{span}(\{x_1, x_2, \dots, x_n\}), \ n \in \mathbb{N},$$

is an n-dimensional subspace of X.

By the *Closedness of Finite-Dimensional Subspaces Theorem* (Theorem 3.12), each subspace X_n, $n \in \mathbb{N}$, is *closed*. Hence, being a closed proper subspace of X, by the *Nowhere Denseness of Closed Proper Subspaces Proposition* (Proposition 3.16, Section 3.5, Problem 11), each subspace X_n, $n \in \mathbb{N}$, is *nowhere dense* in $(X, \| \cdot \|)$.

By the *Baire Category Theorem* (Theorem 2.32), representation (3.5), *contradicts* the completeness of $(X, \| \cdot \|)$, which proves the statement. □

Remark 3.29. Hence, for a Banach space $(X, \| \cdot \|)$,

$$\text{either } \dim X = n, \text{ with some } n \in \mathbb{Z}_+, \text{ or } \dim X \geq \mathfrak{c},$$

where $\mathfrak{c}$ is the cardinality of the continuum, i. e., $\mathfrak{c} = |\mathbb{R}|$ (see Examples 1.1). In the latter case, by the *Dimension-Cardinality Connection Proposition* (Proposition 3.12, Section 3.5, Problem 4), the dimension of X is equal to its *cardinality*:

$$\dim X = |X|.$$

In particular,

$$\dim l_p^{(n)} = n \ (n \in \mathbb{N}, \ 1 \leq p \leq \infty), \ \dim P_n = n + 1 \ (n \in \mathbb{Z}_+)$$

and, by the arithmetic of cardinals (see, e. g., [29, 33, 41, 52]),

$$\dim l_p = \dim c_0 = \dim c = \dim C[a, b] = \mathfrak{c} \ (1 \leq p \leq \infty)$$

(cf. Examples 3.7).

3.4 Riesz's Lemma and Implications

The following celebrated statement has a number of profound implications:

Theorem 3.15 (Riesz's Lemma). *If Y is a closed proper subspace of a normed vector space $(X, \| \cdot \|)$, then*[3]

$$\forall \varepsilon \in (0, 1) \ \exists x_\varepsilon \in Y^c \text{ with } \|x_\varepsilon\| = 1 : \rho(x_\varepsilon, Y) := \inf_{y \in Y} \|x_\varepsilon - y\| > 1 - \varepsilon.$$

---

3 Frigyes Riesz (1880–1956).

Proof. Let $x \in Y^c$ be arbitrary. Then, by the *closedness* of Y, $\rho(x, Y) > 0$ (see Section 2.18, Problem 17), and hence,

$$\forall \varepsilon \in (0, 1) \, \exists y_\varepsilon \in Y : \rho(x, Y) \leq \|x - y_\varepsilon\| < \frac{\rho(x, Y)}{1 - \varepsilon}. \tag{3.6}$$

Setting

$$x_\varepsilon := \frac{1}{\|x - y_\varepsilon\|}(x - y_\varepsilon),$$

we obtain an element $x_\varepsilon \notin Y$ with $\|x_\varepsilon\| = 1$.

Exercise 3.62. Explain.

By *absolute scalability* of norm, we have

$$\forall y \in Y : \|x_\varepsilon - y\| = \left\| \frac{1}{\|x - y_\varepsilon\|}(x - y_\varepsilon) - y \right\| = \frac{1}{\|x - y_\varepsilon\|} \|x - (y_\varepsilon + \|x - y_\varepsilon\| y)\|$$

$$\text{since } y_\varepsilon + \|x - y_\varepsilon\| y \in Y \text{ and in view of (3.6);}$$

$$> \frac{1 - \varepsilon}{\rho(x, Y)} \rho(x, Y) = 1 - \varepsilon,$$

which completes the proof. $\qquad\qquad\qquad\qquad\qquad\qquad\qquad\qquad\qquad\quad$ □

Remark 3.30. The *closedness* condition is essential and cannot be dropped.

Exercise 3.63. Give a corresponding example.

Remark 3.31. As is known (see Remark 2.64), the *unit sphere* $S(0, 1)$, although *closed* and *bounded*, is *not compact* in $(c_{00}, \|\cdot\|_\infty)$ (also in $(c_0, \|\cdot\|_\infty)$, $(c, \|\cdot\|_\infty)$, and l_p ($1 \leq p \leq \infty$)). All the above spaces being *infinite-dimensional*, *Riesz's Lemma* (Theorem 3.15) is instrumental for understanding why this fact is not coincidental.

Corollary 3.6 (Characterization of Finite-Dimensional Normed Vector Spaces). *A normed vector space $(X, \|\cdot\|)$ is finite-dimensional iff the unit sphere*

$$S(0, 1) := \{x \in X \mid \|x\| = 1\}$$

is compact in $(X, \|\cdot\|)$.

Proof. "*Only if*" part. Immediately follows from the *Generalized Heine–Borel Theorem* (Theorem 3.10).

Exercise 3.64. Explain.

"*If*" part. Let us prove this part *by contrapositive*, assuming that X is *infinite-dimensional*.

Let us fix an element $x_1 \in X$ with $\|x_1\| = 1$. Then the *one-dimensional subspace*

$$X_1 := \text{span}(\{x_1\})$$

is a *closed proper subspace* of X, which, by *Riesz's Lemma* (Theorem 3.15) with $\varepsilon = 1/2$, implies that

$$\exists\, x_2 \in X_1^c \text{ with } \|x_1\| = 1 : \|x_2 - x_1\| \geq \rho(x_2, X_1) > \frac{1}{2}.$$

Similarly, since the *two-dimensional subspace*

$$X_2 := \mathrm{span}\{x_1, x_2\}$$

is a *closed proper subspace* of X, by *Riesz's Lemma* with $\varepsilon = 1/2$,

$$\exists\, x_3 \in X_2^c \text{ with } \|x_3\| = 1 : \|x_3 - x_i\| \geq \rho(x_3, X_2) > \frac{1}{2},\ i = 1, 2.$$

Continuing inductively, we obtain a sequence

$$(x_n)_{n \in \mathbb{N}} \subseteq S(0, 1)$$

with the following property:

$$\|x_m - x_n\| > \frac{1}{2},\ m, n \in \mathbb{N}, m \neq n.$$

Thus, not containing a fundamental, and the more so, a convergent subsequence, which, by the *Equivalence of Different Forms of Compactness* (Theorem 2.50), implies that the unit sphere $S(0, 1)$ is not compact in $(X, \|\cdot\|)$, and completes the proof. $\square$

An immediate corollary is as follows.

Corollary 3.7 (Noncompactness of Spheres/Balls). *Any nontrivial sphere/ball in an infinite-dimensional normed vector space is not compact.*

Exercise 3.65. Prove.

Based on this fact, it is not difficult to prove the following quite counterintuitive statement:

Proposition 3.10 (Nowhere Denseness of Compact Sets). *A compact set in an infinite-dimensional normed vector space is nowhere dense.*

Exercise 3.66. Prove.

Hint. Prove *by contradiction*.

3.5 Problems

In the subsequent problems, $\mathbb{F}$ stands for the scalar field of real or complex numbers (i.e., $\mathbb{F} = \mathbb{R}$ or $\mathbb{F} = \mathbb{C}$).

1. (Complexification of Real Vector Spaces)
    The *complexification* of a real vector space X is the complex vector space

    $$X^{\mathbb{C}} := X \times X$$

    with the linear operations

    $$X^{\mathbb{C}} \ni (x_1, y_1), (x_2, y_2) \mapsto (x_1, y_1) + (x_2, y_2) := (x_1 + x_2, y_1 + y_2) \in X^{\mathbb{C}},$$
    $$\mathbb{C} \ni \alpha + i\beta, X^{\mathbb{C}} \ni (x, y) \mapsto (\alpha + i\beta)(x, y) := (\alpha x - \beta y, \beta x + \alpha y) \in X^{\mathbb{C}}.$$

    In particular,

    $$\forall \alpha \in \mathbb{R}, \ \forall (x, y) \in X^{\mathbb{C}} : \alpha(x, y) = (\alpha x, \alpha y),$$
    $$\forall (x, y) \in X^{\mathbb{C}} : i(x, y) = (-y, x),$$

    and hence,

    $$\forall (x, y) \in X^{\mathbb{C}} : (x, y) = (x, 0) + i(y, 0),$$
    $$\forall \alpha + i\beta \in \mathbb{C}, \ \forall (x, y) \in X^{\mathbb{C}} : (\alpha + i\beta)(x, y) = \alpha(x, y) + i\beta(x, y).$$

    The space $X^{\mathbb{C}}$ extends X in the sense of the following *isomorphic embedding*:

    $$X \ni x \mapsto Tx := (x, 0) \in X^{\mathbb{C}}.$$

    Thus, we can identify each vector $x \in X$ with the vector $(x, 0) \in X^{\mathbb{C}}$. Write

    $$\forall (x, y) \in X^{\mathbb{C}} : (x, y) = x + iy.$$

    (a) Show that the linear operations (defined above) satisfy the *vector space axioms*.
    (b) Show that the mapping T *isomorphically embeds* X into the associated with $X^{\mathbb{C}}$ real space $X_{\mathbb{R}}^{\mathbb{C}}$ (see Remarks 3.1).
    (c) Show that, if a set S is linearly independent in X, then the set $TS = \{(x, 0) \mid x \in S\}$ is linearly independent in $X^{\mathbb{C}}$.
    (d) Show that, if a set B is basis for X, then $TB = \{(x, 0) \mid x \in B\}$ is a basis for $X^{\mathbb{C}}$, and hence,

    $$\dim X = \dim X^{\mathbb{C}}.$$

2. Prove

    Proposition 3.11 (Characterization of Isomorphisms). *For vector spaces X and Y, a homomorphism $T : X \to Y$ is an isomorphism iff*

    $$\ker T := \{x \in X \mid Tx = 0\} = \{0\}.$$

3. Show that, if Y is a nonzero *subspace* of a nonzero vector space X, each basis B_Y of Y can be extended to a basis B_X of X.

4. * Prove

   Proposition 3.12 (Dimension-Cardinality Connection). *If for a vector space X, $\dim(X) \ge \mathfrak{c}$ ($\mathfrak{c} := |\mathbb{R}|$), then $\dim X = |X|$.*

   Hint. Use the idea of the proof of the *Dimension Theorem* (Theorem 3.4).

5. Show that the metric

   $$s \ni x := (x_k)_{k \in \mathbb{N}}, y := (y_k)_{k \in \mathbb{N}} \mapsto \rho(x, y) := \sum_{k=1}^{\infty} \frac{1}{2^k} \frac{|x_k - y_k|}{|x_k - y_k| + 1}$$

   (see Section 2.18, Problem 27) is *not* a norm metric.

6. Prove

   Proposition 3.13 (Closure of a Subspace). *If Y is a subspace of a normed vector space $(X, \| \cdot \|)$, then its closure $\overline{Y}$ is also a subspace of $(X, \| \cdot \|)$.*

7. (Cartesian Product of Normed Vector Spaces)
   Let $(X_1, \| \cdot \|_1)$ and $(X_2, \| \cdot \|_2)$ be normed vector spaces over $\mathbb{F}$.
   (a) Show that the Cartesian product $X = X_1 \times X_2$ is a normed vector space relative to the *product norm*

   $$X_1 \times X_2 \ni x = (x_1, x_2) \mapsto \|x\|_{X_1 \times X_2} = \sqrt{\|x_1\|_1^2 + \|x_2\|_2^2}.$$

   (b) Describe convergence in $(X_1 \times X_2, \| \cdot \|_{X_1 \times X_2})$.
   (c) Show that the product space $(X_1 \times X_2, \| \cdot \|_{X_1 \times X_2})$ is Banach space *iff* each space $(X_i, \| \cdot \|_i)$, $i = 1, 2$, is a Banach space.

8. Show that if $(X, \| \cdot \|_X)$ is a *seminormed* vector space, then
   (a) $Y := \{x \in X \mid \|x\| = 0\}$ is a subspace of X, and
   (b) the quotient space X/Y is a normed vector space relative to the norm

   $$X/Y \ni [x] := x + Y \mapsto \|[x]\| = \|x\|_X.$$

   Hint. To show that $\| \cdot \|$ is well defined on X/Y, i. e., is independent of the choice of a representative of the coset $[x]$, use inequality (3.2).

9. Prove

   Proposition 3.14 (Completeness of $BV[a, b]$). *The vector space $BV[a, b]$ ($-\infty < a < b < \infty$) is a Banach space relative to the total-variation norm*

   $$BV[a, b] \ni f \mapsto \|f\| := |f(a)| + V_a^b(f).$$

   Hint. Show first that

   $$\forall f \in BV[a, b] : \sup_{a \le x \le b} |f(x)| \le |f(a)| + V_a^b(f).$$

10. Prove

Proposition 3.15 (Completeness of $C^n[a, b]$). *The vector space* $C^n[a, b]$ *($n \in N$,* $-\infty < a < b < \infty$) *of n-times continuously differentiable on* $[a, b]$ *functions with pointwise linear operations is a Banach space relative to the norm*

$$C^n[a, b] \ni f \mapsto \|f\|_n := \max_{0 \le k \le n} \left[\max_{a \le t \le b} |f^{(k)}(t)| \right].$$

Hint. Use the *Total Change Formula* representation:

$$f^{(k-1)}(t) := f^{(k-1)}(a) + \int_a^t f^{(k)}(s)\, ds, \ k = 1, \dots, n, t \in [a, b].$$

11. (a) Prove

Proposition 3.16 (Nowhere Denseness of Closed Proper Subspaces).
A closed proper subspace Y of a normed vector space $(X, \|\cdot\|)$ is nowhere dense.

   (b) Give an example showing the *closedness requirement* is essential and cannot be dropped.

12. Prove

Theorem 3.16 (Quotient Space Norm). *Let Y be a closed subspace of a normed vector space $(X, \|\cdot\|_X)$. Then*
(1)

$$X/Y \ni [x] = x + Y \mapsto \|[x]\| = \inf_{y \in Y} \|x + y\|_X =: \rho(x, Y)$$

is a norm on X/Y, called the quotient space norm;
(2) *if $(X, \|\cdot\|_X)$ is a Banach space, then $(X/Y, \|\cdot\|)$ is a Banach space.*

Hint. To prove part 2, first describe the convergence in the quotient space $(X/Y, \|\cdot\|)$.

13. (Norm Defining Procedure)
Let X be a vector space, $(Y, \|\cdot\|_Y)$ be a *normed vector space* over $\mathbb{F}$, and $T : X \to Y$ be an *isomorphic embedding* of X into Y.
Show that the mapping

$$X \ni x \mapsto \|x\|_X := \|Tx\|_Y$$

is a *norm* on X, relative to which the isomorphism T is also *isometric*.

14. Show that the mapping

$$P_n \ni p(t) = \sum_{k=0}^{n} a_k t^k \mapsto \|p\| := \sum_{k=0}^{n} |a_k|$$

is a *norm* on the space P_n of all polynomials of degree at most n ($n \in \mathbb{Z}_+$) with real/complex coefficients.

15. (The 60th W. L. Putnam Mathematical Competition, 1999) Prove that there exists a constant $C > 0$ such that, for any polynomial $p(x)$ of degree 1999,

$$|p(0)| \le C \int\limits_{-1}^{1} |p(x)| \, dx.$$

16. In $X = l_1^{(2)}(\mathbb{R})$, i.e., in $\mathbb{R}^2$ with the norm

$$\|(x_1, x_2)\|_1 := |x_1| + |x_2|, \; (x_1, x_2) \in \mathbb{R}^2,$$

determine the *set of all nearest points* to $x := (1, -1)$ in the subspace $Y := \mathrm{span}(\{(1, 1)\}) = \{(\lambda, \lambda) \mid \lambda \in \mathbb{R}\}$.

17. Prove

Proposition 3.17 (Characterization of Finite-Dimensional Banach Spaces). *A Banach space $(X, \| \cdot \|)$ is finite-dimensional iff every subspace of X is closed.*

Hint. Prove the *"if"* part *by contrapositive* applying the *Basis of a Banach Space Theorem* (Theorem 3.14) and the *Characterization of Completeness* (Theorem 2.27).

18. Prove

Proposition 3.18 (Riesz's Lemma in Finite-Dimensional Spaces). *If Y is a proper subspace of a finite-dimensional normed vector space $(X, \| \cdot \|)$, then*

$$\exists x \in Y^C : \|x\| = 1 = \rho(x, Y).$$

19. Prove

Proposition 3.19 (Characterization of Finite-Dimensional Banach Spaces). *A Banach space $(X, \| \cdot \|)$ is finite-dimensional iff it can be represented as a countable union of compact sets.*

Hint. Prove the *"if"* part *by contradiction* applying the *Nowhere Denseness of Compact Sets Proposition* (Proposition 3.10) and the *Baire Category Theorem* (Theorem 2.32).

4 Linear Operators

In this chapter, we discuss an important class of mappings on vector and normed vector spaces, which are closely related to the linear structure of the spaces, or both their linear and topological structures. The major statements of this chapter—the *Uniform Boundedness Principle* and the *Open Mapping Theorem* along with its equivalents: the *Inverse Mapping Theorem* and the *Closed Graph Theorem*—are the two of the so-called *three fundamental principles of linear functional analysis*, which also include *Hahn*[1]–*Banach Theorem* (see, e. g., [8, 10, 16, 45]), whose significance and vast applicability cannot be overstated.

4.1 Linear Operators and Functionals

Linear operators on vector spaces are "married" to the linear structure of such spaces.

4.1.1 Definitions and Examples

Recall that $\mathbb{F}$ stands for the *scalar field* of $\mathbb{R}$ or $\mathbb{C}$.

Definition 4.1 (Linear Operator and Linear Functional). Let X and Y be vector spaces over $\mathbb{F}$.

A *linear operator* (also a *linear transformation* or a *linear mapping*) from X to Y is a mapping

$$A : D(A) \rightarrow Y,$$

where $D(A) \subseteq X$ is a *subspace* of X that preserves linear operations

$$\forall x, y \in D(A), \ \forall \lambda, \mu \in \mathbb{F} : \ A(\lambda x + \mu y) = \lambda A x + \mu A y,$$

i. e., is a *homomorphism* of $D(A)$ to Y.

The subspace $D(A)$ is called the *domain (of definition)* of A.

Notation. $(A, D(A))$.

If $Y = X$, i. e.,

$$A : X \supseteq D(A) \rightarrow X,$$

then A is said to be a linear operator *in* X, or *on* X, provided $D(A) = X$.

---

1 Hans Hahn (1879–1934).

https://doi.org/10.1515/9783110600988-004

If $Y = \mathbb{F}$, i. e.,

$$A : D(A) \to \mathbb{F},$$

then A is called a *linear functional*. These are customarily designated by the lower case letters, and the notation $f(x)$ is used instead of Ax.

Examples 4.1.
1. On a vector space X over $\mathbb{F}$, multiplication by an arbitrary number $\lambda \in \mathbb{F}$,

$$X \ni x \to Ax := \lambda x \in X,$$

is a linear operator (*endomorphism*) (see Examples 3.2).
In particular, we obtain the *zero operator* 0 for $\lambda = 0$ and the *identity operator* I for $\lambda = 1$.

Remark 4.1. Here and henceforth, 0 is used to designate zero operators and functionals, such a connotation being a rather common practice of symbol economy.

2. Recall that, by the *Basis Representation Theorem* (Theorem 3.3), in a vector space X over $\mathbb{F}$ with a *basis* $B := \{x_i\}_{i \in I}$, each element $x \in X$ has a *unique representation* relative to B

$$x = \sum_{i \in I} c_i x_i,$$

in which all but a finite number of the coefficients $c_i \in \mathbb{F}$, $i \in I$, called the *coordinates* of x relative to B, are zero.
For each $j \in I$, the mapping

$$X \ni x = \sum_{i \in I} c_i x_i \mapsto c_j(x) := c_j \in \mathbb{F}$$

is a *linear functional* on X, called the jth *(Hamel) coordinate functional*, relative to the basis B.

3. Multiplication by an $m \times n$ matrix $[a_{ij}]$ $(m, n \in \mathbb{N})$ with entries from $\mathbb{F}$

$$\mathbb{F}^n \ni x \mapsto Ax := [a_{ij}]x \in \mathbb{F}^m$$

is a linear operator from $\mathbb{F}^n$ to $\mathbb{F}^m$ and, provided $m = n$, is a linear operator on $\mathbb{F}^n$ (see Examples 3.2 and Remarks 1.6).
In particular, for $m = 1$, we obtain a *linear functional* on $\mathbb{F}^n$

$$\mathbb{F}^n \ni x := (x_1, \dots, x_n) \mapsto f(x) := \sum_{k=1}^{n} a_k x_k \in \mathbb{F}, \tag{4.1}$$

where $(a_1, \dots, a_n) \in \mathbb{F}^n$.

Conversely, every linear functional on $\mathbb{F}^n$ ($n \in \mathbb{N}$) is of the form given by (4.1) with some $(a_1, \ldots, a_n) \in \mathbb{F}^n$.

4. On c_{00}, the operator of multiplication by a numeric sequence $a := (a_n)_{n \in \mathbb{N}} \in \mathbb{F}^{\mathbb{N}}$,

$$c_{00} \ni x = (x_n)_{n \in \mathbb{N}} \mapsto Ax := (a_n x_n)_{n \in \mathbb{N}} \in c_{00},$$

is a *linear operator*.

5. Let $-\infty < a < b < \infty$.
    (a) $C[a, b] \ni x \to [Ax](t) := \int_a^t x(s)\, ds$, $t \in [a, b]$, is a *linear operator* on $C[a, b]$.
    (b) $C[a, b] \ni x \to f(x) := \int_a^b x(t)\, dt$ is a *linear functional* on $C[a, b]$.
    (c) $C^1[a, b] \ni x \to [Ax](t) := \frac{d}{dt} x(t)$, $t \in [a, b]$, is a *linear operator* from $C^1[a, b]$ to $C[a, b]$ or in $C[a, b]$ with the *domain* $D(A) := C^1[a, b]$.

Exercise 4.1. Verify.

4.1.2 Kernel, Range, and Graph

With every linear operator associated are three important subspaces: its *kernel*, *range*, and *graph*.

Definition 4.2 (Kernel, Range, and Graph). Let X and Y be vector spaces over $\mathbb{F}$ and $(A, D(A))$ be a linear operator from X to Y.
–   The *kernel* (or *null space*) of A is

$$\ker A := \{x \in D(A) \mid Ax = 0\}.$$

–   The *range* of A is

$$R(A) := \{Ax \in Y \mid x \in D(A)\}.$$

–   The *graph* of A is

$$G_A := \{(x, Ax) \in X \times Y \mid x \in D(A)\}.$$

Remarks 4.2. For a linear operator $(A, D(A))$ from X to Y,
– $A0 = 0$,
– $\ker A$ is a *subspace* of $D(A)$ and hence of X,
– $R(A)$ is a *subspace* of Y, and
– G_A is a *subspace* of $X \times Y$.

Exercise 4.2.
(a) Verify.

(b) Show that, for linear operators $(A, D(A))$ and $(B, D(B))$ from a vector space X to a vector space Y over $\mathbb{F}$,

$$A = B, \text{ i. e., } D(A) = D(B) \text{ and } Ax = Bx, \ x \in D(A), \ \Leftrightarrow \ G_A = G_B,$$

i. e., linear operators are equal *iff* their graphs coincide.

4.1.3 Rank-Nullity and Extension Theorems

4.1.3.1 Rank-Nullity Theorem

Definition 4.3 (Rank and Nullity). Let X and Y be vector spaces over $\mathbb{F}$ and $(A, D(A))$ be a linear operator from X to Y. The *rank* and the *nullity* of A are $\dim(R(A))$ and $\dim(\ker A)$, respectively.

The celebrated *Rank-Nullity Theorem* of linear algebra, which states that the *rank* and the *nullity of a matrix* add up to the number of its columns (see, e. g., [54]), allows the following generalization.

Theorem 4.1 (Rank-Nullity Theorem). *Let X and Y be vector spaces over $\mathbb{F}$ and $A : X \to Y$ be a linear operator. Then*

$$\dim(X/\ker A) = \dim R(A),$$

and

$$\dim X = \dim R(A) + \dim \ker A.$$

Proof. Consider the *quotient space* $X/\ker A$ of X modulo $\ker A$ and the *quotient operator*

$$X/\ker A \ni [x] = x + \ker A \mapsto \hat{A}[x] := Ax,$$

which is *well defined*, i. e., for each $[x] \in X/\ker A$, the value $\hat{A}[x]$ is independent of the choice of a representative of the *coset* $[x] = x + \ker A$, and is an *isomorphism* between $X/\ker A$ and $R(A)$.

Exercise 4.3. Verify.

Hint. Apply the *Characterization of Isomorphisms* (Proposition 3.11, Section 3.5, Problem 2).

Hence, by the *Isomorphism Theorem* (Theorem 3.5),

$$\dim(X/\ker A) = \dim R(A).$$

Since, by the *Codimension of a Subspace Corollary* (Corollary 3.4),

$$\dim X = \dim \ker A + \dim(X/\ker A),$$

we conclude that

$$\dim X = \dim \ker A + \dim R(A). \qquad \square$$

Remark 4.3. As immediately follows from the *Rank-Nullity Theorem* and the *Codimension of a Subspace Corollary* (Corollary 3.4), for arbitrary vector spaces X and Y over $\mathbb{F}$ and any linear operator $A : X \to Y$,

$$\operatorname{codim} \ker A = \dim R(A).$$

If furthermore, $Y = X$ and $\dim X = n$ for some $n \in \mathbb{Z}_+$, then

$$\dim \ker A = \operatorname{codim} R(A).$$

Exercise 4.4. Explain the latter.

Using the *Rank-Nullity Theorem*, we can reveal the nature of the kernels of linear functionals.

Proposition 4.1 (Null Space of a Linear Functional). *Let X be a vector space over $\mathbb{F}$ and $f : X \to \mathbb{F}$ be a linear functional on X. Then*

$$\dim(X/\ker f) = \begin{cases} 0, \\ 1, \end{cases}$$

i. e., either $\ker f = X$ for $f = 0$, or $\ker f$ is a hyperplane of X for $f \neq 0$.

Proof. By the *Rank-Nullity Theorem* (Theorem 4.1),

$$\dim(X/\ker f) = \dim R(f) = \begin{cases} 0, \\ 1, \end{cases}$$

since either $R(f) = \{0\}$ for $f = 0$, or $R(f) = \mathbb{F}$ for $f \neq 0$.

Exercise 4.5. Verify. $\qquad \square$

4.1.3.2 Extension Theorem

The following is a fundamental statement concerning the extendability of linear operators. Its proof utilizing *Zorn's Lemma* (Theorem A.5) conceptually foreshadows that of the celebrated *Hahn–Banach Theorem* (the extension form) (see, e. g., [45]).

Theorem 4.2 (Extension Theorem for Linear Operators). *Let X and Y be vector spaces over* $\mathbb{F}$ *and*

$$A : X \supseteq D(A) \to Y,$$

where $D(A)$ is a subspace of X, be a linear operator from X to Y. Then there exists a linear operator $\tilde{A} : X \to Y$ defined on the entire space X such that

$$\forall x \in D(A) : \tilde{A}x = Ax,$$

i. e., there exists a linear extension $\tilde{A}$ of A to the entire space X.

Proof. If $D(A) = X$, then, trivially, $\tilde{A} = A$.

Suppose that $D(A) \neq X$, let $\mathscr{E}$ be the set of all linear extensions of A (in particular, $A \in \mathscr{E}$) partially ordered by *extension*:

$$\forall B, C \in \mathscr{E} : B \leq C \iff C \text{ is an } extension \text{ of } B,$$

i. e.,

$$D(B) \subseteq D(C) \text{ and } \forall x \in D(B) : Cx = Bx,$$

and let $\mathscr{C}$ be an *arbitrary chain* in $(\mathscr{E}, \leq)$.

Then

$$\tilde{D} := \bigcup_{B \in \mathscr{C}} D(B)$$

is a *subspace* of X, and

$$\tilde{D} \ni x \mapsto Cx := Bx,$$

where $B \in \mathscr{C}$ such that $x \in D(B)$ is arbitrary, is a *well-defined linear operator* from X to Y, which is an *upper bound* of $\mathscr{C}$ in $(\mathscr{E}, \leq)$.

Exercise 4.6. Verify.

By *Zorn's Lemma* (Theorem A.5), $(\mathscr{E}, \leq)$ has a *maximal element* $(\tilde{A}, D(\tilde{A}))$, i. e., $(\tilde{A}, D(\tilde{A}))$ is a *maximal linear extension* of $(A, D(A))$.

This implies that $D(\tilde{A}) = X$. Indeed, otherwise, there exists an element $x \in D(\tilde{A})^c$ such that any

$$z \in \text{span}(D(\tilde{A}) \cup \{x\}) = D(\tilde{A}) \oplus \text{span}(\{x\})$$

can be *uniquely* represented as

$$z = u + \lambda x$$

with some $u \in D(\tilde{A})$ and $\lambda \in \mathbb{F}$.

By choosing an arbitrary element $y \in Y$, we can define a linear extension (an "*extension by one dimension*")

$$E : \mathrm{span}(D(\tilde{A}) \cup \{x\}) \to Y$$

of $\tilde{A}$, and hence, of A, as follows:

$$Ez := \tilde{A}u + \lambda y, \; z = u + \lambda x \in \mathrm{span}(D(\tilde{A}) \cup \{x\})$$

(in particular, $Ex := y$).

Exercise 4.7. Verify.

Since $E \in \mathscr{E}$, and $\tilde{A} < E$ (see Section A.2), we obtain a *contradiction* to the maximality of $(\tilde{A}, D(\tilde{A}))$ in $(\mathscr{E}, \leq)$.

Hence, $D(\tilde{A}) = X$, and $\tilde{A}$ is the desired extension of A. □

Remark 4.4. The procedure of "*extension by one dimension*" described in the proof of the prior theorem applies to any linear operator $(A, D(A))$, whose domain is not the entire space $(D(A) \neq X)$. Hence, as readily follows from this procedure, a linear extension of such a linear operator to the whole space, although *existent*, is *not unique* whenever the target space Y is *nonzero* $(Y \neq \{0\})$, being dependent on the choice of the image $y \in Y$ for an $x \in D(\tilde{A})^c$.

4.2 Bounded Linear Operators and Functionals

Bounded linear operators on normed vector spaces are "married" to both their linear and topological structures.

4.2.1 Definitions, Properties, Examples

The following is a remarkable inherent property of linear operators.

Theorem 4.3 (Continuity of a Linear Operator). *Let* $(X, \| \cdot \|_X)$ *and* $(Y, \| \cdot \|_Y)$ *be normed vector spaces over* $\mathbb{F}$. *If a linear operator* $A : X \to Y$ *is continuous at a point* $x \in X$, *then it is Lipschitz continuous on* X.

Proof. Due to the linearity of A, without loss of generality, we can assume that $x = 0$.

Exercise 4.8. Explain.

In view of the fact that $A0 = 0$ (see Remarks 4.2), the continuity of A at 0 implies that

$$\exists \delta > 0 \; \forall x \in X \text{ with } \|x\|_X < \delta : \; \|Ax\|_Y < 1.$$

Then, by the *absolute scalability* of norm, for any *distinct* $x', x'' \in X$,

$$\frac{\delta}{2\|x' - x''\|_X}(x' - x'') \in B_X(0, \delta),$$

and hence,

$$\left\| A\left[\frac{\delta}{2\|x' - x''\|_X}(x' - x'') \right] \right\|_Y < 1.$$

By the *linearity* of A and *absolute scalability* of norm, we infer that

$$\forall x', x'' \in X : \|Ax' - Ax''\|_Y = \|A(x' - x'')\|_Y \le \frac{2}{\delta}\|x' - x''\|_X,$$

which completes the proof. $\qquad\qquad\qquad\qquad\qquad\qquad\qquad\qquad\qquad\quad$ $\square$

Remark 4.5. Thus, a *linear operator* $A : (X, \|\cdot\|_X) \to (Y, \|\cdot\|_Y)$ is either Lipschitz continuous on X, or it is discontinuous at every point of X.

Definition 4.4 (Bounded Linear Operator and Operator Norm). Let $(X, \|\cdot\|_X)$ and $(Y, \|\cdot\|_Y)$ be normed vector spaces over $\mathbb{F}$. A *linear operator* $A : X \to Y$ is called *bounded* if

$$\exists M > 0 \; \forall x \in X : \|Ax\|_Y \le M\|x\|_X,$$

in which case

$$\|A\| := \min\{M > 0 \mid \forall x \in X : \|Ax\|_Y \le M\|x\|_X\} = \sup_{x \in X \setminus \{0\}} \frac{\|Ax\|_Y}{\|x\|_X} = \sup_{\|x\|_X = 1} \|Ax\|_Y$$

is a nonnegative number called the *norm* of the operator A.

Exercise 4.9.
(a) Check the consistency of the above definitions of *operator norm*.
(b) Show that *operator norm* can also be equivalently defined as follows:

$$\|A\| := \sup_{\|x\|_X \le 1} \|Ax\|_Y.$$

(c) Give an example showing that, unless X is finite-dimensional, in the definitions of *operator norm*, sup cannot be replaced with max.

Hint. On $(c_{00}, \|\cdot\|_\infty)$, consider the operator of multiplication by the sequence $(1 - 1/n)_{n \in \mathbb{N}}$:

$$c_{00} \ni x = (x_n)_{n \in \mathbb{N}} \mapsto Ax := ((1 - 1/n)x_n)_{n \in \mathbb{N}} \in c_{00}$$

(see Examples 4.1).

Theorem 4.4 (Characterizations of Bounded Linear Operators). *Let* $(X, \| \cdot \|_X)$ *and* $(Y, \| \cdot \|_Y)$ *be normed vector spaces over* $\mathbb{F}$. *A linear operator* $A : X \to Y$ *is bounded iff any of the following equivalent conditions holds:*

1. *A maps the unit sphere/unit ball of* $(X, \| \cdot \|_X)$ *to a bounded set of* $(Y, \| \cdot \|_Y)$.
2. *A maps an arbitrary bounded set of* $(X, \| \cdot \|_X)$ *to a bounded set of* $(Y, \| \cdot \|_Y)$.
3. *A is Lipschitz continuous on X, with* $\|A\|$ *being its best Lipschitz constant.*
4. *(Sequential Characterization) For any bounded sequence* $(x_n)_{n \in \mathbb{N}}$ *in* $(X, \| \cdot \|_X)$, *the image sequence* $(Ax_n)_{n \in \mathbb{N}}$ *is bounded in* $(Y, \| \cdot \|_Y)$.

Exercise 4.10. Prove.

Remark 4.6. Recall that, in particular, when $Y = \mathbb{F}$, linear operators are called *linear functionals*, and the lower-case letters are used: $f : X \to \mathbb{F}$.

Thus, everything defined/proved for linear operators applies to linear functionals. In particular, the number

$$\|f\| := \sup_{\|x\|_X = 1} |f(x)|$$

is the *norm* of a bounded linear functional $f : (X, \| \cdot \|) \to (\mathbb{F}, |\cdot|)$.

Examples 4.2.

1. On a normed vector space $(X, \| \cdot \|_X)$ over $\mathbb{F}$, multiplication by an arbitrary number $\lambda \in \mathbb{F}$ is a bounded linear operator, and $\|A\| = |\lambda|$.

   In particular, the *zero operator* 0 ($\lambda = 0$) and the *identity operator* I ($\lambda = 1$) are bounded liner operators on X with $\|0\| = 0$ and $\|I\| = 1$.

2. For normed vector spaces $(X, \| \cdot \|_X)$ and $(Y, \| \cdot \|_Y)$ over $\mathbb{F}$ and $n \in \mathbb{N}$, if $f_1, \ldots, f_n$ are *bounded linear functionals* on X and $y_1, \ldots, y_n$ are arbitrary elements in Y, then

$$X \ni x \mapsto Ax := \sum_{k=1}^{n} f_k(x) y_k \in Y$$

   is a *bounded linear operator*, and $\|A\| \leq \sum_{k=1}^{n} \|f_k\| \|y_k\|_Y$.

3. For $m, n \in \mathbb{N}$ and $1 \leq p, p' \leq \infty$, multiplication by an $m \times n$ matrix $[a_{ij}]$ with entries from $\mathbb{F}$,

$$l_p^{(n)} \ni x = (x_1, \ldots, x_n) \mapsto Ax := [a_{ij}]x \in l_{p'}^{(m)},$$

   is a bounded linear operator from $l_p^{(n)}$ to $l_{p'}^{(m)}$ and, provided $m = n$ and $p' = p$, a bounded linear operator on $l_p^{(n)}$.

   By *Hölder's Inequality for n-Tuples* (Theorem 2.2),

$$\|A\| \leq \left\| (\|a_1\|_{p'}, \ldots, \|a_n\|_{p'}) \right\|_q,$$

   where $a_j, j = 1, \ldots, n$, are the *column vectors* of the matrix $[a_{ij}]$, and q is the *conjugate index* to p (see Definition 2.2).

In particular, for $m = 1$, we obtain a *bounded linear functional* on $l_p^{(n)}$:

$$l_p^{(n)} \ni x = (x_1, \ldots, x_n) \mapsto f(x) := \sum_{k=1}^{n} a_k x_k \in \mathbb{F},$$

and (by *Hölder's Inequality for n-Tuples* (Theorem 2.2))

$$\|f\| \le \|a\|_q,$$

where $a := (a_1, \ldots, a_n)$, and q is the *conjugate index* to p (see Definition 2.2).

4. On l_p $(1 \le p \le \infty)$,

   (a) for a *bounded sequence* $a := (a_n)_{n \in \mathbb{N}} \in l_\infty$, the operator of multiplication

   $$l_p \ni x := (x_n)_{n \in \mathbb{N}} \mapsto Ax := (a_n x_n)_{n \in \mathbb{N}} \in l_p$$

   is a *bounded linear operator* with $\|A\| = \|a\|_\infty := \sup_{n \in \mathbb{N}} |a_n|$;

   (b) the *right shift operator*

   $$l_p \ni x := (x_1, x_2, \ldots) \mapsto Ax := (0, x_1, x_2, \ldots) \in l_p$$

   and the *left shift operator*

   $$l_p \ni x := (x_1, x_2, \ldots) \mapsto Bx := (x_2, x_3, x_4, \ldots) \in l_p$$

   are *bounded linear operators* with $\|A\| = \|B\| = 1$, the right shift operator being an *isometry*, i. e.,

   $$\forall x \in l_p : \|Ax\|_p = \|x\|_p;$$

   (c) for a sequence $a := (a_n)_{n \in \mathbb{N}} \in l_q$, where q is the *conjugate index* to p (see Definition 2.2), by *Hölder's inequality for Sequences* (Theorem 2.5),

   $$l_p \ni x := (x_n)_{n \in \mathbb{N}} \mapsto f(x) := \sum_{k=1}^{\infty} a_k x_k \in \mathbb{F}$$

   is a *bounded linear functional* with

   $$\|f\| \le \|a\|_q.$$

Remark 4.7. In fact, more precisely, $\|f\| = \|a\|_q$, and such functionals are the only bounded linear functionals on l_p $(1 \le p \le \infty)$ (see, e. g., [45]).

5. On $(c_{00}, \|\cdot\|_\infty)$, the operator of multiplication by an *unbounded numeric sequence* $a := (a_n)_{n \in \mathbb{N}} \in \mathbb{F}^\mathbb{N}$

   $$c_{00} \ni x := (x_n)_{n \in \mathbb{N}} \mapsto Ax := (a_n x_n)_{n \in \mathbb{N}} \in c_{00}$$

   is an *unbounded linear operator*.

6. On $(c, \| \cdot \|_\infty)$, the *limit functional*

$$c \ni x := (x_n)_{n \in \mathbb{N}} \mapsto l(x) := \lim_{n \to \infty} x_n \in \mathbb{F},$$

assigning to each convergent sequence its limit, is a *bounded linear functional* with $\|l\| = 1$ and $\ker l = c_0$.

7. On l_1, the *sum functional*

$$l_1 \ni x := (x_n)_{n \in \mathbb{N}} \mapsto s(x) := \sum_{n=1}^{\infty} x_n \in \mathbb{F},$$

assigning to each *absolutely summable sequence* the sum of the series composed of its terms, is a *bounded linear functional* with $\|s\| = 1$.

8. On $(C[a, b], \| \cdot \|_\infty)$ $(-\infty < a < b < \infty)$,
   (a) the operator of multiplication by an arbitrary function $m \in C[a, b]$

   $$C[a, b] \ni x \to [Ax](t) := m(t)x(t) \in C[a, b]$$

   is a *bounded linear operator* with $\|A\| = \|m\|_\infty := \max_{a \leq t \leq b} |m(t)|$;
   (b) the *integration operator*

   $$C[a, b] \ni x \to [Ax](t) := \int_a^t x(s)\, ds \in C^1[a, b]$$

   is a *bounded linear operator* with $\|A\| = b - a$;
   (c) the *integration functional*

   $$C[a, b] \ni x \to f(x) := \int_a^b x(t)\, dt \in \mathbb{F}$$

   is a *bounded linear functional* with $\|f\| = b - a$;
   (d) for each $t \in [a, b]$, the *fixed-value functional*

   $$C[a, b] \ni x \to f_t(x) := x(t) \in \mathbb{F}$$

   is a *bounded linear functional* with $\|f_t\| = 1$.

9. The *differentiation operator*

   $$C^1[a, b] \ni x \to [Ax](t) := \frac{d}{dt} x(t) \in C[a, b]$$

   (a) is a *bounded linear operator* from $C^1[a, b]$ with the norm

   $$\|x\| = \max\left[\max_{a \leq t \leq b} |x(t)|, \max_{a \leq t \leq b} |x'(t)| \right]$$

   to $(C[a, b], \| \cdot \|_\infty)$, $\|A\| = 1$ and
   (b) is an *unbounded linear operator* in $(C[a, b], \| \cdot \|_\infty)$ with the *domain* $D(A) := C^1[a, b]$.

Exercise 4.11. Verify.

4.2.2 Kernel

Proposition 4.2 (Kernel of a Bounded Linear Operator). *Let $(X, \| \cdot \|_X)$ and $(Y, \| \cdot \|_Y)$ be normed vector spaces over $\mathbb{F}$. The kernel ker A of an arbitrary bounded linear operator $A : X \to Y$ is a closed subspace of $(X, \| \cdot \|_X)$.*

Exercise 4.12.
(a) Prove.
(b) Give an example showing that a linear operator $A : (X, \| \cdot \|_X) \to (Y, \| \cdot \|_Y)$ with a *closed kernel* need not be bounded.

   Hint. On $(c_{00}, \| \cdot \|_\infty)$, consider the operator of multiplication by a sequence $(n)_{n \in \mathbb{N}}$:

$$c_{00} \ni x = (x_n)_{n \in \mathbb{N}} \mapsto Ax := (n x_n)_{n \in \mathbb{N}} \in c_{00}$$

   (see Examples 4.1).

Remarks 4.8.
– For a *linear functional $f : (X, \| \cdot \|_X) \to \mathbb{F}$* on a normed vector space $(X, \| \cdot \|_X)$, the closedness of the kernel is not only necessary, but also sufficient for its boundedness (see the *Kernel of a Linear Functional Proposition* (Proposition 4.16, Section 4.7, Problem 9)).
– Since, for a bounded linear operator $A : (X, \| \cdot \|_X) \to (Y, \| \cdot \|_Y)$, its kernel ker A is a *closed subspace* of $(X, \| \cdot \|_X)$, the *quotient space norm*

$$X / \ker A \ni [x] = x + \ker A \mapsto \|[x]\| = \inf_{y \in \ker A} \|x + y\| =: \rho(x, \ker A)$$

is well defined, relative to which $X / \ker A$ is a normed vector space, which is a Banach space, provided the space $(X, \| \cdot \|_X)$ is Banach (see the *Quotient Space Norm Theorem* (Theorem 3.16, Section 3.5, Problem 12)).

4.2.3 Space of Bounded Linear Operators, Dual Space

Here, we reveal that bounded linear operators from one normed vector space to another form a normed vector space and consider the convergence of operator sequences in two different senses.

4.2.3.1 Space of Bounded Linear Operators

Theorem 4.5 (Space of Bounded Linear Operators). *Let $(X, \| \cdot \|_X)$ and $(Y, \| \cdot \|_Y)$ be normed vector spaces over $\mathbb{F}$.*

   The set $L(X, Y)$ of all bounded linear operators $A : X \to Y$ is a normed vector space over $\mathbb{F}$ relative to the pointwise linear operations

$$L(X, Y) \ni A, B \mapsto (A + B)x := Ax + Bx, \ x \in X,$$
$$\mathbb{F} \ni \lambda, L(X, Y) \ni A \mapsto (\lambda A)x := \lambda Ax, \ x \in X,$$

and operator norm

$$\|A\| := \sup_{\|x\|_X = 1} \|Ax\|_Y.$$

If $(Y, \| \cdot \|_Y)$ *is a Banach space, then* $(L(X, Y), \| \cdot \|)$ *is also a Banach space.*

Proof. The *vector space axioms* (see Definition 3.1) for $L(X, Y)$ are readily verified.

Exercise 4.13. Verify.

Let us verify the *norm axioms* (see Definition 3.17) for operator norm. *Nonnegativity* is obvious. *Separation* holds as well, since, for an $A \in L(X, Y)$,

$$\|A\| = 0 \ \Leftrightarrow \ \|Ax\|_Y = 0 \text{ for all } x \in X \text{ with } \|x\|_X = 1,$$

the latter being equivalent to the fact that

$$\forall x \in X : \ Ax = 0, \text{ i. e., } A = 0.$$

Exercise 4.14. Explain.

Absolute scalability is easily verified as well.

Exercise 4.15. Verify.

Since, for any $A, B \in L(X, Y)$, by the *subadditivity* of $\| \cdot \|_Y$, for any $x \in X$,

$$\left\|(A + B)x\right\|_Y = \|Ax + Bx\|_Y \leq \|Ax\|_Y + \|Bx\|_Y \leq \|A\|\|x\|_X + \|B\|\|x\|_X$$
$$= [\|A\| + \|B\|]\|x\|_X.$$

Whence, we conclude that

$$\|A + B\| \leq \|A\| + \|B\|,$$

i. e., operator norm is *subadditive*.

Thus, $(L(X, Y), \| \cdot \|)$ is a *normed vector space* over $\mathbb{F}$.

Suppose $(Y, \| \cdot \|_Y)$ is a Banach space and let $(A_n)_{n \in \mathbb{N}}$ be an arbitrary *fundamental sequence* in $(L(X, Y), \| \cdot \|)$. Then, for each $x \in X$,

$$\|A_n x - A_m x\|_Y = \left\|(A_n - A_m)x\right\|_Y = \|A_n - A_m\|\|x\|_X \to 0, \ m, n \to \infty,$$

i. e., $(A_n x)_{n \in \mathbb{N}}$ is a *fundamental sequence* in the space $(Y, \| \cdot \|_Y)$, which, due to the completeness of the latter, converges, and hence, we can define a *linear operator* from X to Y as follows:

$$X \ni x \mapsto Ax := \lim_{n \to \infty} A_n x \in Y.$$

Exercise 4.16. Verify that the operator A is *linear*.

The operator A is *bounded*. Indeed, by the *Properties of Fundamental Sequences* (Theorem 2.22), being fundamental, the sequence $(A_n)_{n \in \mathbb{N}}$ is *bounded* in $(L(X, Y), \| \cdot \|)$, i. e.,

$$\exists M > 0 \, \forall n \in \mathbb{N} : \| A_n \| \le M,$$

and hence, for each $x \in X$ and any $n \in \mathbb{N}$,

$$\| A_n x \|_Y \le \| A_n \| \| x \|_X \le M \| x \|_X.$$

Whence, in view of *continuity* of norm (see Remarks 3.18), passing to the limit as $n \to \infty$ for each fixed $x \in X$, we obtain the estimate

$$\forall x \in X : \| Ax \|_Y \le M \| x \|_X,$$

which implies that $A \in L(X, Y)$.

Since $(A_n)_{n \in \mathbb{N}}$ is a *fundamental sequence* in $(L(X, Y), \| \cdot \|)$,

$$\forall \varepsilon > 0 \, \exists N \in \mathbb{N} \, \forall m, n \ge N : \| A_n - A_m \| < \varepsilon,$$

and hence,

$$\forall x \in X, \ \forall m, n \ge N : \| A_n x - A_m x \|_Y \le \| A_n - A_m \| \| x \|_X \le \varepsilon \| x \|_X.$$

Fixing arbitrary $x \in X$ and $n \ge N$ and passing to the limit as $m \to \infty$, we obtain

$$\| A_n x - Ax \|_Y \le \varepsilon \| x \|_X, \ x \in X,$$

which implies that

$$\forall n \ge N : \| A_n - A \| \le \varepsilon,$$

i. e.,

$$A_n \to A, \ n \to \infty, \ \text{in} \ (L(X, Y), \| \cdot \|),$$

and thus concludes the proof of the *completeness* of the operator space $(L(X, Y), \| \cdot \|)$ when the space $(Y, \| \cdot \|_Y)$ is complete and thus of the entire statement. $\qquad \square$

Remark 4.9. If $X = Y$, we use the notation $L(X)$, $(L(X), \| \cdot \|)$ being a *normed algebra* over $\mathbb{F}$ with *operator multiplication* defined as the composition

$$\forall A, B \in L(X) : (AB)x := A(Bx), \ x \in X,$$

associative and *bilinear* (relative to operator addition and scalar multiplication), the operator norm being *submultiplicative*:

$$\forall A, B \in L(X) : \| AB \| \le \| A \| \| B \|.$$

If $(X, \| \cdot \|_X)$ is a Banach space, then $(L(X), \| \cdot \|)$ is a *Banach algebra* (see, e. g., [3, 4]).

Exercise 4.17. Verify the aforementioned properties of the operator multiplication and *submultiplicativity* of operator norm.

4.2.3.2 Dual Space

Definition 4.5 (Dual Space). For $Y = \mathbb{F}$ with the absolute-value norm, we call the space $L(X, Y)$ of bounded linear functionals on X the *dual space* of X and use the notation X^* for it.

Remark 4.10. The fact that, for any nonzero normed vector space $(X, \|\cdot\|_X)$, $X^* \neq \{0\}$ is guaranteed by the *Hahn–Banach Theorem* (the extension form) (see, e. g., [45]).

From the prior theorem, in view of the completeness of the target space $(\mathbb{F}, |\cdot|)$, we obtain the following:

Corollary 4.1 (Completeness of the Dual Space). *For each normed vector space, its dual is a Banach space.*

4.2.3.3 Uniform and Strong Convergence in $L(X, Y)$

Let us now discuss various forms of convergence of operator sequences.

Definition 4.6 (Uniform Convergence). Let $(X, \|\cdot\|_X)$ and $(Y, \|\cdot\|_Y)$ be normed vector spaces over $\mathbb{F}$.

The convergence of a sequence of operators $(A_n)_{n\in\mathbb{N}}$ to an operator A in the space $(L(X, Y), \|\cdot\|)$:

$$\|A_n - A\| := \sup_{\|x\|_X \leq 1} \|Ax - A_n x\|_Y \to 0, \ n \to \infty,$$

is called *uniform*, and the operator A is called the *uniform limit* of $(A_n)_{n\in\mathbb{N}}$.

Remark 4.11. The name is justified by the fact that such a convergence is equivalent to the uniform convergence of $(A_n)_{n\in\mathbb{N}}$ to A on the closed unit ball

$$B_X(0, 1) := \{x \in X \mid \|x\|_X \leq 1\},$$

of $(X, \|\cdot\|_X)$, or more generally, on any *bounded set* of $(X, \|\cdot\|_X)$.

Example 4.3. Let $(X, \|\cdot\|)$ be a normed vector space. Then the operator sequence $\{\frac{1}{n}I\}_{n=1}^{\infty}$, where I is the *identity operator* on X, *converges uniformly* to the *zero operator* since

$$\left\|\frac{1}{n}I\right\| = \frac{1}{n}\|I\| = \frac{1}{n} \to 0, \ n \to \infty.$$

Definition 4.7 (Strong Convergence). Let $(X, \| \cdot \|_X)$ and $(Y, \| \cdot \|_Y)$ be normed vector spaces over $\mathbb{F}$.

The pointwise convergence of a sequence of operators $(A_n)_{n \in \mathbb{N}}$ to an operator A in the space $(L(X, Y), \| \cdot \|)$:

$$\forall x \in X : A_n x \to Ax, \ n \to \infty, \ \text{in} \ (Y, \| \cdot \|_Y),$$

is called *strong*, and the operator A is called the *strong limit* of $(A_n)_{n \in \mathbb{N}}$.

Remarks 4.12.
- Uniform convergence implies strong convergence.

  Exercise 4.18. Verify.

- However, as the following example shows, the converse is not true, i. e., a strongly convergent operator sequence need not converge uniformly.

Example 4.4. For the *left shift operator* on l_p $(1 \le p < \infty)$

$$l_p \ni x = (x_1, x_2, \dots) \mapsto Bx := (x_2, x_3, x_4, \dots) \in l_p,$$

(see Examples 4.2), the operator sequence $(B^n)_{n \in \mathbb{N}}$ is in $L(l_p)$ with

$$l_p \ni x = (x_1, x_2, \dots) \mapsto B^n x := (x_{n+1}, x_{n+2}, x_{n+3}, \dots) \in l_p.$$

Since

$$\forall x \in l_p : \|B^n x\|_p = \left[\sum_{k=n+1}^{\infty} |x_k|^p \right]^{1/p} \to 0, \ n \to \infty,$$

the operator sequence $(B^n)_{n \in \mathbb{N}}$ strongly converges to the *zero operator*.
However,

$$\forall n \in \mathbb{N} : \|B^n\| = 1.$$

Exercise 4.19. Verify.

Hint. Apply B^n $(n \in \mathbb{N})$ to the sequence $(\delta_{(n+1)k})_{k \in \mathbb{N}} \in l_p$, where δ_{nk} is the *Kronecker delta*.

Therefore, the operator sequence $(B^n)_{n \in \mathbb{N}}$ *does not converge uniformly*.

Exercise 4.20. Explain.

4.3 Uniform Boundedness Principle

The *Uniform Boundedness Principle* was found by Lebesgue[2] in 1908 when studying convergence of Fourier[3] series (see [25]), stated in its general form and published in 1927 by Banach and Steinhaus,[4] is often referred to as the *Banach–Steinhaus Theorem* (for more on the latter, see, e. g., [45]). It states that, for a set of bounded linear operators defined on a Banach space, boundedness in operator norm (*uniform boundedness*) is equivalent to *pointwise boundedness* and has a number of profound implications and far reaching applications.

Theorem 4.6 (Uniform Boundedness Principle). *Let $(X, \| \cdot \|_X)$ and $(Y, \| \cdot \|_Y)$ be normed vector spaces over $\mathbb{F}$ and $\{A_i\}_{i \in I}$ be a set of bounded linear operators in $(L(X, Y), \| \cdot \|)$. Then for*

$$\sup_{i \in I} \|A_i\| < \infty \qquad \text{(uniform boundedness)} \tag{4.2}$$

it is necessary and, provided $(X, \| \cdot \|_X)$ is a Banach space, sufficient that

$$\forall x \in X : \sup_{i \in I} \|A_i x\|_Y < \infty \qquad \text{(pointwise boundedness)}. \tag{4.3}$$

Proof. Necessity. Proving that uniform boundedness (see (4.2)) implies pointwise boundedness (see (4.3)) is straightforward, the implication being true without the additional assumption on the domain space $(X, \| \cdot \|_X)$ to be complete.

Exercise 4.21. Prove.

Sufficiency. Suppose that the domain space $(X, \|\cdot\|_X)$ is Banach and that (4.3) holds. For each $n \in \mathbb{N}$, consider the set

$$X_n := \bigcap_{i \in I} \{x \in X \mid \|A_i x\| \le n\},$$

which, since

$$\forall i \in I, \ \forall n \in \mathbb{N} : \{x \in X \mid \|A_i x\| \le n\} = A_i^{-1}(\overline{B}_Y(0, n)),$$

by the *continuity* of the operators A_i, $i \in I$ (see the *Characterizations of Bounded Linear Operators* (Theorem 4.4)), and in view of the *Characterization of Continuity* (Theorem 2.61, Section 2.18, Problem 15) and the *Properties of Closed Sets* (Proposition 2.20), is *closed* in $(X, \| \cdot \|_X)$.

Exercise 4.22. Explain.

---

2 Henri Lebesgue (1875–1941).
3 Jean-Baptiste Joseph Fourier (1768–1830).
4 Hugo Steinhaus (1887–1972).

Since, by the premise,

$$\forall x \in X \, \exists n(x) \in \mathbb{N} \, \forall i \in I : \|A_i x\|_Y \leq n(x),$$

we infer that

$$X = \bigcup_{n=1}^{\infty} X_n.$$

Whence, by the *Baire Category Theorem* (Theorem 2.32) (see also Corollary 2.5),

$$\exists N \in \mathbb{N} : \text{int}(X_N) \neq \emptyset,$$

i. e., the set X_N is *not nowhere dense* in $(X, \|\cdot\|_X)$, and hence,

$$\exists x_0 \in X, \, \exists r > 0 : \overline{B}(x_0, r) := \{x \in X \mid \|x - x_0\|_X \leq r\} \subseteq X_N.$$

Then

$$\forall i \in I, \, \forall x \in X \text{ with } \|x\|_X \leq r : \|A_i(x_0 + x)\|_Y \leq N,$$

and therefore, by the *linearity of* A_i, $i \in I$, and *subadditivity* of norm,

$$\forall i \in I, \, \forall x \in X \text{ with } \|x\|_X \leq r : \|A_i x\|_Y = \|A_i x + A_i x_0 - A_i x_0\|_Y$$
$$\leq \|A_i x + A_i x_0\|_Y + \|A_i x_0\|_Y = \|A_i(x_0 + x)\|_Y + \|A_i x_0\|_Y \leq 2N.$$

Whence, by the *linearity of* A_i, $i \in I$, and *absolute scalability* of norm, we infer that

$$\forall i \in I, \, \forall x \in X \text{ with } \|x\|_X \leq 1 : \|A_i x\|_Y = \frac{1}{r} \|A_i(rx)\|_Y \qquad \text{since } \|rx\|_X \leq r;$$
$$\leq \frac{2N}{r},$$

which implies that

$$\forall i \in I : \|A_i\| := \sup_{\|x\|_X \leq 1} \|A_i x\|_Y \leq \frac{2N}{r},$$

completing the proof of the sufficiency and the entire statement. $\qquad \Box$

In particular, for $(Y, \|\cdot\|_Y) = (\mathbb{F}, |\cdot|)$, we obtain the following version of the *Uniform Boundedness Principle* for *bounded linear functionals*.

Corollary 4.2 (Uniform Boundedness Principle for Functionals). *Let* $(X, \|\cdot\|_X)$ *be a normed vector space over* $\mathbb{F}$ *and* $\{f_i\}_{i \in I}$ *be a set of bounded linear functionals in the dual space* $(X^*, \|\cdot\|)$. *Then for*

$$\sup_{i \in I} \|f_i\| < \infty$$

it is necessary and, provided $(X, \|\cdot\|_X)$ *is a Banach space, sufficient that*

$$\forall x \in X : \sup_{i \in I} |f_i(x)| < \infty.$$

Remark 4.13. As the following example shows, the condition of *completeness* of the domain space $(X, \|\cdot\|_X)$ in the *sufficiency* of the *Uniform Boundedness Principle* (Theorem 4.6) is essential and cannot be dropped.

Example 4.5. On the *incomplete* normed vector space $(c_{00}, \|\cdot\|_\infty)$ (see Examples 3.13), consider the countable set $\{f_n\}_{n\in\mathbb{N}}$ of linear functionals, defined as follows:

$$c_{00} \ni x := (x_k)_{k\in\mathbb{N}} \mapsto f_n(x) := \sum_{k=1}^{n} x_k, \; n \in \mathbb{N}.$$

Since

$$\forall n \in \mathbb{N}, \; \forall x := (x_k)_{k\in\mathbb{N}} \in X : \left|f_n(x)\right| \le \sum_{k=1}^{n} |x_k| \le n \sup_{k\in\mathbb{N}} |x_k| = n\|x\|_\infty,$$

we infer that

$$\forall n \in \mathbb{N} : f_n \in X^* \text{ with } \|f_n\| \le n.$$

Furthermore, since for $x_n := (\underbrace{1, \dots, 1}_{n \text{ terms}}, 0, 0, \dots) \in c_{00}, \; n \in \mathbb{N},$

$$\|x_n\|_\infty = 1 \quad \text{and} \quad \left|f_n(x_n)\right| = n,$$

we infer that $\|f_n\| = n, \; n \in \mathbb{N}$, and hence,

$$\sup_{n\in\mathbb{N}} \|f_n\| = \infty,$$

i.e., the set $\{f_n\}_{n\in\mathbb{N}}$ is *not uniformly bounded*.

However, for each $x := (x_1, \dots, x_m, 0, 0, \dots) \in c_{00}$ with some $m \in \mathbb{N}$,

$$\left|f_n(x)\right| \le \sum_{k=1}^{m} |x_k| \le m \sup_{k\in\mathbb{N}} |x_k| = m\|x\|_\infty, \; n \in \mathbb{N},$$

and hence,

$$\forall x \in c_{00} : \sup_{n\in\mathbb{N}} \left|f_n(x)\right| < \infty,$$

i.e., the set $\{f_n\}_{n\in\mathbb{N}}$ is *pointwise bounded*.

4.4 Open and Inverse Mapping Theorems

The following three equivalent statements: the *Open Mapping Theorem* (*OMT*), the *Inverse Mapping Theorem* (*IMT*), and the *Closed Graph Theorem* (*CGT*) form another fundamental principle of linear functional analysis.

4.4.1 Open Mapping Theorem

Theorem 4.7 (Open Mapping Theorem). *Let $(X, \|\cdot\|_X)$ and $(Y, \|\cdot\|_Y)$ be Banach spaces over $\mathbb{F}$ and $A : X \to Y$ be a surjective bounded linear operator from X onto Y (i.e., $A \in L(X, Y)$ and $R(A) = Y$). Then A is an open mapping, i.e., the image $A(G)$ under A of each open set G in $(X, \|\cdot\|_X)$ is an open set in $(Y, \|\cdot\|_Y)$.*

Proof. Since

$$X = \bigcup_{n=1}^{\infty} B_X(0, n),$$

by the fact that image preserves unions (Exercise 1.4), and the *surjectivity* and *linearity* of A,

$$Y = R(A) = A\left(\bigcup_{n=1}^{\infty} B_X(0, n) \right) = \bigcup_{n=1}^{\infty} A(B_X(0, n)) = \bigcup_{n=1}^{\infty} nA(B_X(0, 1)).$$

Whence, in view of the *completeness* of $(Y, \|\cdot\|_Y)$, as follows from the *Baire Category Theorem* (Theorem 2.32) (see also Corollary 2.5),

$$\exists N \in \mathbb{N} : \operatorname{int}(\overline{N \cdot A(B_X(0, 1))}) \neq \emptyset,$$

i.e., the set $N \cdot AB_X(0, 1)$ is *not nowhere dense* in $(Y, \|\cdot\|_Y)$. This, since the linear operator of multiplication by a *nonzero* number $\lambda \in \mathbb{F} \setminus \{0\}$,

$$Y \ni y \mapsto \lambda y \in Y,$$

is a *homeomorphism* of $(Y, \|\cdot\|_Y)$ (see Examples 4.2), in view of the *linearity* of A, implies that

$$\forall \delta > 0 : \operatorname{int}(\overline{A(B_X(0, \delta))}) \neq \emptyset.$$

Exercise 4.23. Explain.

Hint. $\forall \delta > 0 : \operatorname{int}(\overline{A(B_X(0, \delta))}) = \frac{\delta}{N} \operatorname{int}(\overline{N \cdot A(B_X(0, 1))}).$

By the *joint continuity* of the difference mapping

$$(x, y) \mapsto x - y,$$

for $(X, \|\cdot\|_X)$,

$$\exists \delta > 0 : B_X(0, 1) \supseteq B_X(0, \delta) - B_X(0, \delta).$$

Hence, by the *linearity* of A, and the *joint continuity* of the difference mapping for $(Y, \|\cdot\|_Y)$, we have

$$\overline{A(B_X(0,1))} \supseteq \overline{A(B_X(0,\delta)) - A(B_X(0,\delta))} \supseteq \overline{A(B_X(0,\delta))} - \overline{A(B_X(0,\delta))}$$
$$\supseteq \mathrm{int}(\overline{A(B_X(0,\delta))}) - \mathrm{int}(\overline{A(B_X(0,\delta))}).$$

Observe that

$$\mathrm{int}(\overline{A(B_X(0,\delta))}) - \mathrm{int}(\overline{A(B_X(0,\delta))}) = \bigcup_{x \in \mathrm{int}(\overline{A(B_X(0,\delta))})} [\mathrm{int}(\overline{A(B_X(0,\delta))}) - x]$$

is an *open set* in $(Y, \|\cdot\|_Y)$ containing 0.

Exercise 4.24. Explain.

Hence, for the set $\overline{A(B_X(0,1))}$, 0 is an *interior point*, i. e.,

$$\exists \delta > 0 : \ B_Y(0,\delta) \subseteq \overline{A(B_X(0,1))}. \tag{4.4}$$

We show that 0 is an *interior point* for the set $AB_X(0,1)$ (without closure!) as well, i. e., we prove the following stronger version of inclusion (4.4):

$$\exists \delta > 0 : \ B_Y(0,\delta) \subseteq A(B_X(0,1)). \tag{4.5}$$

By (4.4), in view of the *linearity* of A,

$$\exists \delta > 0 : \ B_Y(0,\delta) \subseteq \overline{A(B_X(0,1/3))},$$

and hence,

$$\forall n \in \mathbb{Z}_+ : \ B_Y(0,\delta/3^n) \subseteq \overline{A(B_X(0,1/3^{n+1}))}. \tag{4.6}$$

By (4.6) with $n = 0$,

$$\forall y \in B_Y(0,\delta) \ \exists x_0 \in B_X(0,1/3) : \ \|y - Ax_0\|_Y < \delta/3.$$

Exercise 4.25. Explain.

Since, $y - Ax_0 \in B_Y(0,\delta/3)$, by (4.6) with $n = 1$,

$$\exists x_1 \in B_X(0,1/3^2) : \ \|y - Ax_0 - Ax_1\|_Y < \delta/3^2.$$

Continuing inductively, we obtain a sequence of elements $(x_n)_{n \in \mathbb{N}}$ in $(X, \|\cdot\|)$ such that

$$\forall n \in \mathbb{Z}_+ : \ \|x_n\|_X \leq 1/3^{n+1} \tag{4.7}$$

and, by the *linearity* of A,

$$\forall n \in \mathbb{N} : \ \left\| y - A\left[\sum_{k=0}^{n-1} x_k\right] \right\|_Y = \left\| y - \sum_{k=0}^{n-1} Ax_k \right\|_Y < \delta/3^n. \tag{4.8}$$

In view of (4.7), by the *Comparison Test*, the series

$$\sum_{k=0}^{\infty} x_k$$

converges absolutely in $(X, \|\cdot\|_X)$, which, in view of the *completeness* of $(X, \|\cdot\|_X)$, by the *Series Characterization of Banach Spaces* (Theorem 3.8), implies that it converges in $(X, \|\cdot\|_X)$, i. e.,

$$\exists\, x \in X : \; x := \sum_{k=0}^{\infty} x_k = \lim_{n \to \infty} \sum_{k=0}^{n-1} x_k.$$

Then (4.7), by the *subadditivity* of norm, implies that

$$\|x\|_X \le \sum_{k=0}^{\infty} \|x_k\|_X \le \sum_{k=0}^{\infty} \frac{1}{3^{k+1}} = \frac{1/3}{1 - 1/3} = \frac{1}{2} < 1,$$

and hence,

$$x \in B_X(0, 1).$$

Passing to the limit in (4.8) as $n \to \infty$, by the *boundedness*, and hence, the *continuity* of the linear operator A (see the *Characterizations of Bounded Linear Operators* (Theorem 4.4)), *continuity* of norm, and the *Squeeze Theorem*, we infer that

$$\|y - Ax\|_Y = 0,$$

which, by the *separation* norm axiom, implies that

$$y = Ax.$$

Thus, we have shown that

$$\exists\, \delta > 0 \; \forall\, y \in B_Y(0, \delta) \; \exists\, x \in B_X(0, 1) : \; y = Ax,$$

and hence, inclusion (4.5) does hold as desired, i. e., 0 is an *interior point* of the set $A(B_X(0, 1))$. This, by the *linearity* of A, implies that, for an arbitrary $x \in X$ and any $\varepsilon > 0$, Ax is an *interior point* of the set

$$A(B_X(x, \varepsilon)),$$

i. e.,

$$\forall\, x \in X, \; \forall\, \varepsilon > 0 \; \exists\, \delta > 0 : \; B_Y(Ax, \delta) \subseteq A(B_X(x, \varepsilon)). \qquad (4.9)$$

Exercise 4.26. Verify.

Hint. Show that $\forall x \in X, \forall \varepsilon > 0 : A(B_X(x, \varepsilon)) = Ax + \varepsilon A(B_X(0, 1))$.

Now, let G be an arbitrary nonempty *open* set in $(X, \| \cdot \|_X)$. Then,

$$\forall x \in G \, \exists \varepsilon = \varepsilon(x) > 0 : \ B_X(x, \varepsilon) \subseteq G.$$

Whence,

$$A\big(B_X(x, \varepsilon)\big) \subseteq A(G)$$

and

$$\exists \delta = \delta(x, \varepsilon) > 0 : \ B_Y(Ax, \delta) \subseteq A\big(B_X(x, \varepsilon)\big).$$

Therefore, we have the inclusion

$$B_Y(Ax, \delta) \subseteq A\big(B_X(x, \varepsilon)\big) \subseteq A(G),$$

which proves the *openness* of the image $A(G)$ in $(Y, \| \cdot \|_Y)$, and thus, the fact that A is an *open mapping*, completing the proof. $\qquad \square$

Remarks 4.14. *A priori*, we regard the operator $A : X \to Y$ to be *linear*.
- The condition of the *completeness* of the domain space $(X, \| \cdot \|_X)$ in the *Open Mapping Theorem* (*OMT*), provided other conditions hold, is *not essential* and can be dropped (see [45, Remarks 6.19]).
- As the following examples show, the *three* other conditions of the *OMT*:
  (i)   the *completeness* of the target space $(Y, \| \cdot \|_Y)$,
  (ii) the *boundedness* of the linear operator $A : X \to Y$ ($A \in L(X, Y)$), and
  (iii) the *surjectivity* of the linear operator $A : X \to Y$ ($R(A) = Y$),
  are essential and none of them can be dropped.

Examples 4.6.
1. The differentiation operator

$$C^1[a, b] \ni x \to [Ax](t) := \frac{d}{dt} x(t) \in C[a, b]$$

($-\infty < a < b < \infty$) is a *surjective bounded linear operator* from the Banach space $X = C^1[a, b]$ with the norm

$$\|x\| := \max\Big[\max_{a \le t \le b} |x(t)|, \max_{a \le t \le b} |x'(t)|\Big], \ x \in C^1[a, b],$$

onto the incomplete normed vector space $Y = C[a, b]$ with the *integral norm*

$$\|x\|_1 := \int_a^b |x(t)| \, dt, x \in C[a, b]$$

(see Examples 4.2), which is *not an open mapping* (see Section 4.7, Problem 18).

2. Let $(X, \|\cdot\|)$ be an *infinite-dimensional* Banach space with a Hamel basis $\{e_i\}_{i\in I}$, where, without loss of generality, we can regard $\|e_i\| = 1, i \in I$. Observe that, since, by the *Basis of a Banach Space Theorem* (Theorem 3.14), $\{e_i\}_{i\in I}$ is *uncountable*, we can choose a countably infinite subset $J := \{i_n\}_{n\in\mathbb{N}}$ of I and consider the *bijective unbounded linear operator* on X defined on $\{e_i\}_{i\in I}$ as follows:

$$Ae_i := \lambda_i e_i, \ i \in I,$$

    where $\{\lambda_i\}_{i\in I}$ is an *unbounded set of nonzero numbers* with

$$\lambda_{i_n} := 1/n, \ n \in \mathbb{N}.$$

    By the *Characterization of Continuity* (Theorem 2.61, Section 2.18, Problem 15), the operator $A : X \to X$ is *not an open mapping* since its inverse $A^{-1} : X \to X$ is *unbounded*, and hence, is *discontinuous* at every point of X (see Remark 4.5). Indeed,

$$A^{-1}e_i := (1/\lambda_i)e_i, \ i \in I,$$

    with

$$A^{-1}e_{i_n} := ne_{i_n}, \ n \in \mathbb{N}.$$

3. If $(X, \|\cdot\|_X)$ and $(Y, \|\cdot\|_Y)$ are nonzero Banach spaces over $\mathbb{F}$ $(X, Y \neq \{0\})$, then the *zero operator* $0 : X \to Y$ is a *nonsurjective* bounded linear operator that is *not an open mapping*.

Exercise 4.27. Explain and verify.

4.4.2 Inverse Mapping Theorem and Applications

4.4.2.1 Inverse Mapping Theorem
As an immediate corollary of the *Open mapping Theorem* (Theorem 4.7), we obtain the following statement:

Theorem 4.8 (Inverse Mapping Theorem). *Let $(X, \|\cdot\|_X)$ and $(Y, \|\cdot\|_Y)$ be Banach spaces over $\mathbb{F}$ and $A : X \to Y$ be a bijective bounded linear operator from X onto Y ($A \in L(X, Y)$). Then the inverse $A^{-1} : Y \to X$ is a bounded linear operator from Y onto X ($A^{-1} \in L(Y, X)$).*

Exercise 4.28. Prove.

Hint. Apply the *Characterization of Continuity* (Theorem 2.61, Section 2.18, Problem 15) and the *Characterizations of Bounded Linear Operators* (Theorem 4.4).

Remarks 4.15.
- As is shown below (see Section 4.6), the *Open Mapping Theorem* (Theorem 4.7) is equivalent to the *Inverse Mapping Theorem* (Theorem 4.8).
- The *Inverse Mapping Theorem* is also called the *Bounded Inverse Theorem* (see, e. g., [25]).

Let us now consider two profound applications of the *Inverse Mapping Theorem*.

4.4.2.2 Application: Equivalence of Banach Norms

Theorem 4.9 (Equivalence of Banach Norms). *Let a vector space X be a Banach space relative to norms $\|\cdot\|_1$ and $\|\cdot\|_2$.*
  If

$$\exists c > 0 : \ c\|x\|_1 \le \|x\|_2, \ x \in X,$$

then

$$\exists C > 0 : \ \|x\|_2 \le C\|x\|_1, \ x \in X,$$

i. e., if one of two Banach norms on a vector space is stronger than the other, then the norms are equivalent.

Exercise 4.29. Prove.

Hint. Apply the *Inverse Mapping Theorem* (Theorem 4.8) to the *bijective linear operator*

$$(X, \|\cdot\|_2) \ni x \mapsto Ix := x \in (X, \|\cdot\|_1)$$

(see Remarks 3.20).

Remarks 4.16.
- For finite-dimensional Banach spaces, the result is consistent with the *Norm Equivalence Theorem* (Theorem 3.9).

  Exercise 4.30. Explain.

- The requirement of the *completeness* of X relative to both norms is essential. Indeed, the vector space $C[a, b]$ ($-\infty < a < b < \infty$) is *incomplete* relative to the *integral norm*

$$C[a, b] \ni x \mapsto \|x\|_1 = \int_a^b |x(t)| \, dt$$

and is *complete* relative to the *maximum norm*

$$C[a, b] \ni x \mapsto \|x\|_\infty = \max_{a \le t \le b} |x(t)|,$$

the latter being *stronger* than the former, but the norms are *not equivalent* (see Examples 3.13 and Exercise 3.49).

4.4.2.3 Application: Boundedness of the Schauder Coordinate Functionals

Recall that, in a Banach space $(X, \|\cdot\|)$ over $\mathbb{F}$ with a *Schauder basis* $E := \{e_n\}_{n\in\mathbb{N}}$, each element $x \in X$ allows a *unique Schauder expansion*

$$x = \sum_{k=1}^{\infty} c_k e_k$$

with some coefficients $c_n \in \mathbb{F}$, $n \in \mathbb{N}$, called the *coordinates* of x relative to E (see Definition 3.25 and Examples 3.17).

For each $n \in \mathbb{N}$, the mapping

$$X \ni x := \sum_{k=1}^{\infty} c_k e_k \mapsto c_n(x) := c_n \in \mathbb{F}$$

is a well-defined *linear functional* on X, called the nth *Schauder coordinate functional* relative to E.

Exercise 4.31. Verify.

The following statement shows, on a Banach space $(X, \|\cdot\|)$ with a Schauder basis, that all Schauder coordinate functionals are *bounded*, unlike the linear *Hamel coordinate functionals* (cf. the *Unboundedness of Hamel Coordinate Functionals Proposition* (Proposition 4.17, Section 4.7, Problem 10)).

Proposition 4.3 (Boundedness of Schauder Coordinate Functionals). *On a Banach space $(X, \|\cdot\|)$ over $\mathbb{F}$ with a Schauder basis $E := \{e_n\}_{n\in\mathbb{N}}$, all Schauder coordinate functionals*

$$X \ni x := \sum_{k=1}^{\infty} c_k e_k \mapsto c_n(x) := c_n \in \mathbb{F}, \ n \in \mathbb{N},$$

relative to E are bounded, i. e.,

$$\forall n \in \mathbb{N}: \ c_n(\cdot) \in X^*.$$

Proof. Consider the set of $\mathbb{F}$-termed sequences defined as follows:

$$Y := \left\{ y := (c_n)_{n\in\mathbb{N}} \in \mathbb{F}^{\mathbb{N}} \ \middle| \ \sum_{k=1}^{\infty} c_k e_k \text{ converges in } X \right\}.$$

The set Y is a normed vector space relative to the termwise linear operations and the norm

$$Y \ni y := (c_n)_{n\in\mathbb{N}} \mapsto \|y\|_Y := \sup_{n\in\mathbb{N}} \left\| \sum_{k=1}^{n} c_k e_k \right\|.$$

Exercise 4.32. Verify.

Furthermore, the space $(Y, \|\cdot\|_Y)$ is *Banach*. Indeed, for any *fundamental sequence*

$$(y_n := (c_k^{(n)})_{k\in\mathbb{N}})_{n\in\mathbb{N}}$$

in $(Y, \|\cdot\|_Y)$,

$$\forall \varepsilon > 0\, \exists N \in \mathbb{N}\, \forall m, n \geq N: \ \|y_n - y_m\|_Y = \sup_{p\in\mathbb{N}}\left\|\sum_{k=1}^{p}(c_k^{(n)} - c_k^{(m)})e_k\right\| < \varepsilon,$$

and hence,

$$\forall p \in \mathbb{N},\ \forall m, n \geq N: \ \left\|\sum_{k=1}^{p}(c_k^{(n)} - c_k^{(m)})e_k\right\| < \varepsilon.$$

By *subadditivity* of norm, we have the following:

$$\forall j \in \mathbb{N},\ \forall m, n \geq N: \ \|(c_j^{(n)} - c_j^{(m)})e_j\|$$

$$= \left\|\sum_{k=1}^{j}(c_k^{(n)} - c_k^{(m)})e_k - \sum_{k=1}^{j-1}(c_k^{(n)} - c_k^{(m)})e_k\right\|$$

$$\leq \left\|\sum_{k=1}^{j}(c_k^{(n)} - c_k^{(m)})e_k\right\| + \left\|\sum_{k=1}^{j-1}(c_k^{(n)} - c_k^{(m)})e_k\right\| < 2\varepsilon,$$

which, by the *absolute scalability* of norm, implies that

$$\forall j \in \mathbb{N},\ \forall m, n \geq N: \ |c_j^{(n)} - c_j^{(m)}| = \frac{1}{\|e_j\|}\|(c_j^{(n)} - c_j^{(m)})e_j\| < \frac{2\varepsilon}{\|e_j\|}.$$

Whence, we conclude that, for each $j \in \mathbb{N}$, the numeric sequence $(c_j^{(n)})_{n\in\mathbb{N}}$ is *fundamental* in $(\mathbb{F}, |\cdot|)$, and thus, *converges*, i. e.,

$$\forall j \in \mathbb{N}\, \exists c_j \in \mathbb{F}: \ c_j^{(n)} \to c_j,\ j \to \infty.$$

Whereby, we obtain a numeric sequence $y := (c_n)_{n\in\mathbb{N}}$.
It can be shown that $y \in Y$ and

$$y_n \to y,\ n \to \infty,\ \text{in } (Y, \|\cdot\|_Y)$$

(see, e. g., [25, 43]), which proves the *completeness* of $(Y, \|\cdot\|_Y)$.
Since $E := \{e_n\}_{n\in\mathbb{N}}$ is a *Schauder basis* of $(X, \|\cdot\|)$, the operator

$$Y \ni y := (c_n)_{n\in\mathbb{N}} \mapsto Ay := \sum_{k=1}^{\infty} c_k e_k \in X$$

is a *bijective linear operator* from $(Y, \|\cdot\|_Y)$ *onto* $(X, \|\cdot\|)$, which is *bounded*:

$$\forall y \in Y : \|Ay\| = \left\|\sum_{k=1}^{\infty} c_k e_k\right\| = \lim_{n \to \infty}\left\|\sum_{k=1}^{n} c_k e_k\right\| \leq \sup_{n \in \mathbb{N}}\left\|\sum_{k=1}^{n} c_k e_k\right\| = \|y\|_Y.$$

Hence, by the *Inverse Mapping Theorem* (Theorem 4.8), the inverse operator

$$A^{-1} : X \to Y$$

is *bounded*, and, for each $x \in X$,

$$x = \sum_{k=1}^{\infty} c_k e_k = Ay$$

with some $y := (c_n)_{n \in \mathbb{N}} \in Y$ and any fixed $j \in \mathbb{N}$, by the *norm axioms*,

$$\begin{aligned}
|c_j(x)| = |c_j| &= \frac{\|c_j e_j\|}{\|e_j\|} = \frac{1}{\|e_j\|}\left\|\sum_{k=1}^{j} c_k e_k - \sum_{k=1}^{j-1} c_k e_k\right\| \\
&\leq \frac{1}{\|e_j\|}\left[\left\|\sum_{k=1}^{j} c_k e_k\right\| + \left\|\sum_{k=1}^{j-1} c_k e_k\right\|\right] \leq \frac{1}{\|e_j\|} 2 \sup_{n \in \mathbb{N}}\left\|\sum_{k=1}^{n} c_k e_k\right\| \\
&= \frac{2}{\|e_j\|}\|y\|_Y = \frac{2}{\|e_j\|}\|A^{-1}x\|_Y \leq \frac{2\|A^{-1}\|}{\|e_j\|}\|x\|,
\end{aligned}$$

which proves that each Schauder coordinate functional $c_j(\cdot)$, $j \in \mathbb{N}$, is *bounded*, i.e.,
$c_j(\cdot) \in X^*, j \in \mathbb{N}$. $\quad\square$

Remark 4.17. A Schauder basis $E := \{e_n\}_{n \in \mathbb{N}}$, and the set of the Schauder coordinate functionals $\{c_n(\cdot)\}_{n \in \mathbb{N}}$ relative to E are *biorthogonal*, i.e.,

$$c_i(e_j) = \delta_{ij}, \quad i, j \in \mathbb{N},$$

where δ_{ij} is the *Kronecker delta*, which, in particular, implies that the Schauder coordinate functionals are *linearly independent*.

Exercise 4.33. Verify.

4.5 Closed Linear Operators, Closed Graph Theorem

4.5.1 Definition, Characterizations, Examples

The following defines a very important class of linear operators, which need not be bounded.

Definition 4.8 (Closed Linear Operator). Let $(X, \|\cdot\|_X)$ and $(Y, \|\cdot\|_Y)$ be normed vector spaces over $\mathbb{F}$. A linear operator $(A, D(A))$ from X to Y is called *closed* if its *graph*

$$G_A := \{(x, Ax) | x \in D(A)\}$$

is a *closed subspace* in the product space $X \times Y$ relative to the product norm

$$X \times Y \ni (x, y) \mapsto \|(x, y)\|_{X \times Y} := \sqrt{\|x\|_X^2 + \|y\|_Y^2}$$

(see Section 3.5, Problem 7).

Remark 4.18. The product norm $\|\cdot\|_{X \times Y}$ on $X \times Y$ can be replaced with the *equivalent* one

$$X \times Y \ni (x, y) \mapsto \|(x, y)\| := \|x\|_X + \|y\|_Y,$$

which may be a little easier to handle.

Exercise 4.34. Verify the *norm axioms* for the latter and the equivalence of $\|\cdot\|_{X \times Y}$ and $\|\cdot\|$ on $X \times Y$.

In view of the *componentwise* nature of convergence in a product space (see the *Characterization of Convergence in Product Space* (Proposition 2.14, Section 2.18, Problem 6)), by the *Sequential Characterization of Closed Sets* (Theorem 2.19) we obtain the following statement:

Proposition 4.4 (Sequential Characterization of Closed Linear Operators).
Let $(X, \|\cdot\|_X)$ and $(Y, \|\cdot\|_Y)$ be normed vector spaces over $\mathbb{F}$. A linear operator $(A, D(A))$ from X to Y is closed iff, for any sequence $(x_n)_{n \in \mathbb{N}}$ in the domain $D(A)$ such that

$$\lim_{n \to \infty} x_n = x \text{ in } (X, \|\cdot\|_X) \quad and \quad \lim_{n \to \infty} Ax_n = y \text{ in } (Y, \|\cdot\|_Y),$$

the following is true:

$$x \in D(A) \quad and \quad y = Ax.$$

Exercise 4.35. Prove.

Remark 4.19. Provided $D(A) = X$, the condition $x \in D(A)$ holds automatically.

Using the prior sequential characterization, one can prove the following fact:

Proposition 4.5 (Characterization of Closedness for Bounded Linear Operators). *Let $(X, \|\cdot\|_X)$ and $(Y, \|\cdot\|_Y)$ be normed vector spaces. For a bounded linear operator $A : X \supseteq D(A) \to Y$ to be closed it is sufficient and, provided $(Y, \|\cdot\|_Y)$ is a Banach space, necessary that the domain $D(A)$ be a closed subspace in $(X, \|\cdot\|_X)$.*
In particular, each $A \in L(X, Y)$ is a closed operator.

Exercise 4.36. Prove.

Examples 4.7.

1. By the prior proposition, all bounded linear operators from Examples 4.2 are closed.

2. In the Banach space l_p $(1 \le p \le \infty)$, the linear operator A of multiplication by a numeric sequence $a := (a_n)_{n \in \mathbb{N}} \in \mathbb{F}^{\mathbb{N}}$,

$$(x_n)_{n \in \mathbb{N}} \mapsto Ax := (a_n x_n)_{n \in \mathbb{N}},$$

with the maximal domain

$$D(A) := \{(x_n)_{n \in \mathbb{N}} \in l_p \mid (a_n x_n)_{n \in \mathbb{N}} \in l_p\}$$

is *closed* and is *bounded*, and $D(A) = l_p$ *iff* $a \in l_\infty$ (see Examples 4.2).

Remark 4.20. Observe that $c_{00} \subseteq D(A)$, which, for $1 \le p < \infty$, immediately implies that the domain $D(A)$ is *dense* in l_p, and hence, makes the operator A to be *densely defined*.

3. In the Banach space $(C[a,b], \|\cdot\|_\infty)$ $(-\infty < a < b < \infty)$, the *unbounded* linear differentiation operator

$$C^1[a,b] =: D(A) \ni x \to Ax := \frac{dx}{dt} \in C[a,b]$$

(see Examples 4.2) is *closed*.

4. In the Banach space $(c_0, \|\cdot\|_\infty)$, the *unbounded* linear operator of multiplication

$$c_{00} =: D(A) \ni x = (x_n)_{n \in \mathbb{N}} \mapsto Ax := (nx_n)_{n \in \mathbb{N}} \in c_{00}$$

(see Examples 4.2) is *not closed*.

Exercise 4.37. Verify.

Hints.

- Apply the *Sequential Characterization of Closed Linear Operators* (Proposition 4.4).
- For (c), use the *Total Change Formula*

$$x(t) = x(a) + \int_a^t x'(s)\, ds, \ t \in [a,b],$$

valid for every $x \in C^1[a,b]$.

- For (d), in the domain $D(A) = c_{00}$, consider the sequence

$$(x_n := (1, 1/2^2, \ldots, 1/n^2, 0, 0, \ldots))_{n=1}^\infty.$$

Remark 4.21. Thus, a closed linear operator need not be bounded, and a linear operator need not be closed.

4.5.2 Kernel

Proposition 4.6 (Kernel of a Closed Linear Operator). *Let $(X, \| \cdot \|_X)$ and $(Y, \| \cdot \|_Y)$ be normed vector spaces over $\mathbb{F}$. The kernel* $\ker A$ *of an arbitrary closed linear operator* $(A, D(A))$ *from X to Y is a closed subspace of* $(X, \| \cdot \|_X)$.

Exercise 4.38. Prove.

Hint. Apply the *Sequential Characterization of Closed Linear Operators* (Proposition 4.4).

Remarks 4.22.
- The *Kernel of a Bounded Linear Operator Proposition* (Proposition 4.2) is now a direct corollary of the prior proposition and the *Characterization of Closedness for Bounded Linear Operators* (Proposition 4.5).
- For a linear operator, which is not a linear functional, the closedness of the kernel is necessary, but *not sufficient* for its boundedness (see the *Kernel of a Linear Functional Proposition* (Proposition 4.16, Section 4.7, Problem 9) and Remarks 4.8).
- For a linear operator, which is not a linear functional, the kernel need not be a hyperplane.

  Exercise 4.39. Give a corresponding example.

- Since, for a closed linear operator $A : (X, \| \cdot \|_X) \to (Y, \| \cdot \|_Y)$, its kernel $\ker A$ is a *closed subspace* of $(X, \| \cdot \|_X)$, the *quotient space norm*

$$X / \ker A \ni [x] = x + \ker A \mapsto \|[x]\| = \inf_{y \in \ker A} \|x + y\| =: \rho(x, \ker A)$$

  is well defined, relative to which $X / \ker A$ is a normed vector space, which is a Banach space, provided the space $(X, \| \cdot \|_X)$ is Banach (see the *Quotient Space Norm Theorem* (Theorem 3.16, Section 3.5, Problem 12), cf. Remarks 4.8).

4.5.3 Closed Graph Theorem

The *Inverse Mapping Theorem* (Theorem 4.8) underlies the proof of the following important statement:

Theorem 4.10 (Closed Graph Theorem). *Let $(X, \| \cdot \|_X)$ and $(Y, \| \cdot \|_Y)$ be Banach spaces over $\mathbb{F}$ and $A : X \to Y$ be a closed linear operator. Then A is bounded $(A \in L(X, Y))$.*

Proof. By the *completeness* of the spaces $(X, \| \cdot \|_X)$ and $(Y, \| \cdot \|_Y)$, the product space $X \times Y$ is a *Banach space* relative to the norm

$$X \times Y \ni (x, y) \mapsto \|(x, y)\| := \|x\|_X + \|y\|_Y$$

(see Section 3.5, Problem 7, Remark 4.18, and Remarks 3.20), and (by the *Characterization of Completeness* (Theorem 2.27)) so is the *graph* G_A of A, being a *closed subspace* in $(X \times Y, \| \cdot \|)$.

The mapping

$$G_A \ni (x, Ax) \mapsto P(x, Ax) := x \in X$$

is a *bijective linear operator* from $(G_A, \| \cdot \|)$ *onto* $(X, \| \cdot \|_X)$.

Exercise 4.40. Explain.

The operator P is *bounded* since

$$\forall x \in X : \ \|P(x, Ax)\|_X = \|x\|_X \le \|x\|_X + \|Ax\|_Y = \|(x, Ax)\|.$$

By the *Inverse Mapping Theorem* (Theorem 4.8), the *inverse operator*

$$X \ni x \to P^{-1}x = (x, Ax) \in G_A$$

is *bounded*, and hence,

$$\forall x \in X : \ \|Ax\|_Y \le \|x\|_X + \|Ax\|_Y = \|(x, Ax)\| = \|P^{-1}x\| \le \|P^{-1}\|\|x\|_X,$$

which implies the *boundedness* for A, completing the proof. □

Remarks 4.23.

- The condition of the *completeness* of the target space $(Y, \| \cdot \|_Y)$ in the *Closed Graph Theorem* (*CGT*), provided other conditions hold, is *not essential* and can be dropped (see [45, Remarks 6.23]).
- The condition of the *completeness* of the domain space $(X, \| \cdot \|_X)$ in the *CGT* is essential and cannot be dropped. Indeed, as is known (see Examples 4.7), the differentiation operator

$$C^1[a, b] \ni x \to [Ax](t) := \frac{d}{dt}x(t) \in C[a, b]$$

$(-\infty < a < b < \infty)$ is a *closed unbounded linear operator* from the *incomplete* normed vector space $(C^1[a, b], \| \cdot \|_\infty)$ onto the Banach space $(C[a, b], \| \cdot \|_\infty)$.
- As is shown below (see Section 4.6), the *Inverse Mapping Theorem* (Theorem 4.8) is equivalent to the *Closed Graph Theorem* (Theorem 4.10).
- Thus far, we have the following chain of implications

$$OMT \Rightarrow IMT \Rightarrow CGT$$

(see Section 4.4.2.1).

4.5.4 Application: Projection Operators

4.5.4.1 Projection Operators on Vector Spaces

We study the important class of linear operators called *projections*.

Recall that each subspace Y in a vector space X has a *complementary subspace* Z:

$$X = Y \oplus Z,$$

i. e., every $x \in X$ allows a *unique decomposition*

$$x = y + z$$

with $y \in Y$ and $z \in Z$ (see the *Existence of a Complementary Subspace Theorem* (Theorem 3.6) and the *Direct Sum Decompositions Proposition* (Proposition 3.5)).

Remarks 4.24.
- The complementary subspaces Y and Z are necessarily *disjoint*, i. e.,

$$Y \cap Z = \{0\}$$

  (see Definition 3.14).
- Except when $Y = \{0\}$, the complementary subspace Z need not be unique (see Remark 3.15).

Exercise 4.41. In the space $C[-a, a]$ $(0 < a < \infty)$,
(a) show that the subspace

$$Y := \{y \in C[-a, a] \mid y(-t) = -y(t)\}$$

of all *odd* continuous on $[-a, a]$ functions and the subspace

$$Z := \{y \in C[-a, a] \mid y(-t) = y(t)\}$$

of all *even* continuous on $[-a, a]$ functions are *complementary*;
(b) for each $x \in C[-a, a]$, find the *unique decomposition*

$$x = y + z$$

with $y \in Y$ and $z \in Z$.

With every *decomposition* of a vector space X into a direct sum of complementary subspaces Y and Z associated is a linear operator called the *projection onto Y along Z*.

Definition 4.9 (Projection Operator on a Vector Space). Let Y and Z be complementary subspaces in a vector space X over $\mathbb{F}$. The linear operator P on X, defined as

$$X \ni x = y + z, \; y \in Y, z \in Z \mapsto Px := y \in Y,$$

is called the *projection operator* (or *projection*) *onto Y along Z*.

Exercise 4.42. Verify that P is *well defined* and *linear*.

Example 4.8. For the direct product $X := \prod_{i \in I} X_i$ of a nonempty collection $\{X_i\}_{i \in I}$ of vector spaces (see Definition 3.10) and each $j \in I$, the linear operator

$$X \ni x = (x_i)_{i \in I} \mapsto P_j x := (\delta_{ij} x_i)_{i \in I},$$

where δ_{ij} is the *Kronecker delta*, is a *projection* onto the subspace

$$Y := \{x = (x_i)_{i \in I} \in X \mid x_i = 0,\ i \in I, i \neq j\}$$

along the subspace

$$Z := \{x = (x_i)_{i \in I} \in X \mid x_j = 0\}.$$

Proposition 4.7 (Properties of Projection Operators). *Let Y and Z be complementary subspaces in a vector space X. The projection P onto Y along Z has the following properties:*
1. $Px = x \Leftrightarrow x \in Y$,
2. $Px = 0 \Leftrightarrow x \in Z$,
3. $P^2 = P$, i.e., the operator P is idempotent.

Exercise 4.43. Prove.

We immediately obtain the following corollary.

Corollary 4.3 (Properties of Projection Operators). *Let Y and Z be complementary subspaces in a vector space X. For the projection P onto Y along Z,*
(1) $R(P) = Y$,
(2) $\ker P = Z$,
(3) $I - P$ *(I is the identity operator on X) is the projection onto Z along Y with*

$$R(I - P) = Z = \ker P \quad and \quad \ker(I - P) = Y = R(P).$$

Exercise 4.44.
(a) Prove.
(b) Prove that, on a vector space X, P is a projection operator *iff* $I - P$ is a projection operator.

Remarks 4.25.
– Hence, projection operators on a vector space X occur in complementary pairs, P, $I - P$, adding up to the identity operator I.
– There are always at least two *complementary projections* on a nonzero vector space X: the *zero operator* 0 and the *identity operator* I.

Thus, a projection operator on a vector space X is an *idempotent linear operator* (see Proposition 4.7). The converse is true as well.

Proposition 4.8 (Characterization of Projections on a Vector Space). *A linear operator P on a vector space X is a projection iff P is idempotent, in which case*

$$X = Y \oplus Z$$

with $Y = R(P)$ and $Z = \ker P$.

Exercise 4.45. Prove.

Remark 4.26. Thus, one can define a projection operator P on a vector space X as an idempotent linear operator on X. To define a projection operator P on a normed vector space $(X, \| \cdot \|)$, we add the boundedness condition.

4.5.4.2 Projection Operators on Normed Vector Spaces

Definition 4.10 (Projection Operator on a Normed Vector Space). Let $(X, \| \cdot \|)$ be a normed vector space. A *projection operator* (or *projection*) P on $(X, \|\cdot\|)$ is an *idempotent bounded linear operator* on $(X, \| \cdot \|)$ (i. e., $P^2 = P$ and $P \in L(X)$).

Exercise 4.46. Prove that, on a normed vector space $(X, \|\cdot\|)$, P is a projection operator *iff $I - P$ is a projection operator*.

Remarks 4.27.
- Hence, projection operators on a normed vector space $(X, \| \cdot \|)$ occur in complementary pairs, $P, I - P$, adding up to the identity operator I.
- There are always at least two *complementary projections* on a nonzero normed vector space $(X, \| \cdot \|)$: the *zero operator* 0 and the *identity operator* I.

Examples 4.9.
1. In the space $l_p^{(2)}(\mathbb{R})$ $(1 \le p \le \infty)$,

    $$P(x, y) = (x, 0), \ (x, y) \in \mathbb{R}^2,$$

    and

    $$(I - P)(x, y) = (0, y), \ (x, y) \in \mathbb{R}^2,$$

    is a complementary pair of *projection operators* (*orthogonal* ones for $p = 2$ (see Examples 6.10)).
2. In the space $l_p^{(3)}(\mathbb{R})$ $(1 \le p \le \infty)$,

    $$P(x, y, z) = (x, y, 0), \ (x, y, z) \in \mathbb{R}^3,$$

and

$$(I - P)(x, y, z) = (0, 0, z), \quad (x, y, z) \in \mathbb{R}^3,$$

is a complementary pair of *projection operators* (*orthogonal* ones for $p = 2$ (see Examples 6.10)).
3. On l_p $(1 \le l \le \infty)$,

$$P(x_n)_{n \in \mathbb{N}} := (x_1, 0, x_3, 0, \dots), \quad (x_n)_{n \in \mathbb{N}} \in l_p,$$

and

$$(I - P)(x_n)_{n \in \mathbb{N}} := (0, x_2, 0, x_4, 0, \dots), \quad (x_n)_{n \in \mathbb{N}} \in l_p,$$

is a complementary pair of *projection operators* (*orthogonal* ones for $p = 2$ (see Examples 6.10)).

Exercise 4.47. Verify.

Proposition 4.9 (Norm of a Projection Operator). *For a nontrivial projection $P \ne 0$ on a normed vector space $(X, \| \cdot \|)$,*

$$\|P\| \ge 1.$$

Exercise 4.48. Prove (see the *Characterization of Orthogonal Projections* (Proposition 6.24, Section 6.12, Problem 14)).

Example 4.10. Let $X := l_2^{(2)}(\mathbb{R})$, $Y := \{(x, 0) \mid x \in \mathbb{R}\}$, $Z := \{(x, -x) \mid x \in \mathbb{R}\}$, and P be the projection operator onto Y along Z. Then, since

$$P(1, 1) = (2, 0),$$

we have:

$$\sqrt{2}\|P\| = \|P\|\|(1, 1)\|_2 \ge \|P(1, 1)\|_2 = \|(2, 0)\|_2 = 2,$$

which implies that

$$\|P\| \ge \sqrt{2} > 1.$$

Exercise 4.49. Verify.

Theorem 4.11 (Projections on a Normed Vector Space).
1. *If $(X, \|\cdot\|)$ is a normed vector space and P is a projection operator on X, then $Y := R(P)$ and $Z := \ker P$ are closed complementary subspaces:*

$$X = Y \oplus Z.$$

2. *Conversely, if $(X, \| \cdot \|)$ is a Banach space, and Y and Z are closed complementary subspaces:*

$$X = Y \oplus Z, \tag{4.10}$$

then the projection P onto Y along Z in the vector space sense is a projection in the normed vector space sense, i. e., $P \in L(X)$.

Proof.
1. The proof of this part immediately follows by the *Kernel of a Bounded Linear Operator Proposition* (Proposition 4.2) from the fact that

$$Z := \ker P \quad \text{and} \quad Y := R(P) = \ker(I - P).$$

2. To prove this part, let us to show that the *projection operator* $P : X \rightarrow X$ onto Y along Z is a *closed* linear operator.
   Indeed, let $(x_n)_{n \in \mathbb{N}}$ be an arbitrary sequence in $(X, \| \cdot \|)$ such that

$$\lim_{n \to \infty} x_n = x \in X \quad \text{and} \quad \lim_{n \to \infty} Px_n = y \in X.$$

Then, by the *closedness* of Y in $(X, \| \cdot \|)$, $y \in Y$ and, since

$$x_n = Px_n + (I - P)x_n, \ n \in \mathbb{N}, \tag{4.11}$$

in view of direct sum decomposition (4.10) and the *closedness* of Z in $(X, \| \cdot \|)$,

$$Z \ni (I - P)x_n = x_n - Px_n \rightarrow x - y =: z \in Z, \ n \to \infty.$$

Hence, passing to the limit in (4.11) as $n \to \infty$, we arrive at

$$x = y + z$$

with $y \in Y$ and $z \in Z$, which implies that $y = Px$.
By the *Sequential Characterization of Closed Linear Operators* (Proposition 4.4), we infer that the operator P is *closed*, which, considering that $(X, \| \cdot \|)$ is a Banach space, by the *Closed Graph Theorem* (Theorem 4.10), implies that $P \in L(X)$ and completes the proof. □

Remarks 4.28.

– The condition of the *completeness* of the space $(X, \| \cdot \|)$ in part 2 of the prior theorem, provided other conditions hold, is *not essential* and can be dropped (see [45, Remarks 6.28]).

- Thus, similarly to the case for projections on vector spaces, every *decomposition* of a normed vector space $(X, \| \cdot \|)$ into a direct sum of complementary closed subspaces Y and Z generates a *projection operator* on $(X, \| \cdot \|)$ and vice versa.
- For more on closed subspaces of normed vector spaces, which allow *closed complementary subspaces*, and thus, are called *closed complemented subspaces*, see, e. g., [45, Section 6.2.6].

  Generally, a closed subspace in a normed vector space need not be closed complemented. A counterexample built in l_1 can be found in [62, Section 5.7, Problem 9].

4.6 Equivalence of *OMT*, *IMT*, and *CGT*

Here, we prove the equivalence of the *Open Mapping Theorem* (*OMT*), the *Inverse Mapping Theorem* (*IMT*), and the *Closed Graph Theorem* (*CGT*).

Theorem 4.12 (Equivalence Theorem). *The Open Mapping Theorem (OMT), the Inverse Mapping Theorem (IMT), and the Closed Graph Theorem (CGT) are equivalent statements.*

Proof. Let us prove the following closed chain of implications:

$$OMT \Rightarrow IMT \Rightarrow CGT \Rightarrow OMT.$$

Observe that we already have

$$OMT \Rightarrow IMT \Rightarrow CGT$$

(see Remarks 4.23), and hence, it remains to prove the last implication

$$CGT \Rightarrow OMT$$

in the above chain.

Suppose that *CGT* holds and let

$$A : (X, \| \cdot \|_X) \to (Y, \| \cdot \|_Y)$$

be a surjective bounded linear operator from a Banach space $(X, \| \cdot \|_X)$ onto a Banach space $(Y, \| \cdot \|_Y)$ over $\mathbb{F}$.

Since, by *Kernel of a Bounded Linear Operator Proposition* (Proposition 4.2), the kernel $\ker A$ of the bounded linear operator A is a closed subspace of $(X, \| \cdot \|_X)$, the *quotient space norm*

$$X/\ker A \ni [x] = x + \ker A \mapsto \|[x]\| = \inf_{y \in \ker A} \|x + y\| =: \rho(x, \ker A)$$

is well defined, relative to which $X/\ker A$ is a *Banach space* (see the *Quotient Space Norm Theorem* (Theorem 3.16, Section 3.5, Problem 12), see Remarks 4.8), and the *bijective* linear operator

$$(X/\ker A, \|\cdot\|) \ni [x] = x + \ker A \mapsto \hat{A}[x] := Ax \in (Y, \|\cdot\|_Y)$$

(see the proof of the *Rank-Nullity Theorem* (Theorem 4.1)) is *bounded* (see Section 4.7, Problem 13).

By the *Characterization of Closedness for Bounded Linear Operators* (Proposition 4.5), the operator $\hat{A}$ is *closed*, and hence, by the *Closedness of Inverse Operator Proposition* (Proposition 4.23, Section 4.7, Problem 20), so is the inverse operator

$$\hat{A}^{-1} : (Y, \|\cdot\|_Y) \to (X/\ker A, \|\cdot\|).$$

This, by the *Closed Graph Theorem* (Theorem 4.10), implies that the inverse operator $\hat{A}^{-1}$ is *bounded*, and hence, by the *Characterizations of Bounded Linear Operators* (Theorem 4.4), is *continuous* on $(Y, \|\cdot\|_Y)$.

By the *Characterization of Continuity* (Theorem 2.61, Section 2.18, Problem 15), for each *open set* G in $(X, \|\cdot\|_X)$, the image $A(G)$ is *open* in $(Y, \|\cdot\|_Y)$ since

$$A(G) = \left(\hat{A}^{-1}\right)^{-1}(T(G)),$$

where

$$(X, \|\cdot\|) \ni x \mapsto Tx := [x] = x + Y \in (X/\ker A, \|\cdot\|),$$

is the *canonical homomorphism*, which is an *open mapping* (see the *Canonical Homomorphism Proposition* (Proposition 4.22, Problem 16)), i.e., the image $T(G)$ is *open* in the quotient space $(X/\ker A, \|\cdot\|)$ (see the *Quotient Space Norm Theorem* (Theorem 3.16, Section 3.5, Problem 12)).

Therefore, $A : X \to Y$ is an *open mapping*, which completes the proof. □

4.7 Problems

In the subsequent problems, $\mathbb{F}$ stands for the scalar field of real or complex numbers (i.e., $\mathbb{F} = \mathbb{R}$ or $\mathbb{F} = \mathbb{C}$).

1. (Complexification of Linear Operators in Real Vector Spaces)

   Let X be a real vector space and $X^{\mathbb{C}}$ be its *complexification* (see Section 3.5, Problem 1).

   The *complexification* $(\tilde{A}, D(\tilde{A}))$ of an arbitrary linear operator $(A, D(A))$ in X is its natural linear extension into the complex vector space $X^{\mathbb{C}}$ defined as follows:

   $$D(\tilde{A}) := D(A) \times D(A) \ni (x, y) \mapsto \tilde{A}(x, y) := (Ax, Ay) = Ax + iAy.$$

In the matrix form, the operator $(\tilde{A}, D(\tilde{A}))$ can be written as

$$\begin{bmatrix} A & 0 \\ 0 & A \end{bmatrix}.$$

(a) Show that the mapping $(\tilde{A}, D(\tilde{A}))$ is a *linear operator* in $X_{\mathbb{C}}$.
(b) Show that $(\tilde{A}, D(\tilde{A}))$ is an extension of $(A, D(A))$, in the following sense:

$$D(A) \times \{0\} = \{(x, 0) \mid x \in D(A)\} \subseteq D(\tilde{A}) \text{ and } \tilde{A}(x, 0) = (Ax, 0), \ x \in D(A).$$

2. Prove

Proposition 4.10 (Characterization of the Graph of a Linear Operator). *Let X and Y be vector spaces over $\mathbb{F}$. A subspace G of the product space $X \times Y$ is the graph of a linear operator $(A, D(A))$ from X to Y iff it does not contain points of the form $(0, y)$ with $y \neq 0$.*

3. **Definition 4.11** (Inverse Operator). Let X and Y be vector spaces over $\mathbb{F}$ and

$$X \supseteq D(A) \ni x \mapsto Ax \in R(A) \subseteq Y$$

be a *one-to-one (injective)* linear operator, i. e.,

$$\forall y \in R(A) \ \exists! \, x \in D(A) : \ y = Ax.$$

The *inverse* to A is a linear operator defined as follows:

$$X \supseteq R(A) =: D(A^{-1}) \ni y \mapsto A^{-1}y = x \in R(A^{-1}) =: D(A) \subseteq Y,$$

where $x \in D(A)$ is the unique element such that

$$Ax = y.$$

The operator A is said to be *invertible*.

Verify that the inverse A^{-1} is a *linear operator* and

$$\forall x \in D(A) : A^{-1}(Ax) = x,$$
$$\forall y \in R(A) : A(A^{-1}y) = y,$$

which implies that $\left(A^{-1}\right)^{-1} = A.$

4. Prove

Proposition 4.11 (Kernel Characterization of Invertibility). *Let X and Y be vector spaces over $\mathbb{F}$. A linear operator $(A, D(A))$ from X to Y is invertible, i. e., there exists an inverse operator $A^{-1} : R(A) \rightarrow D(A)$, iff*

$$\ker A := \{x \in D(A) \mid Ax = 0\} = \{0\}.$$

See the *Characterization of Isomorphisms* (Proposition 3.11, Problems 3.5, Problem 2).

5. Prove

Proposition 4.12 (Graph Characterization of Invertibility). *Let X and Y be vector spaces over $\mathbb{F}$. A linear operator $(A, D(A))$ from X to Y is invertible, i.e., there exists an inverse operator $A^{-1} : R(A) \to D(A)$ iff the subspace*

$$G_A^{-1} := \{(Ax, x) \in Y \times X \mid x \in D(A)\}$$

of $Y \times X$ is the graph of a linear operator from Y to X, in which case

$$G_{A^{-1}} = G_A^{-1}.$$

6. Prove

Proposition 4.13 (Hyperplane Characterization). *A subspace Y of a nontrivial vector space X over $\mathbb{F}$ is a hyperplane iff there exists a nontrivial linear functional $f : X \to \mathbb{F}$ such that $Y = \ker f$.*

7. Prove

Proposition 4.14 (Boundedness of Linear Operators from Finite-Dimensional Spaces). *Let $(X, \| \cdot \|_X)$ and $(Y, \| \cdot \|_Y)$ be normed vector spaces over $\mathbb{F}$, the space X being finite-dimensional. Then each linear operator $A : X \to Y$ is bounded.*

Remark 4.29. In particular, for $Y = \mathbb{F}$ with the absolute-value norm, we conclude that all linear functionals on a *finite-dimensional* normed vector space $(X, \| \cdot \|_X)$ are *bounded*.

8. Prove

Proposition 4.15 (Existence of Unbounded Linear Operators). *Let $(X, \| \cdot \|_X)$ be an infinite-dimensional normed vector space and $(Y, \| \cdot \|_Y)$ be a nonzero normed vector space $(Y \neq \{0\})$ over $\mathbb{F}$. Then there exists an unbounded linear operator $A : X \to Y$.*

Hint. Use the *Extension Theorem for Linear Operators* (Theorem 4.2).

Remark 4.30. In particular, for $Y = \mathbb{F}$ with the absolute-value norm, we conclude that, on every *infinite-dimensional* normed vector space $(X, \| \cdot \|_X)$, there exists an *unbounded linear functional*.

9. * Prove

Proposition 4.16 (Kernel of a Linear Functional). *A linear functional f on a normed vector space $(X, \| \cdot \|)$ is bounded iff its kernel $\ker f$ is a closed subspace of $(X, \| \cdot \|_X)$.*

Hint. Prove the *"if"* part *by contrapositive*.

10. (a) Prove

> **Proposition 4.17** (Unboundedness of Hamel Coordinate Functionals).
> Let $(X, \| \cdot \|)$ be an infinite-dimensional Banach space over $\mathbb{F}$ with a Hamel basis $B := \{x_i\}_{i \in I}$. Then all but a finite number of the Hamel coordinate functionals
>
> $$X \ni x = \sum_{i \in I} c_i x_i \mapsto c_j(x) := c_j \in \mathbb{F}, \ j \in I,$$
>
> relative to B are unbounded.
>
> **Hint.** Prove the "*if*" part *by contradiction*, applying the prior proposition and the *Baire Category Theorem* (Theorem 2.32).

(b) Give an example showing that the *completeness* requirement for the space is essential and cannot be dropped.

11. Prove

> **Proposition 4.18** (Characterization of Bounded Linear Operators). *Let* $(X, \| \cdot \|_X)$ *and* $(Y, \| \cdot \|_Y)$ *be normed vector spaces over* $\mathbb{F}$. *A linear operator* $A : X \to Y$ *is bounded iff the preimage* $A^{-1} B_Y(0, 1)$ *under A of the open unit ball* $B_Y(0, 1) := \{y \in Y \mid \|y\|_Y < 1\}$ *of* $(Y, \| \cdot \|_Y)$ *has a nonempty interior in* $(X, \| \cdot \|_X)$.

12. (a) Prove

> **Proposition 4.19** (Completeness of the Range). *Let* $(X, \| \cdot \|_X)$ *be a Banach space,* $(Y, \| \cdot \|_Y)$ *be a normed vector space over* $\mathbb{F}$, *and an operator* $A \in L(X, Y)$ *have a bounded inverse* $A^{-1} : R(A) \to X$. *Then* $(R(A), \| \cdot \|_Y)$ *is a Banach space.*

(b) Give an example showing that the *completeness* requirement for $(X, \| \cdot \|_X)$ is essential and cannot be dropped.

13. Let $(X, \| \cdot \|_X)$ and $(Y, \| \cdot \|_Y)$ be normed vector spaces over $\mathbb{F}$ and $A : X \to Y$ be a linear operator. Show that the *injective* linear operator

$$(X/ \ker A, \| \cdot \|) \ni [x] = x + \ker A \mapsto \hat{A}[x] := Ax \in (Y, \| \cdot \|_Y),$$

where $\| \cdot \|$ is the *quotient space norm* (see the proof of the *Rank-Nullity Theorem* (Theorem 4.1) and the *Quotient Space Norm Theorem* (Theorem 3.16, Section 3.5, Problem 12)), is *bounded*, provided the operator A is *bounded*.

14. Prove

> **Proposition 4.20** (Boundedness of Strong Limit). *Let* $(X, \| \cdot \|_X)$ *be a Banach space,* $(Y, \| \cdot \|_Y)$ *be normed vector space over* $\mathbb{F}$, *and* $(A_n)_{n \in \mathbb{N}}$ *be a sequence of bounded linear operators in* $(L(X, Y), \| \cdot \|)$. *If, for all* $x \in X$, *there exists the limit*
>
> $$Ax := \lim_{n \to \infty} A_n x \text{ in } (Y, \| \cdot \|_Y),$$
>
> *then the limit mapping* $A : X \to Y$ *defined above is a bounded linear operator (i. e.,* $A \in L(X, Y)$).

Corollary 4.4 (Boundedness of Strong Limit). *Let* $(X, \|\cdot\|_X)$ *be a Banach space over* $\mathbb{F}$ *and* $(f_n)_{n\in\mathbb{N}}$ *be a sequence of bounded linear functionals in the dual space* $(X^*, \|\cdot\|)$. *If, for all* $x \in X$, *there exists the limit*

$$f(x) := \lim_{n\to\infty} f_n(x),$$

then the limit mapping $f : X \to \mathbb{F}$ *defined above is a bounded linear functional (i. e.,* $f \in X^*$).

Hint. Apply the *Uniform Boundedness Principle* (Theorem 4.6).

15. * Prove

Proposition 4.21 (Continuity of a Bilinear Functional). *Let* $(X_1, \|\cdot\|_1)$ *and* $(X_2, \|\cdot\|_2)$ *be Banach spaces over* $\mathbb{F}$, $X = X_1 \times X_2$ *be their Cartesian product, which is also a Banach space relative to the product norm*

$$X \ni (x,y) \mapsto \|(x,y)\|_{X_1 \times X_2} = \sqrt{\|x\|_1^2 + \|y\|_2^2}$$

(see Section 3.5, Problem 7), and $B(\cdot,\cdot) : X_1 \times X_2 \to \mathbb{F}$ *be a bilinear functional, i. e.,*

$$\forall x, y \in X_1, z \in X_2 \; \forall \lambda, \mu \in \mathbb{F} : \; B(\lambda x + \mu y, z) = \lambda B(x,z) + \mu B(y,z),$$
$$\forall x \in X_1, y, z \in X_2, \; \forall \lambda, \mu \in \mathbb{F} : \; B(x, \lambda y + \mu z) = \lambda B(x,y) + \mu B(x,z),$$

continuous relative to each argument. Then $B(\cdot,\cdot)$ *is jointly continuous on* $X_1 \times X_2$, *i. e.,*

$$\forall (x,y) \in X \; \forall ((x_n, y_n))_{n=1}^\infty \subseteq X, \; (x_n, y_n) \to (x,y), \; n \to \infty, \; in \; X :$$
$$B(x_n, y_n) \to B(x,y), \; n \to \infty.$$

Hint. Apply the *Uniform Boundedness Principle* (Theorem 4.6).

16. Prove *with* and *without* the *Open Mapping Theorem* (Theorem 4.7).

Proposition 4.22 (Canonical Homomorphism). *Let* Y *be a closed subspace of a Banach space* $(X, \|\cdot\|_X)$. *Then the canonical homomorphism*

$$(X, \|\cdot\|) \ni x \mapsto Tx := [x] = x + Y \in (X/Y, \|\cdot\|),$$

where

$$X/Y \ni [x] = x + Y \mapsto \|[x]\| = \inf_{y \in Y} \|x + y\|_X =: \rho(x, Y)$$

is the quotient space norm (see the Quotient Space Norm Theorem *(Theorem 3.16, Section 3.5, Problem 12)),*

(1) is a surjective bounded linear operator;

(2) is an open mapping, i. e., the image $T(G)$ under T of each open set G in $(X, \| \cdot \|_X)$ is an open set in $(X/Y, \| \cdot \|)$.

Hint. To prove part 2 without the *Open Mapping Theorem* (Theorem 4.7), take advantage of the nature of convergence in the quotient space $(X/Y, \| \cdot \|)$ (see Section 3.5, Problem 12) and apply the *Sequential Characterizations of Open Sets* (Theorem 2.15).

17. Let $(X, \| \cdot \|_X)$ be a Banach space, $(Y, \| \cdot \|_Y)$ be a normed vector space over $\mathbb{F}$, and $A \in L(X, Y)$. Show that either $R(A) = Y$ or $R(A)$ is of the *first category* in $(Y, \| \cdot \|_Y)$.

Hint. Use some ideas of the proof of the *Open Mapping Theorem* (Theorem 4.7).

18. Let $(X, \| \cdot \|_X)$ be a Banach space, $(Y, \| \cdot \|_Y)$ be a normed vector space over $\mathbb{F}$, and $A : X \to Y$ be a surjective bounded linear operator from X onto Y (i. e., $A \in L(X, Y)$ and $R(A) = Y$). Prove that, if A is an open mapping, then $(Y, \| \cdot \|_Y)$ is a Banach space.

Hint. Show that, if A is an open mapping, then the *bijective* bounded linear operator

$$(X/ \ker A, \| \cdot \|) \ni [x] = x + \ker A \mapsto \hat{A}[x] := Ax \in (Y, \| \cdot \|_Y),$$

where $\| \cdot \|$ is the quotient space norm (see Problem 13), is also an *open mapping* (considering that the *canonical homomorphism* $X \ni x \mapsto Tx := [x] \in X/ \ker A$ is an *open mapping* (see the *Canonical Homomorphism Proposition* (Proposition 4.22, Problem 16)) has a bounded inverse $\hat{A}^{-1} \in L(Y, X/ \ker A)$, which, by the *Completeness of the Range Proposition* (Proposition 4.19, Problem 12)), makes $(Y, \| \cdot \|_Y)$ to be a Banach space.

19. Let Y and Z be closed disjoint subspaces in a Banach space $(X, \| \cdot \|)$. Prove that the subspace $Y \oplus Z$ is closed in $(X, \| \cdot \|)$ *iff*

$$\exists c > 0 \, \forall y \in Y, \, \forall z \in Z : \, c[\|y\| + \|z\|] \leq \|y + z\|.$$

Hint. Apply the *Equivalence of Banach Norms Theorem* (Theorem 4.9).

20. Prove

Proposition 4.23 (Closedness of Inverse Operator). *Let $(X, \| \cdot \|_X)$ and $(Y, \| \cdot \|_Y)$ be normed vector spaces over $\mathbb{F}$ and $A : X \supseteq D(A) \to Y$ be a closed linear operator. Then if the inverse operator $A^{-1} : Y \supseteq R(A) \to X$ exists, then it is a closed linear operator.*

21. Prove

Proposition 4.24 (Closed Linear Operators and Series). *Let* $(X, \| \cdot \|_X)$ *and* $(Y, \| \cdot \|_Y)$ *be normed vector spaces over* $\mathbb{F}$ *and* $(A, D(A))$ *be a closed linear operator from* X *to* Y. *If a sequence* $(x_n)_{n \in \mathbb{N}}$ *in* $D(A)$ *is such that*

$$\sum_{k=1}^{\infty} x_k = x \text{ in } (X, \| \cdot \|_X) \quad and \quad \sum_{k=1}^{\infty} A x_k = y \text{ in } (Y, \| \cdot \|_Y),$$

then

$$x \in D(A) \quad and \quad \sum_{k=1}^{\infty} A x_k = y = Ax = A\left(\sum_{k=1}^{\infty} x_k \right).$$

Hint. Apply the *Sequential Characterization of Closed Linear Operators* (Proposition 4.4).

22. Let $(X, \| \cdot \|_X)$ be a Banach space, $(Y, \| \cdot \|_Y)$ be a normed vector space over $\mathbb{F}$, and $A : X \to Y$ be a linear operator. Show that the *injective* linear operator

$$(X/ \ker A, \| \cdot \|) \ni [x] := x + \ker A \mapsto \hat{A}[x] := Ax \in (Y, \| \cdot \|_Y),$$

where $\| \cdot \|$ is the *quotient space norm* (see Problem 13), is *closed*, provided the operator A is *closed*.

23. Prove

Proposition 4.25 (Sufficient Condition of Boundedness). *Let* $\quad (X, \| \cdot \|_X) \quad$ *and* $(Y, \| \cdot \|_Y)$ *be Banach spaces over* $\mathbb{F}$ *and* $A : X \to Y$ *be a linear operator. If, for any sequence* $(x_n)_{n \in \mathbb{N}} \subset X$ *such that*

$$\lim_{n \to \infty} x_n = 0 \text{ in } (X, \| \cdot \|_X) \text{ and } \lim_{n \to \infty} A x_n = y \text{ in } (Y, \| \cdot \|_Y),$$

we have

$$y = 0,$$

then the operator A *is bounded (i. e.,* $A \in L(X, Y)$*).*

Hint. Apply the *Sequential Characterization of Closed Linear Operators* (Proposition 4.4) and the *Closed Graph Theorem* (Theorem 4.10).

5 Elements of Spectral Theory in a Banach Space Setting

The concepts, approaches, and applications discussed in this chapter gravitate to the concept of the *spectrum* of a closed linear operator, a natural generalization of the set of eigenvalues of a matrix, well-known from linear algebra (see, e. g., [34, 54]).

5.1 Algebraic Operations on Linear Operators

Recall that, in Section 4.2.3.1, we define linear operations on bounded linear operators from a normed vector space $(X, \| \cdot \|_X)$ to a normed vector space $(Y, \| \cdot \|_Y)$ as pointwise. We also defined *multiplication* for bounded linear operators on a normed vector space $(X, \| \cdot \|)$ via composition (see Remark 4.9).

In the same fashion, we can define algebraic operations on linear operators, bounded or not, defined in the entire space or not, as follows:

Definition 5.1 (Algebraic Operations on Linear Operators). For linear operators $(A, D(A))$ and $(B, D(B))$ from a vector space X to a vector space Y over $\mathbb{F}$ ($\mathbb{F} := \mathbb{R}$ or $\mathbb{F} := \mathbb{C}$), and an arbitrary number $\lambda \in \mathbb{F}$,

$$(A + B)x := Ax + Bx, \ x \in D(A + B) := D(A) \cap D(B),$$
$$(\lambda A)x := \lambda Ax, \ x \in D(\lambda A) := D(A).$$

Provided $Y = X$, one can also define the product:

$$(AB)x := A(Bx), \ x \in D(AB) := \{x \in D(B) \mid Bx \in D(A)\},$$

and, in particular, natural powers:

$$A^1 := A, \ A^n := AA^{n-1} \text{ with } D(A^n) := \{x \in D(A^{n-1}) \mid A^{n-1}x \in D(A)\}.$$

Remark 5.1. Such linear operations turn the set of all linear operators $A : X \to Y$ into a *vector space* over $\mathbb{F}$, and, provided $Y = X$, jointly with the operator multiplication, into an *algebra* over F (see, e. g., [4]).

Examples 5.1. Let $(A, D(A))$ be a linear operator in a (real or complex) vector space X.
1. $A + (-A)$ and $0A$ (0 is understood as the zero of the scalar field, or the zero operator on X) are the *zero operator* in X, whose domain coincides with $D(A)$.
2. $A0 = 0$ (0 is the zero operator on X).
3. For the *identity operator* I on X, $AI = IA = A$.
4. In the (real or complex) Banach space $(C[a, b], \| \cdot \|_\infty)$ ($-\infty < a < b < \infty$), for the *unbounded* linear differentiation operator

$$C^1[a, b] =: D(A) \ni x \to Ax := \frac{dx}{dt} \in C[a, b]$$

https://doi.org/10.1515/9783110600988-005

(see Examples 4.2), A^n ($n \in \mathbb{N}$) is the *nth derivative operator*:

$$A^n x := \frac{d^n x}{dt^n} \quad \text{with} \quad D(A^n) = C^n[a, b],$$

where $C^n[a, b]$ is the subspace of all n times continuously differentiable on $[a, b]$ real- or complex-valued functions.

Exercise 5.1. Give examples showing that
(a) the sum of two closed linear operators in a normed vector space $(X, \| \cdot \|)$ need not be closed;
(b) the product of two closed linear operators in a normed vector space $(X, \| \cdot \|)$ need not be closed.

Cf. Section 5.8, Problem 1.

5.2 Resolvent Set and Spectrum of a Closed Linear Operator

5.2.1 Definitions

For the purposes of the spectral theory, it is essential that the underlying space be *complex*.

Remark 5.2. Observe that $(A, D(A))$ be a *closed linear operator* in a complex Banach space $(X, \| \cdot \|)$. Then, for an arbitrary $\lambda \in \mathbb{C}$, the linear operator

$$A - \lambda I$$

in $(X, \| \cdot \|)$ with the domain $D(A - \lambda I) = D(A) \cap X = D(A)$ is also *closed* (see Section 5.8, Problem 1).

Definition 5.2 (Resolvent Set and Spectrum of a Closed Linear Operator). Let $(A, D(A))$ be a *closed linear operator* in a complex Banach space $(X, \| \cdot \|)$.
– A number $\lambda \in \mathbb{C}$ is said to be a *regular point* of A if the closed linear operator $A - \lambda I : D(A) \to X$ is *one-to-one* (*injective*) and *onto* (*surjective*), i. e., $R(A - \lambda I) = X$.
– The set of all regular points of A

$$\rho(A) := \{\lambda \in \mathbb{C} \mid A - \lambda I \text{ is } \textit{one-to-one} \text{ and } R(A - \lambda I) = X\},$$

is called the *resolvent set* of A.
– The *spectrum* $\sigma(A)$ of A is the complement of its resolvent set:

$$\sigma(A) := \rho(A)^c = \mathbb{C} \setminus \rho(A).$$

Remarks 5.3.

– Thus, the fact that $\lambda \in \rho(A)$ is *equivalent* to any of the following:

(1) The equation

$$(A - \lambda I)x = y$$

has a *unique solution for each* $y \in X$.

(2) The inverse operator $(A - \lambda I)^{-1}$ exists and is defined in the whole Banach space $(X, \| \cdot \|)$, i. e.,

$$\forall x \in D(A) : \ (A - \lambda I)^{-1}(A - \lambda I)x = x,$$

and

$$\forall y \in X : \ (A - \lambda I)(A - \lambda I)^{-1}y = y.$$

– For each $\lambda \in \rho(A)$, the inverse $(A - \lambda I)^{-1}$ being a *closed operator* (see the *Closedness of Inverse Operator Proposition* (Proposition 4.23, Section 4.7, Problem 20)), by the *Closed Graph Theorem* (Theorem 4.10), we conclude that it is *bounded*, i. e.,

$$\forall \lambda \in \rho(A) : \ (A - \lambda I)^{-1} \in L(X).$$

– By the *Characterization of Closedness for Bounded Linear Operators* (Proposition 4.5), the prior definition covers bounded linear operators defined on X, i. e., all operators from $L(X)$.

In this case, showing that $\lambda \in \rho(A)$ has purely algebraic meaning, i. e., is equivalent to showing that the operator $A - \lambda I$, as an element of the operator algebra $L(X)$ (see Remark 4.9), has an *inverse element*:

$$\exists (A - \lambda I)^{-1} \in L(X) : \ (A - \lambda I)(A - \lambda I)^{-1} = (A - \lambda I)^{-1}(A - \lambda I) = I.$$

5.2.2 Spectrum Classification, Geometric Multiplicity

Definition 5.3 (Spectrum Classification). The spectrum $\sigma(A)$ of a *closed linear operator* A in a complex Banach space $(X, \| \cdot \|)$ is naturally partitioned into the following pairwise disjoint subsets:

$$\sigma_p(A) := \{\lambda \in \mathbb{C} \mid A - \lambda I \text{ is } not \text{ one-to-one}\},$$

$$\sigma_c(A) := \{\lambda \in \mathbb{C} \mid A - \lambda I \text{ is } one\text{-}to\text{-}one \text{ and } R(A - \lambda I) \neq X, \text{ but } \overline{R(A - \lambda I)} = X\},$$

$$\sigma_r(A) := \{\lambda \in \mathbb{C} \mid A - \lambda I \text{ is } one\text{-}to\text{-}one \text{ and } \overline{R(A - \lambda I)} \neq X\},$$

called the *point*, *continuous*, and *residual spectrum* of A, respectively.

Remarks 5.4.

- Thus, the fact that $\lambda \in \sigma_p(A)$ is *equivalent* to any of the following:

  (1) The operator $A - \lambda I$ is *not invertible*, i. e., there does not exist an inverse operator $(A - \lambda I)^{-1} : R(A) \to D(A)$.

  (2)
  $$\ker(A - \lambda I) \neq \{0\}$$

  (see the *Kernel Characterization of Invertibility* (Proposition 4.11, Problems 4.7, Problem 4)), i. e., λ is an *eigenvalue* of A with the associated *eigenspace* $\ker(A - \lambda I)$.

- For $\lambda \in \sigma_c(A) \cup \sigma_r(A)$, the operator $A - \lambda I$ is *invertible*.

- For $\lambda \in \sigma_c(A)$, the range $R(A - \lambda I)$ not being a closed subspace in $(X, \|\cdot\|)$, by the *Characterization of Closedness for Bounded Linear Operators* (Proposition 4.5), the closed inverse operator $(A - \lambda I)^{-1}$ is *unbounded*.

- For $\lambda \in \sigma_r(A)$, the closed inverse operator $(A - \lambda I)^{-1}$ can be bounded or unbounded (see Examples 5.2, 4 (d) and Section 5.8, Problem 9), which by the *Characterization of Closedness for Bounded Linear Operators* (Proposition 4.5) is equivalent to the closedness of the range $R(A - \lambda I)$ in $(X, \|\cdot\|)$.

Definition 5.4 (Geometric Multiplicity of an Eigenvalue). Let X be a vector space over $\mathbb{F}$, $A : X \to X$ be a linear operator on X, and let $\lambda \in \mathbb{F}$ be an *eigenvalue* of A, i. e.,

$$\exists x \in X \setminus \{0\} : Ax = \lambda x.$$

The *geometric multiplicity* of λ is the dimension of the *eigenspace* $\ker(A - \lambda I)$ corresponding to λ is

$$\dim \ker(A - \lambda I).$$

5.2.3 Examples

Now, let us consider some examples.

Examples 5.2.

1. Let $(X, \|\cdot\|)$ be a complex Banach space.

    (a) For the bounded linear operator A of multiplication by a number $\lambda_0 \in \mathbb{C}$ (see Examples 4.2),

    $$\sigma(A) = \sigma_p(A) = \{\lambda_0\},$$

    the corresponding *eigenspace* being the whole X.
    Hence, in this case, $\sigma_c(A) = \sigma_r(A) = \emptyset$.

    Exercise 5.2. Verify.

In particular, for the *zero operator* 0 ($\lambda_0 = 0$),

$$\sigma(0) = \sigma_p(0) = \{0\},$$

and, for the *identity operator* I ($\lambda_0 = 1$),

$$\sigma(I) = \sigma_p(I) = \{1\}.$$

(b) For a *projection operator* $P : X \to X$, except when $P = 0$ or $P = I$,

$$\sigma(P) = \sigma_p(P) = \{0, 1\},$$

the corresponding *eigenspaces* being $\ker P$ and $R(P)$, respectively.
Hence, $\sigma_c(P) = \sigma_r(P) = \emptyset$.
Indeed, in this case, for any $\lambda \neq 0, 1$,

$$(P - \lambda I)^{-1} = -\frac{1}{\lambda}(I - P) + \frac{1}{1 - \lambda}P \in L(X),$$

Exercise 5.3. Verify by showing that, for any $\lambda \neq 0, 1$,

$$(P - \lambda I)\left[-\frac{1}{\lambda}(I - P) + \frac{1}{1 - \lambda}P\right] = \left[-\frac{1}{\lambda}(I - P) + \frac{1}{1 - \lambda}P\right](P - \lambda I) = I$$

(see the *Left/Right Inverse Proposition* (Proposition 5.9, Section 5.8, Problem 11)).

2. On the complex space $l_p^{(n)}$ ($n \in \mathbb{N}, 1 \leq p \leq \infty$), for the bounded linear operator A of multiplication by an $n \times n$ matrix $[a_{ij}]$ with complex entries (see Examples 4.2),

$$\sigma(A) = \sigma_p(A) = \{\lambda_1, \dots, \lambda_m\} \ (1 \leq m \leq n),$$

where the latter is the *set of all eigenvalues* of the multiplier matrix $[a_{ij}]$.
Hence, $\sigma_c(A) = \sigma_r(A) = \emptyset$. Indeed, since, for any $\lambda \in \mathbb{C}$,

$$(A - \lambda I)x = [a_{ij} - \lambda \delta_{ij}]x,$$

where δ_{ij} is the *Kronecker delta*. Hence, $\lambda \in \rho(A)$ iff the matrix $[a_{ij} - \lambda \delta_{ij}]$ is *invertible*, i. e., iff λ is *not an eigenvalue* of the matrix $[a_{ij}]$, which is equivalent to

$$\det([a_{ij} - \lambda \delta_{ij}]) \neq 0$$

(see, e. g., [34, 54]).
E. g., on the complex space $l_p^{(2)}$ ($1 \leq p \leq \infty$), for the operators A, B, and C of multiplication by the matrices

$$\begin{bmatrix} 1 & 0 \\ 0 & 2 \end{bmatrix}, \begin{bmatrix} 0 & 1 \\ 0 & 0 \end{bmatrix}, \text{ and } \begin{bmatrix} 0 & 1 \\ -1 & 0 \end{bmatrix},$$

respectively,

$$\sigma(A) = \sigma_p(A) = \{1, 2\}, \quad \sigma(B) = \sigma_p(B) = \{0\}, \quad \text{and} \quad \sigma(C) = \sigma_p(C) = \{-i, i\},$$

where i is the *imaginary unit* ($i^2 = -1$).

3. In the complex sequence space l_p $(1 \le p < \infty)$, for the closed linear operator A of multiplication by a complex-termed sequence $a := (a_n)_{n\in\mathbb{N}}$,

$$(x_n)_{n\in\mathbb{N}} \mapsto Ax := (a_n x_n)_{n\in\mathbb{N}}$$

with the maximal domain

$$D(A) := \{(x_n)_{n\in\mathbb{N}} \in l_p \mid (a_n x_n)_{n\in\mathbb{N}} \in l_p\}$$

(see Examples 4.7),

$$\sigma(A) = \overline{\{a_n\}_{n\in\mathbb{N}}} \text{ with } \sigma_p(A) = \{a_n\}_{n\in\mathbb{N}} \text{ and } \sigma_c(A) = \sigma(A) \setminus \sigma_p(A),$$

where $\{a_n\}_{n\in\mathbb{N}}$ is the set of values of $(a_n)_{n\in\mathbb{N}}$.
Hence, $\sigma_r(A) = \emptyset$.

Exercise 5.4. Verify.

E. g., for the operators A and B of multiplication by the sequences

$$(a_n)_{n\in\mathbb{N}} := (1/n)_{n\in\mathbb{N}} \quad \text{and} \quad (b_n)_{n\in\mathbb{N}} := (n)_{n\in\mathbb{N}},$$

respectively,

$$\sigma(A) = \{1/n\}_{n\in\mathbb{N}} \cup \{0\} \text{ with } \sigma_p(A) = \{1/n\}_{n\in\mathbb{N}} \text{ and } \sigma_c(A) = \{0\},$$

and

$$\sigma(B) = \sigma_p(B) = \{n\}_{n\in\mathbb{N}} = \mathbb{N},$$

each eigenvalue being of *geometric multiplicity* 1.

4. All the subsequent examples are in the complex function space $(C[a,b], \|\cdot\|_\infty)$ $(-\infty < a < b < \infty)$.
   (a) For the bounded linear operator A of multiplication by a function $m \in C[a,b]$,

$$C[a,b] \ni x \to [Ax](t) := m(t)x(t) \in C[a,b]$$

(see Examples 4.2),

$$\sigma(A) = \{\lambda \in \mathbb{C} \mid \exists t \in [a,b] : \lambda = m(t)\} = m([a,b]),$$

i. e., $\sigma(A)$ is the *range* of the multiplier function m, with

$$\lambda \in \begin{cases} \sigma_p(A) & \text{if } m(t) - \lambda = 0 \text{ on a subinterval of } [a,b], \\ \sigma_r(A) & \text{otherwise}, \end{cases}$$

which implies that $\sigma_c(A) = \emptyset$.

Exercise 5.5. Verify.

E. g., in $C[0,1]$, for the operator A and B of multiplication by the functions

$$m_1(t) = t, \ t \in [0,1], \ \text{and} \ m_2(t) = \begin{cases} 0, & 0 \le t \le 1/2, \\ (t-1/2)^2, & 1/2 \le t \le 1, \end{cases}$$

respectively,

$$\sigma(A) = \sigma_r(A) = [0,1] \ \text{and}$$
$$\sigma(B) = [0,1/4] \ \text{with} \ \sigma_p(B) = \{0\} \ \text{and} \ \sigma_r(B) = (0,1/4].$$

(b) For the bounded linear *integration operator*

$$C[a,b] \ni x \to [Ax](t) := \int_a^t x(s)\, ds \in C[a,b]$$

(see Examples 4.2),

$$\sigma(A) = \sigma_r(A) = \{0\},$$

which implies that $\sigma_p(A) = \sigma_c(A) = \emptyset$.
Indeed, A has the *inverse*

$$[A^{-1}y](t) := y'(t),$$

but

$$R(A) = \{y \in C^1[a,b] \mid y(a) = 0\},$$

is not a dense subspace in $(C[a,b], \|\cdot\|_\infty)$.
Hence, $0 \in \sigma_r(A)$.
As can be shown *inductively*, for any $n \in \mathbb{N}$ and arbitrary $x \in C[a,b]$,

$$\left|[A^n x](t)\right| \le \frac{(t-a)^n}{n!} \|x\|_\infty, \ t \in [a,b].$$

Exercise 5.6. Show.

This implies that, for any $n \in \mathbb{N}$ and an arbitrary $x \in C[a,b]$,

$$\|A^n x\|_\infty \le \frac{(b-a)^n}{n!} \|x\|_\infty.$$

Whence, we conclude that

$$\|A^n\| \le \frac{(b-a)^n}{n!}, \ n \in \mathbb{N},$$

where $\| \cdot \|$ is the operator norm of the space $L(C[a, b])$, and therefore, by absolute scalability of norm, for any $\lambda \in \mathbb{C} \setminus \{0\}$,

$$\left\| \left[\frac{A}{\lambda} \right]^n \right\| = \left\| \frac{A^n}{\lambda^n} \right\| = \frac{\|A^n\|}{|\lambda|^n} \leq \frac{[(b-a)|\lambda|^{-1}]^n}{n!}, \quad n \in \mathbb{N},$$

which, in view of the fact that, by the *Space of Bounded Linear Operators Theorem* (Theorem 4.5), $(L(C[a, b]), \| \cdot \|)$ is a Banach space, by the *Comparison Test* and the *Series Characterization of Banach Spaces* (Theorem 3.8), implies that the operator series

$$\sum_{n=0}^{\infty} \left[\frac{A}{\lambda} \right]^n,$$

where $A^0 := I$ (I is the *identity operator*), converges in the operator space $(L(C[a, b]), \| \cdot \|)$, i. e., *uniformly* (see Section 4.2.3.3).

Then, for any $\lambda \in \mathbb{C} \setminus \{0\}$, by the *Invertibility of Bounded Linear Operators Proposition* (Proposition 5.10, Section 5.8, Problem 12),

$$(A - \lambda I)^{-1} = \left[-\lambda \left(I - \frac{A}{\lambda} \right) \right]^{-1} = -\frac{1}{\lambda} \left(I - \frac{A}{\lambda} \right)^{-1} = -\frac{1}{\lambda} \sum_{n=0}^{\infty} \left[\frac{A}{\lambda} \right]^n$$

$$= -\sum_{n=0}^{\infty} \frac{A^n}{\lambda^{n+1}}.$$

Exercise 5.7. Verify by showing that, for any $\lambda \in \mathbb{C} \setminus \{0\}$,

$$\left(I - \frac{A}{\lambda} \right) \sum_{n=0}^{\infty} \left[\frac{A}{\lambda} \right]^n = \left[\sum_{n=0}^{\infty} \left[\frac{A}{\lambda} \right]^n \right] \left(I - \frac{A}{\lambda} \right) = I,$$

the *Left/Right Inverse Proposition* (Proposition 5.9, Section 5.8, Problem 11).

Remark 5.5. Similarly, one can show that, all the above remains in place in a more general case of the *Volterra*[1] *integral operator*

$$C[a, b] \ni x \to [Ax](t) := \int_a^t K(t, s)x(s)\, ds \in C[a, b],$$

with some $K \in C([a, b] \times [a, b])$ (see, e. g., [26]).

---

1 Vito Volterra (1860–1940).

(c) For the closed unbounded linear *differentiation operator*

$$Ax := \frac{dx}{dt}, \ x \in D(A) := C^1[a, b]$$

(see Examples 4.7),

$$\sigma(A) = \sigma_p(A) = \mathbb{C},$$

since, for all $\lambda \in \mathbb{C}$, the equation

$$(A - \lambda I)x = 0,$$

equivalent to the exponential differential equation

$$x'(t) = \lambda x(t),$$

whose *general solution* is

$$x(t) = ce^{\lambda t}, \ t \in [a, b] \ (c \in \mathbb{C})$$

(see, e. g., [18]).

Exercise 5.8. Verify.

This implies that, each $\lambda \in \mathbb{C}$ is an *eigenvalue* of A with the associated one-dimensional *eigenspace*

$$\{ce^{\lambda t} \mid t \in [a, b] \ (c \in \mathbb{C})\} = \mathrm{span}(\{e^{\lambda t} \mid t \in [a, b]\}).$$

Hence, $\sigma_c(A) = \sigma_r(A) = \emptyset$, and $\rho(A) = \emptyset$.

(d) For the closed unbounded linear *differentiation operator*

$$Ax := \frac{dx}{dt}, \ x \in D(A) := \{x \in C^1[a, b] \mid x(a) = 0\},$$

$\sigma(A) = \emptyset$, which implies that $\rho(A) = \mathbb{C}$.

Indeed, for any $\lambda \in \mathbb{C}$ and $y \in C[a, b]$, the equation

$$(A - \lambda I)x = y$$

is equivalent to the *initial-value problem*

$$x'(t) - \lambda x(t) = y(t), \ x(a) = 0,$$

which, as can be shown via the *method of integrating factor* (see, e. g., [18]), has the *unique solution*

$$x(t) = e^{\lambda t} \int_a^t e^{-\lambda s} y(s) \, ds = \int_a^t e^{\lambda(t-s)} y(s) \, ds.$$

Exercise 5.9. Verify that the operator A is *closed* and the latter.

Hence, for any $\lambda \in \mathbb{C}$ and an arbitrary $y \in C[a,b]$,

$$[(A - \lambda I)^{-1}y](t) = \int_a^t e^{\lambda(t-s)}y(s)\, ds, \ t \in [a,b],$$

which is a bounded linear *Volterra integral operator* (see Remark 5.5). In particular,

$$[A^{-1}y](t) = \int_a^t y(s)\, ds, \ y \in D(A^{-1}) = C[a,b], t \in [a,b],$$

i. e., the inverse to the differentiation operator A of the present example is the integration operator considered in 4(b). Conversely (see Section 4.7, Problem 3), the inverse to the integration operator considered in 4(b), for which 0 is the point of its *residual spectrum*, is the *unbounded* linear differentiation operator A of this example (see Remarks 5.4).

5.3 Spectral Radius

Definition 5.5 (Spectral Radius). Let $(A, D(A))$ be a *closed linear operator* in a complex Banach space $(X, \|\cdot\|)$ with a *nonempty* spectrum $\sigma(A)$. The *spectral radius* of A is

$$r(A) := \sup_{\lambda \in \sigma(A)} |\lambda| \in [0, \infty].$$

Remark 5.6. If $\sigma(A) = \emptyset$, $r(A)$ is *undefined*.

Examples 5.3.
1. In a complex Banach space $(X, \|\cdot\|)$,
   (a) for the bounded linear operator A of multiplication by a number $\lambda_0 \in \mathbb{C}$ with $\sigma(A) = \{\lambda_0\}$ (see Examples 5.2),

$$r(A) = \max\{|\lambda_0|\} = |\lambda_0|;$$

   (b) for a *projection operator* $P : X \to X$,

$$r(P) = \begin{cases} 1 & \text{if } P \neq 0, \\ 0 & \text{if } P = 0 \end{cases}$$

(see Examples 5.2).

2. In the complex space $l_p^{(n)}$ ($n \in \mathbb{N}$, $1 \le p \le \infty$), for the a bounded linear operator A of multiplication by an $n \times n$ matrix $[a_{ij}]$ with complex entries (see Examples 4.2),

$$\sigma(A) = \sigma_p(A) = \{\lambda_1, \ldots, \lambda_m\} \ (1 \le m \le n),$$

where the latter is the *set of all eigenvalues* of the multiplier matrix $[a_{ij}]$ (see Examples 5.2), and hence,

$$r(A) = \max_{1 \le k \le m} |\lambda_k|.$$

In particular, in the complex space $l_p^{(2)}$ ($1 \le p \le \infty$), for the operators A, B, and C of multiplication by the matrices

$$\begin{bmatrix} 1 & 0 \\ 0 & 2 \end{bmatrix}, \begin{bmatrix} 0 & 1 \\ 0 & 0 \end{bmatrix}, \text{ and } \begin{bmatrix} 0 & 1 \\ -1 & 0 \end{bmatrix},$$

respectively,

$$r(A) = \max(1, 2) = 2, \ r(B) = 0, \text{ and } r(C) = \max(|-i|, |i|) = 1$$

(see Examples 5.2).

3. In the complex sequence space l_p ($1 \le p < \infty$), for the closed linear operator A of multiplication by a complex-termed sequence $a := (a_n)_{n \in \mathbb{N}}$:

$$(x_n)_{n \in \mathbb{N}} \mapsto Ax := (a_n x_n)_{n \in \mathbb{N}}$$

with the maximal domain

$$D(A) := \{(x_n)_{n \in \mathbb{N}} \in l_p \mid (a_n x_n)_{n \in \mathbb{N}} \in l_p\},$$
$$\sigma(A) = \overline{\{a_n\}}_{n \in \mathbb{N}} \text{ with } \sigma_p(A) = \{a_n\}_{n \in \mathbb{N}} \text{ and } \sigma_c(A) = \sigma(A) \setminus \sigma_p(A),$$

where $\{a_n\}_{n \in \mathbb{N}}$ is the set of values of $(a_n)_{n \in \mathbb{N}}$ (see Examples 5.2), and hence,

$$r(A) = \sup_{n \in \mathbb{N}} |a_n|.$$

In particular, for the operators A and B of multiplication by the sequences

$$(a_n)_{n \in \mathbb{N}} := (1/n)_{n \in \mathbb{N}}, \text{ and } (b_n)_{n \in \mathbb{N}} := (n)_{n \in \mathbb{N}},$$

respectively,

$$\sigma(A) = \{0\} \cup \{1/n\}_{n \in \mathbb{N}}, \text{ and } \sigma(B) = \sigma_p(B) = \{n\}_{n \in \mathbb{N}} = \mathbb{N}$$

(see Examples 5.2), and hence,

$$r(A) = 1, \text{ and } r(B) = \infty.$$

4. All the subsequent examples are in the complex function space $(C[a, b], \| \cdot \|_\infty)$ $(-\infty < a < b < \infty)$.

   (a) For the bounded linear operator A of multiplication by a function $m \in C[a, b]$,

   $$\sigma(A) = m([a, b]),$$

   (see Examples 5.2), and hence,

   $$r(A) = \max_{a \le t \le b} |m(t)|.$$

   (b) For the bounded linear *integration operator*

   $$C[a, b] \ni x \to [Ax](t) := \int_a^t x(s)\, ds \in C^1[a, b],$$

   $\sigma(A) = \{0\}$ (see Examples 5.2), and hence, $r(A) = 0$.

   (c) For the closed unbounded linear *differentiation operator*

   $$C^1[a, b] =: D(A) \ni x \to Ax := \frac{dx}{dt} \in C[a, b],$$

   $\sigma(A) = \mathbb{C}$ (see Examples 5.2), and hence, $r(A) = \sup_{\lambda \in \mathbb{C}} |\lambda| = \infty$.

   (d) For the closed unbounded linear *differentiation operator*

   $$Ax := \frac{dx}{dt}$$

   with the domain

   $$D(A) := \{x \in C^1[a, b] \mid x(a) = 0\},$$

   $\sigma(A) = \emptyset$ (see Examples 5.2), and hence, $r(A)$ is *undefined*.

5.4 Resolvent of a Closed Linear Operator

5.4.1 Definition and Examples

Definition 5.6 (Resolvent Function of a Closed Linear Operator). Let $(A, D(A))$ be a *closed linear operator* in a complex Banach space $(X, \| \cdot \|)$ with a *nonempty* resolvent set $\rho(A)$.

The operator-valued function

$$\rho(A) \ni \lambda \mapsto R(\lambda, A) := (A - \lambda I)^{-1} \in L(X)$$

is called the *resolvent function* (or *resolvent*) of A.

Remark 5.7. If $\rho(A) = \emptyset$, the resolvent function of A is *undefined*.

Examples 5.4.
1. Let $(X, \|\cdot\|)$ be a complex Banach space.
   (a) For the bounded linear operator A of multiplication by a number $\lambda_0 \in \mathbb{C}$,

$$\rho(A) = \{\lambda_0\}^c$$

(see Examples 5.2) and, for each $\lambda \in \rho(A)$,

$$R(\lambda, A)x = (A - \lambda I)^{-1}x = \frac{1}{\lambda_0 - \lambda}x, \ x \in X,$$

is the operator of multiplication by the number $1/(\lambda_0 - \lambda)$.

Example 5.5. Verify.

   (b) For a *projection operator* $P : X \to X$, except when $P = 0$ or $P = I$,

$$\rho(A) = \{0, 1\}^c$$

and, for each $\lambda \in \rho(A)$,

$$R(\lambda, P) = (P - \lambda I)^{-1} = -\frac{1}{\lambda}(I - P) + \frac{1}{1 - \lambda}P$$

(see Examples 5.2).
2. On the complex space $l_p^{(n)}$ ($n \in \mathbb{N}$, $1 \le p \le \infty$), for the a bounded linear operator A of multiplication by an $n \times n$ matrix $[a_{ij}]$ with complex entries,

$$\rho(A) = \{\lambda_1, \ldots, \lambda_m\}^c,$$

where the latter is the *set of all eigenvalues* of $[a_{ij}]$, and, for each $\lambda \in \rho(A)$,

$$R(\lambda, A)x = (A - \lambda I)^{-1}x = [a_{ij} - \lambda\delta_{ij}]^{-1}x, \ x \in l_p^{(n)},$$

where δ_{ij} is the *Kronecker delta*, is the operator of multiplication by the inverse matrix $[a_{ij} - \lambda\delta_{ij}]^{-1}$ (see Examples 5.2).

Exercise 5.10. Verify.

E. g., on the complex space $l_p^{(2)}$ ($1 \le p \le \infty$), for the operators A, B, and C of multiplication by the matrices

$$\begin{bmatrix} 1 & 0 \\ 0 & 2 \end{bmatrix}, \begin{bmatrix} 0 & 1 \\ 0 & 0 \end{bmatrix}, \text{ and } \begin{bmatrix} 0 & 1 \\ -1 & 0 \end{bmatrix},$$

respectively,

$$\rho(A) = \{1, 2\}^c, \ \rho(B) = \{0\}^c, \text{ and } \sigma(C) = \{-i, i\}^c$$

(see Examples 5.2) and

$$R(A,\lambda)x = \begin{bmatrix} 1-\lambda & 0 \\ 0 & 2-\lambda \end{bmatrix}^{-1} x = \begin{bmatrix} 1/(1-\lambda) & 0 \\ 0 & 1/(2-\lambda) \end{bmatrix} x, \; \lambda \in \rho(A), x \in l_p^{(2)},$$

$$R(B,\lambda)x = \begin{bmatrix} -\lambda & 1 \\ 0 & -\lambda \end{bmatrix}^{-1} x = \frac{1}{\lambda^2} \begin{bmatrix} -\lambda & -1 \\ 0 & -\lambda \end{bmatrix} x, \; \lambda \in \rho(B), x \in l_p^{(2)},$$

$$R(C,\lambda)x = \begin{bmatrix} -\lambda & 1 \\ -1 & -\lambda \end{bmatrix}^{-1} x = \frac{1}{\lambda^2+1} \begin{bmatrix} -\lambda & -1 \\ 1 & -\lambda \end{bmatrix} x, \; \lambda \in \rho(C), x \in l_p^{(2)}.$$

Exercise 5.11. Verify.

3. In the complex space l_p $(1 \le p < \infty)$, for the closed linear operator A of multiplication by a complex-termed sequence $a = (a_n)_{n\in\mathbb{N}}$,

$$(x_n)_{n\in\mathbb{N}} \mapsto Ax := (a_n x_n)_{n\in\mathbb{N}}$$

with the domain

$$D(A) := \{(x_n)_{n\in\mathbb{N}} \in l_p \mid (a_n x_n)_{n\in\mathbb{N}} \in l_p\}$$

(see Examples 4.7),

$$\rho(A) = \left(\overline{\{a_n\}_{n\in\mathbb{N}}}\right)^c$$

(see Examples 5.2) and, for each $\lambda \in \rho(A)$,

$$R(\lambda, A)x = (A - \lambda I)^{-1}x = (x_n/(a_n - \lambda))_{n\in\mathbb{N}}, \; x \in l_p,$$

is the operator of multiplication by the *bounded* sequence $(1/(a_n - \lambda))_{n\in\mathbb{N}}$. E. g., for the operators A and B of multiplication by the sequences

$$(a_n)_{n\in\mathbb{N}} := (1/n)_{n\in\mathbb{N}} \quad \text{and} \quad (b_n)_{n\in\mathbb{N}} := (n)_{n\in\mathbb{N}},$$

respectively,

$$\rho(A) = \left(\{0\} \cup \{1/n\}_{n\in\mathbb{N}}\right)^c, \quad \text{and} \quad \rho(B) = \mathbb{N}^c$$

(see Examples 5.2), and

$$R(\lambda, A)x = (x_n/(1/n - \lambda))_{n\in\mathbb{N}}, \; \lambda \in \rho(A), x \in l_p,$$
$$R(\lambda, B)x = (x_n/(n - \lambda))_{n\in\mathbb{N}}, \; \lambda \in \rho(B), x \in l_p.$$

4. All the subsequent examples are in the complex space $(C[a,b], \|\cdot\|_\infty)$ $(-\infty < a < b < \infty)$.

(a) For the bounded linear operator A of multiplication by a function $m \in C[a, b]$,

$$\rho(A) = (m([a, b]))^c$$

(see Examples 5.2), and

$$[R(\lambda, A)x](t) = [(A - \lambda I)^{-1}x](t) = \frac{x(t)}{m(t) - \lambda}, \quad \lambda \in \rho(A), x \in C[a, b],$$

is the operator of multiplication by the function $1/(m(\cdot) - \lambda) \in C[a, b]$.

(b) For the bounded linear *integration operator*

$$C[a, b] \ni x \rightarrow [Ax](t) := \int_a^t x(s)\, ds \in C[a, b],$$

$\rho(A) = \{0\}^c$, and, for any $\lambda \neq 0$,

$$R(\lambda, A) = (A - \lambda I)^{-1} = -\sum_{n=0}^{\infty} \frac{A^n}{\lambda^{n+1}},$$

the operator series converging *uniformly* (see Examples 5.2).

Considering the operator's unsophisticated structure, we can also explicitly find that, for any $\lambda \in \rho(A) = \{0\}^c$ and an arbitrary $y \in C[a, b]$,

$$[R(\lambda, A)y](t) = -\frac{1}{\lambda}\left[\frac{1}{\lambda}\int_a^t e^{\frac{t-s}{\lambda}} y(s)\, ds + y(t)\right], \quad t \in [a, b].$$

Exercise 5.12. Verify by solving the equation

$$(A - \lambda I)x = y,$$

i. e., the integral equation

$$\int_a^t x(s)\, ds - \lambda x(t) = y(t)$$

with arbitrary $\lambda \neq 0$ and $y \in C[a, b]$.

Hint. Via the substitution

$$z(t) := \int_a^t x(s)\, ds, \quad t \in [a, b],$$

the integral equation is transformed into the *equivalent* initial value problem

$$z'(t) - \frac{1}{\lambda}z(t) = -\frac{1}{\lambda}y(t), \quad z(a) = 0,$$

which can be solved via the *method of integrating factor* (see Examples 5.2, 4(d)).

(c) For the closed unbounded linear *differentiation operator*

$$C^1[a,b] =: D(A) \ni x \to Ax := \frac{dx}{dt} \in C[a,b],$$

$\rho(A) = \emptyset$ (see Examples 5.2), and hence, the resolvent function is undefined.

(d) For the closed unbounded linear *differentiation operator*

$$Ax := \frac{dx}{dt}$$

with the domain

$$D(A) := \{x \in C^1[a,b] \mid x(a) = 0\},$$

$\rho(A) = \mathbb{C}$, and hence, for any $\lambda \in \mathbb{C}$ and an arbitrary $y \in C[a,b]$,

$$[R(\lambda,A)y](t) = \int_a^t e^{\lambda(t-s)} y(s) \, ds, \ t \in [a,b],$$

(see Examples 5.2).

5.4.2 Resolvent Identity

Theorem 5.1 (Resolvent Identity). *For a closed linear operator $(A, D(A))$ in a complex Banach space $(X, \|\cdot\|)$ with $\rho(A) \neq \emptyset$, the following identity holds:*

$$\forall \lambda, \mu \in \rho(A) : \ R(\lambda,A) - R(\mu,A) = (\lambda - \mu)R(\lambda,A)R(\mu,A).$$

Proof. For arbitrary $\lambda, \mu \in \rho(A)$,

$$(\lambda - \mu)R(\lambda,A)R(\mu,A) = R(\lambda,A)(\lambda - \mu)R(\mu,A)$$
$$= R(\lambda,A)[(A - \mu I) - (A - \lambda I)]R(\mu,A)$$
$$= R(\lambda,A)(A - \mu I)R(\mu,A) - R(\lambda,A)(A - \lambda I)R(\mu,A) = R(\lambda,A) - R(\mu,A). \qquad \square$$

Remark 5.8. The *resolvent identity* is also called *Hilbert's identity* after David Hilbert,[2] who also coined the very term *"resolvent"*.

---

2 David Hilbert (1862–1943).

5.4.3 Analytic Vector Functions

To proceed, we need to make some observations on *vector functions* (also called *vector-valued functions*), i. e., functions with values in vector spaces, when such functions of one complex variable take values in a complex Banach space and are *analytic*.

Definition 5.7 (Analytic Vector Function). Let D be an open set in the complex plane $\mathbb{C}$ and $(X, \|\cdot\|)$ be a complex Banach space. A vector function $f : D \to X$ is called *analytic on D* if, for any $\lambda_0 \in D$, the *derivative of f at λ_0*

$$f'(\lambda_0) := \lim_{\lambda \to \lambda_0} \frac{f(\lambda) - f(\lambda_0)}{\lambda - \lambda_0} \ \text{in} \ (X, \|\cdot\|)$$

exists, i. e.,

$$\forall \varepsilon > 0 \ \exists \delta > 0 \ \forall \lambda \in D \setminus \{\lambda_0\} \ \text{with} \ |\lambda - \lambda_0| < \delta : \ \left\| \frac{f(\lambda) - f(\lambda_0)}{\lambda - \lambda_0} - f'(\lambda_0) \right\| < \varepsilon.$$

An analytic vector function $f : D \to X$, provided $D = \mathbb{C}$, is called *entire*.

A vector function f with values in $(X, \|\cdot\|)$ is called *analytic at a point $\lambda_0 \in \mathbb{C}$* if it is analytic on an open disk

$$\{\lambda \in \mathbb{C} \mid |\lambda - \lambda_0| < R_0\}$$

centered at λ_0 with some $0 < R_0 < \infty$. A point $\lambda_0 \in \mathbb{C}$, at which a vector function f with values in $(X, \|\cdot\|)$ is not analytic, is called a *singular point of f*.

Remarks 5.9.
- One can also define the above limit sequentially as follows:

$$\forall (\lambda_n)_{n \in \mathbb{N}} \subset D \setminus \{\lambda_0\} : \ \frac{f(\lambda_n) - f(\lambda_0)}{\lambda_n - \lambda_0} \to f'(\lambda_0), \ n \to \infty, \ \text{in} \ (X, \|\cdot\|).$$

- A vector function f with values in a complex Banach space $(X, \|\cdot\|)$ analytic at a point $\lambda_0 \in \mathbb{C}$ is *infinite differentiable at λ_0*, i. e., has derivatives $f^{(n)}(\lambda_0)$ of all orders $n \in \mathbb{Z}_+$ ($f^{(0)}(\lambda_0) := f(\lambda_0)$) (see, e. g., [65]).

Examples 5.6. Let D be an open set in the complex plane $\mathbb{C}$ and $(X, \|\cdot\|)$ be a complex Banach space.
1. For any $x \in X$, the constant vector function

$$f(\lambda) := x, \ \lambda \in D,$$

is analytic on D with $f'(\lambda) = 0, \lambda \in D$.

2. For any $x_0, \ldots, x_n \in X$ ($n \in \mathbb{N}$), the vector polynomial

$$f(\lambda) := \sum_{k=0}^{n} \lambda^k x_k, \ \lambda \in D,$$

is analytic on D with $f'(\lambda) = \sum_{k=1}^{n} k\lambda^{k-1} x_k$, $\lambda \in D$.

3. For any $x \in X$, and an arbitrary analytic complex-valued function $f : D \to \mathbb{C}$, the vector function

$$F(\lambda) := f(\lambda)x, \ \lambda \in D,$$

is analytic on D with $F'(\lambda) = f'(\lambda)x$, $\lambda \in D$.

4. For any $x \in X \setminus \{0\}$, the vector function

$$f(\lambda) := \begin{cases} \frac{x}{\lambda}, & \lambda \in \mathbb{C} \setminus \{0\}, \\ 0, & \lambda = 0, \end{cases}$$

is analytic on $\mathbb{C} \setminus \{0\}$ with $f'(\lambda) = -\frac{x}{\lambda^2}$, $\lambda \in \mathbb{C} \setminus \{0\}$, 0 being the only *singular point* of f.

Exercise 5.13. Verify.

The following generalized versions of the well-known classical results (see, e. g., [32, 61]) hold for analytic vector functions.

Theorem 5.2 (Taylor Series Expansion). *Let a vector function f with values in a complex Banach space $(X, \|\cdot\|)$ be analytic on an open disk*[3]

$$D := \{\lambda \in \mathbb{C} \mid |\lambda - \lambda_0| < R_0\}$$

centered at a point $\lambda_0 \in \mathbb{C}$ with a radius $0 < R_0 \le \infty$.
    Then

$$f(\lambda) = \sum_{n=0}^{\infty} (\lambda - \lambda_0)^n x_n, \ \lambda \in D, \quad \text{with} \quad x_n = \frac{f^{(n)}(\lambda_0)}{n!} \in X, \ n \in \mathbb{Z}_+,$$

the power series, called the Taylor series *of f centered at λ_0, converging on the open disk*

$$\{\lambda \in \mathbb{C} \mid |\lambda - \lambda_0| < R\},$$

called the disk of convergence, *where*

$$R_0 \le R = \frac{1}{\overline{\lim}_{n \to \infty} \|x_n\|^{1/n}} \le \infty,$$

with $R := \infty$ when $\overline{\lim}_{n \to \infty} \|x_n\|^{1/n} = 0$, called the radius of convergence.

---

3 Brook Taylor (1685–1731).

The Taylor series may converge at some or all points of its circle of convergence

$$\{\lambda \in \mathbb{C} \mid |\lambda - \lambda_0| = R\};$$

its sum is analytic on the disk of convergence and, provided the radius of convergence is finite (i. e., $0 < R < \infty$), has at least one singular point on the circle of convergence.

See, e. g., [65].

Theorem 5.3 (Laurent Series Expansion). *Let a vector function f with values in a complex Banach space $(X, \|\cdot\|)$ be analytic on an open annulus*[4]

$$D := \{\lambda \in \mathbb{C} \mid r_0 < |\lambda - \lambda_0| < R_0\}$$

centered at a singular point $\lambda_0 \in \mathbb{C}$ with inner and outer radii r_0 and R_0, respectively, $0 \le r_0 < R_0 \le \infty$.
    Then

$$f(\lambda) = \sum_{n=-\infty}^{\infty} (\lambda - \lambda_0)^n x_n, \ \lambda \in D,$$

with some $x_n \in X$, $n \in \mathbb{Z}$, and the power series, called the Laurent series *of f centered at λ_0, converging on the open annulus*

$$\{\lambda \in \mathbb{C} \mid r < |\lambda - \lambda_0| < R\},$$

called the annulus of convergence, *where*

$$0 \le r = \varlimsup_{n \to \infty} \|x_{-n}\|^{1/n} \le r_0 \quad and \quad R_0 \le R = \frac{1}{\varlimsup_{n \to \infty} \|x_n\|^{1/n}} \le \infty,$$

with $R := \infty$ when $\varlimsup_{n \to \infty} \|x_n\|^{1/n} = 0$, called the inner *and* outer radius of convergence, *respectively.*
    The Laurent series may converge at some or all points of its inner circle of convergence

$$\{\lambda \in \mathbb{C} \mid |\lambda - \lambda_0| = r\}$$

and outer circle of convergence

$$\{\lambda \in \mathbb{C} \mid |\lambda - \lambda_0| = R\};$$

its sum is analytic on the annulus of convergence, and has at least one singular point on the inner circle of convergence and, provided the outer radius of convergence is finite (i. e., $0 < R < \infty$), at least one singular point on the outer circle of convergence.

---

4 Pierre Alphonse Laurent (1813–1854).

See, e. g., [65].

Theorem 5.4 (Liouville's Theorem for Vector Functions). *If an entire vector function f with values in a complex Banach space $(X, \| \cdot \|)$ is bounded, i. e.,*

$$\exists C > 0 \, \forall \lambda \in \mathbb{C} : \left| f(\lambda) \right| \leq C,$$

then f is constant.

See, e. g., [4].

5.4.4 Neumann Expansion Theorem

The following theorem is a fundamental statement for our subsequent discourse:

Theorem 5.5 (Neumann Expansion Theorem). *Let $(X, \| \cdot \|)$ be a (real or complex) Banach space. Then, for each $A \in L(X)$ with $\|A\| < 1$, where the same notation $\| \cdot \|$ is used to designate the operator norm,*[5]

$$\exists (I - A)^{-1} = \sum_{n=0}^{\infty} A^n \in L(X),$$

where $A^0 := I$, with the operator series converging uniformly.
  Furthermore,

$$\left\| (I - A)^{-1} \right\| \leq \frac{1}{1 - \|A\|}.$$

Proof. First, observe that, since $(X, \| \cdot \|)$ is Banach space, by the *Space of Bounded Linear Operators Theorem* (Theorem 4.5), $(L(X), \| \cdot \|)$, where the same notation $\| \cdot \|$ is used to designate the operator norm, is a Banach space.
  By *submultiplicativity* of operator norm (see Remark 4.9),

$$\left\| A^n \right\| \leq \|A\|^n, \; n \in \mathbb{Z}_+.$$

This, in view of $\|A\| < 1$, by the *Comparison Test*, implies that the operator series

$$\sum_{n=0}^{\infty} A^n$$

converges absolutely, and hence, in view of the fact that $(L(X), \| \cdot \|)$ is a Banach space, by the *Series Characterization of Banach Spaces* (Theorem 3.8), implies that it converges in the operator space $(L(X), \| \cdot \|)$, i. e., *uniformly* (see Section 4.2.3.3).

---

5 Carl Gottfried Neumann (1832–1925).

Therefore, by the continuity of operator multiplication, following directly from *submultiplicativity* of operator norm (see Remark 4.9) and the *Left/Right Inverse Proposition* (Proposition 5.9, Section 5.8, Problem 11), we have

$$(I - A) \sum_{n=0}^{\infty} A^n = \left[\sum_{n=0}^{\infty} A^n \right] (I - A) = \sum_{n=0}^{\infty} A^k - \sum_{n=0}^{\infty} A^{n+1} = \sum_{n=0}^{\infty} A^n - \sum_{n=1}^{\infty} A^n = I,$$

which implies that there exists $(I - A)^{-1} \in L(X)$, and

$$(I - A)^{-1} = \sum_{n=0}^{\infty} A^n.$$

By *continuity*, *subadditivity*, and *submultiplicativity* of operator norm (see Remark 4.9),

$$\left\| (I - A)^{-1} \right\| = \left\| \sum_{n=0}^{\infty} A^n \right\| \leq \sum_{n=0}^{\infty} \|A^n\| \leq \sum_{n=0}^{\infty} \|A\|^n = \frac{1}{1 - \|A\|},$$

which completes the proof. ☐

Remarks 5.10.
-   The prior proposition is a particular case of a more general statement for unital Banach algebras (see, e. g., [3, 4]).
-   For a bounded linear operator $A \in L(X)$, its *Neumann series*

$$\sum_{k=0}^{\infty} A^n$$

converges uniformly when the power series with operator coefficients

$$\sum_{n=0}^{\infty} \lambda^n A^n$$

converges for $\lambda = 1$ in $(L(X), \|\cdot\|)$, which may happen when its *radius of convergence*

$$R = \frac{1}{\lim_{n \to \infty} \|A^n\|^{1/n}} \geq 1 \Leftrightarrow \overline{\lim_{n \to \infty}} \|A^n\|^{1/n} \leq 1$$

(see the *Taylor Series Expansion Theorem* (Theorem 5.2)), i. e., under a less stringent condition than $\|A\| < 1$.

5.4.5 Properties of Resolvent Set and Resolvent Function

Theorem 5.6 (Properties of Resolvent Set and Resolvent Function). *Let $(A, D(A))$ be a closed linear operator in a complex Banach space $(X, \| \cdot \|)$. Then*

(1) the resolvent set $\rho(A)$ of A is an open subset of the complex plane $\mathbb{C}$;
(2) provided $\rho(A) \neq \emptyset$, the resolvent function

$$\rho(A) \ni \lambda \mapsto R(\lambda, A) \in (L(X), \|\cdot\|),$$

where the same notation $\|\cdot\|$ is used to designate the operator norm, is analytic on $\rho(A)$;
(3) for each $\lambda_0 \in \rho(A)$,

$$\left\{ \lambda \in \mathbb{C} \mid |\lambda - \lambda_0| < \frac{1}{\|R(\lambda_0, A)\|} \right\} \subseteq \rho(A)$$

with

$$R(\lambda, A) = \sum_{n=0}^{\infty} (\lambda - \lambda_0)^n R(\lambda_0, A)^{n+1} \tag{5.1}$$

whenever $|\lambda - \lambda_0| < 1/\|R(\lambda_0, A)\|$, the operator series converging uniformly;
(4) furthermore, for each $\lambda_0 \in \rho(A)$, provided $\sigma(A) \neq \emptyset$,

$$\|R(\lambda_0, A)\| \geq \frac{1}{\rho(\lambda_0, \sigma(A))},$$

where

$$\rho(\lambda_0, \sigma(A)) := \inf_{\mu \in \sigma(A)} |\lambda - \mu|$$

is the distance from the point λ_0 to the spectrum $\sigma(A)$, which implies that

$$\|R(\lambda, A)\| \to \infty, \quad \rho(\lambda, \sigma(A)) \to 0.$$

Proof. (1–3) When $\rho(A) = \emptyset$, the statement of part (1) is trivially true.
Suppose that $\rho(A) \neq \emptyset$ and let us fix an arbitrary $\lambda_0 \in \rho(A)$.
Since, for all $\lambda \in \mathbb{C}$,

$$A - \lambda I = A - \lambda_0 I - (\lambda - \lambda_0)I = [I - (\lambda - \lambda_0)R(\lambda_0, A)](A - \lambda_0 I),$$

if further $|\lambda - \lambda_0| < 1/\|R(\lambda_0, A)\|$, in view of

$$\|(\lambda - \lambda_0)R(\lambda_0, A)\| = |\lambda - \lambda_0|\|R(\lambda_0, A)\| < 1,$$

by the *Neumann Expansion Theorem* (Theorem 5.5) and *Invertibility of Linear Operators Proposition* (Proposition 5.5, Section 5.8, Problem 2), we infer that

$$(A - \lambda I)^{-1} = [[I - (\lambda - \lambda_0)R(\lambda_0, A)](A - \lambda_0 I)]^{-1}$$

$$= (A - \lambda_0 I)^{-1}[I - (\lambda - \lambda_0)R(\lambda_0, I)]^{-1} = R(\lambda_0, A)\sum_{n=0}^{\infty}[(\lambda - \lambda_0)R(\lambda_0, A)]^n$$

$$= \sum_{n=0}^{\infty} (\lambda - \lambda_0)^n R(\lambda_0, A)^{n+1},$$

the operator series converging *uniformly*.

Hence, if $\rho(A) \neq \emptyset$, along with each point $\lambda_0 \in \rho(A)$, the resolvent set $\rho(A)$ contains the *open disk*

$$\left\{ \lambda \in \mathbb{C} \,\middle|\, |\lambda - \lambda_0| < \frac{1}{\|R(\lambda_0, A)\|} \right\} \subseteq \rho(A)$$

centered at λ_0 of radius $1/\|R(\lambda_0, A)\|$, on which the Taylor series expansion of $R(\cdot, A)$ centered at λ_0 given by (5.1) holds, the operator series converging uniformly, which, by the *Taylor Series Expansion Theorem* (Theorem 5.2), implies that the resolvent function is *analytic* on $\rho(A)$.

This completes the proof of parts (1–3).

(4) It follows that, for each $\lambda_0 \in \rho(A)$, provided $\sigma(A) \neq \emptyset$,

$$\frac{1}{\|R(\lambda_0, A)\|} \leq \rho(\lambda_0, \sigma(A)),$$

or equivalently,

$$\|R(\lambda_0, A)\| \geq \frac{1}{\rho(\lambda_0, \sigma(A))},$$

which implies that

$$\|R(\lambda, A)\| \to \infty, \ \rho(\lambda, \sigma(A)) \to 0,$$

completing the proof of part (4) and the entire statement. □

Remark 5.11. By the *Taylor Series Expansion Theorem* (Theorem 5.2),

$$\forall n \in \mathbb{Z}_+, \lambda \in \rho(A) : \ \frac{d^n}{d\lambda^n} R(\lambda, A) = n! R(\lambda, A)^{n+1}.$$

5.4.6 Spectrum

Theorem 5.7 (Spectrum of a Closed Linear Operator). *The spectrum $\sigma(A)$ of a closed linear operator $(A, D(A))$ in a complex Banach space is a closed set subset of the complex plane $\mathbb{C}$. Conversely, any closed subset σ of $\mathbb{C}$ is the spectrum of a closed linear operator.*

Proof. The first part of the statement follows directly from the prior theorem, and the fact that $\sigma(A) = \rho(A)^c$.

The example of a *differentiation operator* in the complex space $(C[a, b], \|\cdot\|_\infty)$ demonstrates that the spectrum of a closed operator can be $\emptyset$ (see Examples 5.2). For an arbitrary nonempty closed subset σ of $\mathbb{C}$, considering that σ is a *separable space* as a subspace of $\mathbb{C}$ (see the *Subspace of Separable Metric Space Proposition* (Proposition 2.19, Section 2.18, Problem 19)), one can choose a complex-termed sequence $(a_n)_{n\in\mathbb{N}}$ such that

$$\overline{\{a_n\}}_{n\in\mathbb{N}} = \sigma,$$

where $\{a_n\}_{n\in\mathbb{N}}$ is the set of values of $(a_n)_{n\in\mathbb{N}}$.

Hence, in the complex sequence space l_p $(1 \le p < \infty)$, for the closed linear operator A of multiplication by the sequence $(a_n)_{n\in\mathbb{N}}$:

$$(x_n)_{n\in\mathbb{N}} \mapsto Ax := (a_n x_n)_{n\in\mathbb{N}}$$

with the maximal domain

$$D(A) := \{(x_n)_{n\in\mathbb{N}} \in l_p \mid (a_n x_n)_{n\in\mathbb{N}} \in l_p\},$$
$$\sigma(A) = \overline{\{a_n\}}_{n\in\mathbb{N}} = \sigma$$

(see Examples 5.2), which completes the proof of the second part and the entire statement. $\qquad\square$

5.5 Spectral Theory of Bounded Linear Operators

In this section, we study spectral features inherent to bounded linear operators, and introduce a functional calculus for such operators.

5.5.1 Gelfand's Spectral Radius Theorem

Examples 5.2 suggest that, for a bounded linear operator A on a complex Banach space, both the resolvents set $\rho(A)$ and spectrum $\sigma(A)$ are *nonempty* subsets of $\mathbb{C}$, the spectrum $\sigma(A)$ being *compact* in $\mathbb{C}$. *Gelfand's*[6] *Spectral Radius Theorem* (Theorem 5.8), the central statement in the spectral theory of bounded linear operators, explains, in particular, that all the above inferences are not coincidental.

5.5.1.1 Limit Lemma and Corollary
We need the following lemma for our discourse:

---

6 Israel Gelfand (1913–2009).

Lemma 5.1 (Limit Lemma). *Let* $(x_n)_{n \in \mathbb{N}}$ *be a sequence of nonnegative numbers such that*

$$\forall\, m, n \in \mathbb{N} : x_{m+n} \leq x_m x_n. \tag{5.2}$$

Then

$$\lim_{n \to \infty} x_n^{1/n} \in [0, \infty]$$

exists.

Exercise 5.14.
(a) Give two examples of nonnegative a sequences satisfying condition (5.2).
(b) Do sequences $(n!)_{n \in \mathbb{N}}$ and $(e^{n^2})_{n \in \mathbb{N}}$ satisfy condition (5.2)?

Proof. Observe that condition (5.2) immediately implies that

$$\forall\, k, m \in \mathbb{N} : x_{km} \leq x_m^k \tag{5.3}$$

and, if $x_m = 0$ for some $m \in \mathbb{N}$, then $x_n = 0$ for all $n \geq m$.

Exercise 5.15. Verify.

Since

$$\forall\, m \in \mathbb{N}\ \forall\, n \in \mathbb{N},\ n \geq m\ \exists\, k \in \mathbb{N}\ \exists\, l \in \mathbb{Z}_+,\ 0 \leq l \leq m - 1 : n = km + l, \tag{5.4}$$

by (5.2) and (5.3),

$$x_n = x_{km+l} \leq x_{km} x_l \leq x_m^k x_l.$$

Hence, for arbitrary $m \in \mathbb{N}$ and $n \in \mathbb{N}$, $n \geq m$, and certain $k \in \mathbb{N}$ and $l \in \mathbb{Z}_+$, $0 \leq l \leq m - 1$, satisfying (5.4),

$$x_n^{1/n} = x_{km+l}^{1/n} \leq \left(x_m^k x_l \right)^{1/n} = x_m^{k/n} x_l^{1/n} = x_m^{1/m - l/(mn)} x_l^{1/n}.$$

Fixing an arbitrary $m \in \mathbb{N}$ and letting $n \to \infty$, we arrive at

$$\overline{\lim_{n \to \infty}}\, x_n^{1/n} \leq \overline{\lim_{n \to \infty}}\, x_m^{1/m - l/(mn)} \overline{\lim_{n \to \infty}} \left[\max_{0 \leq l \leq m-1} x_l \right]^{1/n} = x_m^{1/m}$$

(see Section 1.3).

Exercise 5.16. Explain.

Now, letting $m \to \infty$, we obtain the estimate

$$\overline{\lim_{n \to \infty}}\, x_n^{1/n} \leq \underline{\lim_{m \to \infty}}\, x_m^{1/m}$$

(see Section 1.3).

Whence, considering that

$$\varliminf_{n\to\infty} x_n^{1/n} \leq \varlimsup_{n\to\infty} x_n^{1/n}$$

(see Exercise 1.5), we conclude that

$$\varliminf_{n\to\infty} x_n^{1/n} = \varlimsup_{n\to\infty} x_n^{1/n},$$

which, by the *Characterization of Limit Existence* (Proposition 1.3), implies that

$$\exists \lim_{n\to\infty} x_n^{1/n} = \varliminf_{n\to\infty} x_n^{1/n} = \varlimsup_{n\to\infty} x_n^{1/n}. \qquad \square$$

From the prior lemma, by *submultiplicativity* of operator norm (see Remark 4.9), we immediately obtain

Corollary 5.1 (Limit Corollary). *Let $(X, \| \cdot \|)$ be a Banach space and $A \in L(X)$. Then*

$$\lim_{n\to\infty} \|A^n\|^{1/n},$$

exists, where the same notation $\| \cdot \|$ is used to designate the operator norm, and

$$0 \leq \lim_{n\to\infty} \|A^n\|^{1/n} \leq \|A\|.$$

Exercise 5.17. Prove.

5.5.1.2 Gelfand's Spectral Radius Theorem

Theorem 5.8 (Gelfand's Spectral Radius Theorem). *Let $(X, \| \cdot \|)$ be a complex Banach space and $A \in L(X)$. Then*
1. *the resolvent set $\rho(A)$ is a nonempty subset of the complex plane $\mathbb{C}$ with*

$$\left\{ \lambda \in \mathbb{C} \mid |\lambda| > \lim_{n\to\infty} \|A^n\|^{1/n} \right\} \subseteq \rho(A),$$

where the same notation $\| \cdot \|$ is used to designate the operator norm, and

$$R(\lambda, A) = -\sum_{n=0}^{\infty} \frac{A^n}{\lambda^{n+1}}$$

whenever $|\lambda| > \lim_{n\to\infty} \|A^n\|^{1/n}$, the operator series converging uniformly;
2. *the estimate*

$$\|R(\lambda, A)\| \leq \frac{1}{|\lambda| - \|A\|}$$

holds whenever $|\lambda| > \|A\|$, which implies that

$$\|R(\lambda, A)\| \to 0, \ |\lambda| \to \infty; \qquad (5.5)$$

3. $\rho(A) \neq \mathbb{C}$, *the spectrum $\sigma(A)$ is a nonempty compact subset of $\mathbb{C}$, and*

$$r(A) = \max_{\lambda \in \sigma(A)} |\lambda| = \lim_{n \to \infty} \|A^n\|^{1/n} \quad \text{(Spectral Radius Formula)}.$$

Proof. Observe that, $A \in L(X)$ being a *closed linear operator* by the *Characterization of Closedness for Bounded Linear Operators* (Proposition 4.5), the *Properties of Resolvent Set and Resolvent Function* (Theorem 5.6) and the *Spectrum of a Closed Linear Operator Theorem* (Theorem 5.7) apply.

1.   For an arbitrary $\lambda \in \mathbb{C} \setminus \{0\}$,

$$A - \lambda I = -\lambda\left(I - \frac{A}{\lambda}\right),$$

and hence, whenever $|\lambda| > \|A\|$, since, by absolute scalability of operator norm,

$$\left\|\frac{A}{\lambda}\right\| = \frac{\|A\|}{|\lambda|} < 1,$$

by the *Neumann Expansion Theorem* (Theorem 5.5) and in view of the *Invertibility of Bounded Linear Operators Proposition* (Proposition 5.10, Section 5.8, Problem 12),

$$\exists\,(A - \lambda I)^{-1} = \left[-\lambda\left(I - \frac{A}{\lambda}\right)\right]^{-1} = -\frac{1}{\lambda}\left(I - \frac{A}{\lambda}\right)^{-1} = -\frac{1}{\lambda}\sum_{n=0}^{\infty}\left[\frac{A}{\lambda}\right]^n$$

$$= -\sum_{n=0}^{\infty} \frac{A^n}{\lambda^{n+1}} \in L(X),$$

the operator series converging uniformly. Whence, we infer that

$$\{\lambda \in \mathbb{C} \mid |\lambda| > \|A\|\} \subseteq \rho(A),$$

which implies that $\rho(A) \neq \emptyset$, and

$$R(\lambda, A) = -\sum_{n=0}^{\infty} \frac{A^n}{\lambda^{n+1}} = -\sum_{n=0}^{\infty} \lambda^{-(n+1)} A^n \text{ whenever } |\lambda| > \|A\|,$$

the operator series converging uniformly.

The above operator series, being the *Laurent series* of the resolvent function $R(\cdot, A)$ with the *inner radius of convergence*

$$r = \overline{\lim_{n \to \infty}} \|A^n\|^{1/n}$$

and the *outer radius of convergence $R = \infty$* (see the *Laurent Series Expansion Theorem* (Theorem 5.3)), since, in view of the *Limit Corollary* (Corollary 5.1),

$$r = \overline{\lim_{n \to \infty}} \|A^n\|^{1/n} = \lim_{n \to \infty} \|A^n\|^{1/n} \leq \|A\|,$$

converges uniformly, and represents the resolvent function wherever $|\lambda| > r$ (see Examples 5.4, 4 (b)).

Thus,

$$\left\{\lambda \in \mathbb{C} \mid |\lambda| > \lim_{n \to \infty} \|A^n\|^{1/n}\right\} \subseteq \rho(A)$$

and

$$R(\lambda, A) = -\sum_{n=0}^{\infty} \frac{A^n}{\lambda^{n+1}} = -\sum_{n=0}^{\infty} \lambda^{-(n+1)} A^n$$

whenever $|\lambda| > \lim_{n \to \infty} \|A^n\|^{1/n}$ and, possibly, for some or all $\lambda \in \mathbb{C}$ with $|\lambda| = \lim_{n \to \infty} \|A^n\|^{1/n}$, the operator series converging uniformly (see the *Laurent Series Expansion Theorem* (Theorem 5.3)).

2. Whenever $|\lambda| > \|A\|$, i. e.,

$$|\lambda|^{-1} \|A\| < 1,$$

by *continuity*, *subadditivity*, *absolute scalability*, and *submultiplicativity* of operator norm (see Remark 4.9), we have the following estimate:

$$\|R(\lambda, A)\| = \left\| -\sum_{n=0}^{\infty} \lambda^{-(n+1)} A^n \right\| \leq \sum_{n=0}^{\infty} |\lambda|^{-(n+1)} \|A\|^n = |\lambda|^{-1} \sum_{n=0}^{\infty} \left[|\lambda|^{-1} \|A\|\right]^n$$

$$= \frac{|\lambda|^{-1}}{1 - |\lambda|^{-1} \|A\|} = \frac{1}{|\lambda| - \|A\|},$$

which implies (5.5).

3. Let us prove that $\rho(A) \neq \mathbb{C}$ by contradiction. Assume that $\rho(A) = \mathbb{C}$. Then, by the *Properties of Resolvent Set and Resolvent Function* (Theorem 5.6), the resolvent

$$\mathbb{C} \ni \lambda \mapsto R(\lambda, A) \in (L(X), \|\cdot\|)$$

is an *entire vector function*, i. e., is analytic on $\mathbb{C}$. This, in view of (5.5), by *Liouville's Theorem for Vector Functions* (Theorem 5.4), implies that

$$R(\lambda, A) = 0, \ \lambda \in \mathbb{C},$$

which is a *contradiction* implying that $\rho(A) \neq \mathbb{C}$.

Hence, $\sigma(A) = \rho(A)^c \neq \emptyset$. By the *Spectrum of a Closed Linear Operator Theorem* (Theorem 5.7), the spectrum $\sigma(A)$ is a *closed* subset of $\mathbb{C}$. Since, as follows from part 1 and the *Limit Corollary* (Corollary 5.1),

$$\sigma(A) \subseteq \left\{\lambda \in \mathbb{C} \mid |\lambda| > \lim_{n \to \infty} \|A^n\|^{1/n}\right\}^c = \left\{\lambda \in \mathbb{C} \mid |\lambda| \leq \lim_{n \to \infty} \|A^n\|^{1/n}\right\},$$

the spectrum $\sigma(A)$ is also *bounded* in $\mathbb{C}$, by the *Heine–Borel Theorem* (Theorem 2.47), we infer that the spectrum $\sigma(A)$ is a *compact* subset of $\mathbb{C}$.

Considering the prior inclusion and the *compactness* of $\sigma(A)$, by the *Extreme Value Theorem for Modulus* (Theorem 2.10), we have

$$r(A) = \max_{\lambda \in \sigma(A)} |\lambda| \le \lim_{n \to \infty} \|A^n\|^{1/n}.$$

Since on the *inner circle of convergence*

$$\left\{ \lambda \in \mathbb{C} \mid |\lambda| = \lim_{n \to \infty} \|A^n\|^{1/n} \right\}$$

of the resolvent's *Laurent series*, there must be at least one *singular point* of $R(\cdot, A)$, i. e., a point where $R(\cdot, A)$ is *not* analytic (see the *Laurent Series Expansion Theorem*, Theorem 5.3), which, by the *Properties of Resolvent Set and Resolvent Function* (Theorem 5.6), can only be a point of the spectrum $\sigma(A)$, we conclude that

$$r(A) = \lim_{n \to \infty} \|A^n\|^{1/n}.$$

This completes the proof of part 3 and the entire statement. □

Remarks 5.12.
- *Gelfand's Spectral Radius Theorem* is a particular case of a more general eponymous statement for unital Banach algebras [22, 23] (see also [3, 4]).
- An arbitrary nonempty compact set in $\mathbb{C}$ can be the spectrum of a *bounded* linear operator.

  Exercise 5.18. Prove (see the proof of the *Spectrum of a Closed Linear Operator Theorem* (Theorem 5.7)).

- As is seen from the following examples, for a bounded linear operator A on a complex Banach space, the inequality

$$r(A) = \lim_{n \to \infty} \|A^n\|^{1/n} \le \|A\|$$

  may be strict, i. e., the spectral radius $r(A)$ of A need not be equal its norm $\|A\|$.

Examples 5.7.
1. In a complex Banach space $(X, \| \cdot \|)$,
   (a) for the bounded linear operator A of multiplication by a number $\lambda_0 \in \mathbb{C}$ with $\|A\| = \lambda_0$ and $\sigma(A) = \{\lambda_0\}$ (see Examples 4.2 and 5.3),

$$r(A) = \max\{|\lambda_0|\} = |\lambda_0| = \|A\|;$$

   (b) for a *projection operator* $P : X \to X$, in view of

$$P^n = P, \ n \in \mathbb{N},$$

by the *Gelfand's Spectral Radius Theorem* (Theorem 5.8), we have the following:

$$r(P) = \begin{cases} 1 & \text{if } P \neq 0, \\ 0 & \text{if } P = 0 \end{cases} = \lim_{n \to \infty} \|P^n\|^{1/n} \leq \|P\|$$

(see Examples 5.3 and the *Norm of a Projection Operator Proposition* (Proposition 4.9)).

2. In the complex space l_p ($1 \leq p < \infty$), for the bounded linear operators of *right shift*

$$l_p \ni x = (x_1, x_2, \dots) \mapsto Ax := (0, x_1, x_2, \dots) \in l_p$$

and *left shift*

$$l_p \ni x = (x_1, x_2, \dots) \mapsto Bx := (x_2, x_3, x_4, \dots) \in l_p$$

(see Examples 4.2),

$$\|A^n\| = \|B^n\| = 1, \; n \in \mathbb{N}.$$

Exercise 5.19. Verify (see Example 4.4).

Hence, as follows from the *Gelfand's Spectral Radius Theorem* (Theorem 5.8),

$$r(A) = r(B) = 1,$$

and both spectra $\sigma(A)$ and $\sigma(B)$ are subsets of the closed unit ball

$$\overline{B}(0, 1) = \{\lambda \in \mathbb{C} \mid |\lambda| \leq 1\}.$$

In fact, it can be shown that

$$\sigma(A) = \sigma(B) = \overline{B}(0, 1),$$

with

$$\sigma_r(A) = \sigma_p(B) = B(0, 1) \text{ and } \sigma_c(A) = \sigma_c(B) = S(0, 1)$$

(see, e. g., [16], cf. Section 5.8, Problem 9 and Examples 6.18, 3 (b)).

3. On the complex space $(C[a, b], \| \cdot \|_\infty)$ ($-\infty < a < b < \infty$), for the bounded linear integration operator

$$C[a, b] \ni x \to [Ax](t) := \int_a^t x(s) \, ds \in C^1[a, b],$$

$\sigma(A) = \{0\}$ and, for any $\lambda \in \rho(A) = \{0\}^C$, as is consistent with the *Gelfand's Spectral Radius Theorem* (Theorem 5.8),

$$R(\lambda, A) = - \sum_{n=0}^{\infty} \frac{A^n}{\lambda^{n+1}},$$

the operator series converging *uniformly* (see Examples 5.4, 4 (b)).

Also, by the *Gelfand's Spectral Radius Theorem*,

$$\lim_{n \to \infty} \|A^n\|^{1/n} = r(A) = \max\{0\} = 0 < b - a = \|A\|$$

(see Examples 4.2, 8 (b)).

5.5.2 Quasinilpotent Operators

The latter is an example of a bounded linear operator in a complex Banach space with zero spectral radius. Such operators enjoy a special name and "reputation".

Definition 5.8 (Quasinilpotent Operator). A bounded linear operator A in a complex Banach space $(X, \| \cdot \|)$ with $\sigma(A) = \{0\}$, which is equivalent to $r(A) = 0$, is called *quasinilpotent*.

Remarks 5.13.
- Any *nilpotent operator* $A \in L(X)$, i. e., such that

$$\exists N \in \mathbb{N} : A^N = 0;$$

  in particular, the *zero operator* is quasinilpotent.
- Any *Volterra integral operator* (see Remark 5.5), in particular, the *integration operator* in the latter example is quasinilpotent.

Exercise 5.20. Give another example of a nonzero *quasinilpotent* operator.

5.5.3 Operator Polynomials

Here, we study naturally emerging *operator polynomials* and, in particular, prove the *Spectral Mapping Theorem* for them.

Definition 5.9 (Operator Polynomial). For a bounded linear operator A on a Banach space $(X, \| \cdot \|)$ over $\mathbb{F}$ ($\mathbb{F} := \mathbb{R}$ or $\mathbb{F} := \mathbb{C}$) and

$$p(\lambda) = \sum_{k=0}^{n} a_k \lambda^k, \ \lambda \in \mathbb{F},$$

a polynomial of degree $n \in \mathbb{Z}_+$ with coefficients $a_k \in \mathbb{F}$, $k = 0, \ldots, n$,

$$p(A) := \sum_{k=0}^{n} a_k A^k \in L(X)$$

with $A^0 := I$.

Examples 5.8.
1. $\forall \lambda \in \mathbb{F}: A \pm \lambda I = p(A)$ with $p(\mu) := \mu \pm \lambda$, $\mu \in \mathbb{F}$.
2. $\forall \lambda \in \mathbb{F}: \lambda A = p(A)$ with $p(\mu) := \lambda \mu$, $\mu \in \mathbb{F}$.
3. $\forall n \in \mathbb{Z}_+: A^n = p(A)$ with $p(\lambda) := \lambda^n$, $\lambda \in \mathbb{F}$ ($A^0 := I$).

Exercise 5.21. Let A be a bounded linear operator A on a (real or complex) Banach space $(X, \|\cdot\|)$. Show that, for any polynomials p and q,

$$p(A)q(A) = q(A)p(A),$$

i. e., the operators $p(A)$ and $q(A)$ *commute*.

Provided the underlying space is complex, what is the relationship between the spectra $\sigma(A)$ and $\sigma(p(A))$? The following statement, naturally referred to as a *spectral mapping theorem* (see the *Spectral Mapping Theorem for Inverse Operators* (Theorem 5.22, Section 5.8, Problem 8)), gives an intuitively predictable answer.

Theorem 5.9 (Spectral Mapping Theorem for Operator Polynomials). *Let A be a bounded linear operator A on a complex Banach space $(X, \|\cdot\|)$ (i. e., $A \in L(X)$) and*

$$p(\lambda) = \sum_{k=0}^{n} a_k \lambda^k, \ \lambda \in \mathbb{C},$$

be an arbitrary polynomial of degree $n \in \mathbb{Z}_+$ with complex coefficients $a_k \in \mathbb{C}$, $k = 0, \ldots, n$. Then, for the operator polynomial

$$p(A) := \sum_{k=0}^{n} a_k A^k,$$

$$\sigma(p(A)) = p(\sigma(A)) := \{p(\lambda) \mid \lambda \in \sigma(A)\}.$$

Proof. Observe that the statement is trivially true for polynomials of degree $n = 0$ that are merely complex numbers. For any such polynomial

$$p(\lambda) = a_0 \lambda^0$$

with some $a_0 \in \mathbb{C}$,

$$p(A) = a_0 I,$$

which implies that

$$\sigma(p(A)) = \sigma(a_0 I) = \{a_0\} = p(\sigma(A)).$$

Suppose that

$$p(\lambda) = \sum_{k=0}^{n} a_k \lambda^k, \ \lambda \in \mathbb{C},$$

is an arbitrary polynomial of degree $n \in \mathbb{N}$ with $a_n \neq 0$.

First, let us prove the inclusion

$$p(\sigma(A)) \subseteq \sigma(p(A)). \tag{5.6}$$

For an arbitrary $\mu \in \mathbb{C}$, since μ is a root of the polynomial

$$p(\lambda) - p(\mu), \ \lambda \in \mathbb{C},$$

we have

$$p(\lambda) - p(\mu) = (\lambda - \mu)q(\lambda), \ \lambda \in \mathbb{C},$$

where q is a polynomial of degree $n - 1$.

Hence,

$$p(A) - p(\mu)I = (A - \mu I)q(A) = q(A)(A - \mu I).$$

If $p(\mu) \in \rho(p(A))$, then we infer from the latter that

$$\ker(A - \mu I) = \{0\} \quad \text{and} \quad R(A - \mu I) = X,$$

and hence, $\mu \in \rho(A)$.

Exercise 5.22. Explain.

Therefore, *by contrapositive*, if $\mu \in \sigma(A)$, then $p(\mu) \in \sigma(p(A))$, which proves inclusion (5.6).

Example 5.9. Explain.

To prove the inverse inclusion

$$\sigma(p(A)) \subseteq p(\sigma(A)), \tag{5.7}$$

let $\mu \in \mathbb{C}$ be arbitrary and, by the *Fundamental Theorem of Algebra*, suppose that $\mu_1, \ldots, \mu_n \in \mathbb{C}$ are the roots of the polynomial

$$p(\lambda) - \mu,$$

each one counted as many times as its *multiplicity*.

Then

$$p(\lambda) - \mu = a_n(\lambda - \mu_1) \cdots (\lambda - \mu_n),$$

and hence,

$$p(A) - \mu I = a_n(A - \mu_1 I) \cdots (A - \mu_n I).$$

If $\mu_1, \ldots, \mu_n \in \rho(A)$, then

$$\forall k = 1, \ldots, n \exists (A - \mu_k I)^{-1} \in L(X),$$

and hence, by the *Invertibility of Bounded Linear Operators Proposition* (Proposition 5.10, Section 5.8, Problem 12),

$$\exists (p(A) - \mu I)^{-1} = [a_n(A - \mu_1 I) \cdots (A - \mu_n I)]^{-1}$$
$$= \frac{1}{a_n}(A - \mu_n I)^{-1} \cdots (A - \mu_1 I)^{-1} \in L(X),$$

which implies that $\mu \in \rho(p(A))$.

Example 5.10. Verify.

Therefore, *by contrapositive*, if $\mu \in \sigma(p(A))$, then

$$\exists k = 1, \ldots, n : \mu_k \in \sigma(A),$$

and hence,

$$\mu = p(\mu_k) \in p(\sigma(A)).$$

Exercise 5.23. Explain the latter.

This proves inclusion (5.7), completing the proof. □

We immediately obtain the following:

Corollary 5.2 (Spectra of Shifts, Constant Multiples, and Powers). *Let A be a bounded linear operator A on a complex Banach space $(X, \|\cdot\|)$ (i. e., $A \in L(X)$). Then*
1. $\forall \lambda \in \mathbb{C} : \sigma(A \pm \lambda I) = \sigma(A) \pm \lambda := \{\mu \pm \lambda \mid \mu \in \sigma(A)\};$
2. $\forall \lambda \in \mathbb{C} : \sigma(\lambda A) = \lambda \sigma(A) := \{\lambda \mu \mid \mu \in \sigma(A)\}.$
3. $\forall n \in \mathbb{Z}_+ : \sigma(A^n) = \sigma(A)^n := \{\lambda^n \mid \lambda \in \sigma(A)\} (A^0 := I).$
   In particular, if $A \in L(X)$ is quasinilpotent, then, for each $n \in \mathbb{N}$, A^n is also quasinilpotent.

Exercise 5.24.
(a) Apply the prior statement to find the spectrum of the polynomial operator

$$A^2 + A + I,$$

where A is the bounded linear operator on the complex space $l_p^{(2)}$ $(1 \le p \le \infty)$ of multiplication by each of the 2×2 matrices of Examples 5.2, 2.
(b) Apply the prior statement to find the spectrum of the integral operator in the complex space $(C[a, b], \| \cdot \|_\infty)$ $(-\infty < a < b < \infty)$:

$$\int_a^t \left[\int_a^s x(r)\, dr \right] ds + \int_a^t x(s)\, ds - \lambda x(t), \; t \in [a, b],$$

where $\lambda \ne 0$.

Remark 5.14. Provided $0 \in \rho(A)$, i. e., $\exists A^{-1} \in L(X)$, part 3 can be extended to negative integer exponents $n = -1, -2, \ldots$, with

$$A^{-n} := [A^{-1}]^n, \; n \in \mathbb{N}.$$

Exercise 5.25. Verify by using the prior corollary and the *Spectral Mapping Theorem for Inverse Operators* (Theorem 5.22, Section 5.8, Problem 8).

Corollary 5.3 (Solvability of Operator Polynomial Equations). *Let A be a bounded linear operator A on a complex Banach space $(X, \|\cdot\|)$ (i. e., $A \in L(X)$) and p be a polynomial with complex coefficients. Then the equation*

$$p(A)x = y$$

has a unique solution for each $y \in X$ iff

$$\forall \lambda \in \sigma(A) : \; p(\lambda) \ne 0.$$

Exercise 5.26.
(a) Prove.
(b) Apply the prior statement to show that, for arbitrary $y \in \mathbb{C}^2$, the polynomial operator equation

$$A^2 x + Ax + x = y,$$

where A is the bounded linear operator on the complex space $l_p^{(2)}$ $(1 \le p \le \infty)$ of multiplication by any of the 2×2 matrices of Examples 5.2, 2, has a unique solution $x \in \mathbb{C}^2$ without solving the equation.
(c) Apply the prior statement to show that, for any $\lambda \ne 0$ and arbitrary $y \in C[a, b]$ $(-\infty < a < b < \infty)$, the integral equation

$$\int_a^t \left[\int_a^s x(r)\, dr \right] ds + \int_a^t x(s)\, ds - \lambda x(t) = y(t), \; t \in [a, b],$$

has a unique solution $x \in C[a, b]$ without solving the equation.

5.5.4 Operator Exponentials

One can also naturally define *operator exponentials*, which are instrumental for numerous applications.

Definition 5.10 (Operator Exponentials). Let A be a bounded linear operator on a Banach space $(X, \| \cdot \|)$ over $\mathbb{F}$ (i. e., $A \in L(X)$). The *operator exponentials* are defined as follows:

$$e^{tA} := \sum_{k=0}^{\infty} \frac{t^k A^k}{k!} \text{ in } (L(X), \| \cdot \|), \ t \in \mathbb{F},$$

where the notation $\| \cdot \|$ is used to designate the operator norm.

Exercise 5.27. Show that, for any $t \in \mathbb{F}$, the operator $e^{tA} \in L(X)$ is *well defined*, i. e., the above operator series *converges uniformly*.

Proposition 5.1 (Operator Exponentials). *Let A be a bounded linear operator on a Banach space $(X, \| \cdot \|)$ over $\mathbb{F}$ (i. e., $A \in L(X)$). Then, for any $t \in \mathbb{F}$,*

$$\frac{d}{dt} e^{tA} = A e^{tA} \text{ in } (L(X), \| \cdot \|).$$

Exercise 5.28. Prove via the termwise differentiation of the operator series.

Remark 5.15. Hence, for a bounded linear operator on a (real or complex) Banach space and $(X, \| \cdot \|)$ (i. e., $A \in L(X)$), the *abstract Cauchy problem*

$$\begin{cases} y'(t) = Ay(t), \ t \in \mathbb{R}, \\ y(0) = y_0, \end{cases}$$

has the *unique solution*

$$y(t) = e^{tA} y_0 = \left[\sum_{k=0}^{\infty} \frac{t^k A^k}{k!} \right] y_0, \ t \in \mathbb{R},$$

for an arbitrary *initial value* $y_0 \in X$.

As a particular case, when A is the operator of multiplication by an $n \times n$ ($n \in \mathbb{N}$) matrix $[a_{ij}]$ on the Banach space $l_p^{(n)}$ ($1 \le p \le \infty$), we obtain the *Global Existence and Uniqueness Theorem for Constant-Coefficient Homogeneous Linear Systems of Differential Equations* (Theorem 2.42).

5.6 Spectral Theory in a Finite-Dimensional Setting

Throughout this section, unless specified otherwise, A is regarded as a bounded linear operator on a complex finite-dimensional Banach space $(X, \| \cdot \|)$ (i. e., $A \in L(X)$).

It is noteworthy that the prior sentence has a certain two-fold redundancy. Indeed,

- each finite-dimensional normed vector space $(X, \|\cdot\|)$ is automatically *Banach* (see the *Completeness of Finite-Dimensional Spaces Theorem* (Theorem 3.11)) and
- each linear operator $A : X \to X$ on a finite-dimensional Banach space is necessarily *bounded* (see *Boundedness of Linear Operators from Finite-Dimensional Spaces Proposition* (Proposition 4.14, Section 4.7, Problem 7)).

Thus, in a finite-dimensional setting, one need not specify that a linear operator $A : X \to X$ is bounded. For such operators, all the facts of the spectral theory of bounded linear operators, established in Section 5.5, remain in place. In particular, by *Gelfand's Spectral Radius Theorem* (Theorem 5.8), the *spectrum* of any such operator is a *nonempty compact set* in the complex plane $\mathbb{C}$. However, the finite-dimensionality of the underlying space gives rise to *matrix representations* (see Section 1.4), and hence, allows us to say much more about the spectral structure of (bounded) linear operator, and derive some profound applications.

5.6.1 Basics

By the *Matrix Representation Theorem* (Theorem 1.5), every (bounded) linear operator A on an n-dimensional ($n \in \mathbb{N}$) vector space X over $\mathbb{F}$ with an (ordered) basis $B := \{x_1, \dots, x_n\}$ can be represented by the $n \times n$ matrix

$$[A]_B := [[Ax_1]_B \quad [Ax_2]_B \quad \cdots \quad [Ax_n]_B]$$

with entries from $\mathbb{F}$, whose columns are the *coordinate vectors* $[Ax_j]_B$, $j = 1, \dots, n$, of Ax_j, $j = 1, \dots, n$, relative to basis B (see Section 1.4) as follows:

$$[Ax]_B = [A]_B[x]_B, \ x \in X.$$

The matrix $[A]_B$ is called the *matrix representation of A relative to basis B* (see Section 1.4).

For an arbitrary vector

$$x = \sum_{k=1}^{n} c_k x_k \in X,$$

where $c_j \in \mathbb{F}$, $j = 1, \dots, n$, are the *coordinates of x relative to basis B* (see Section 3.1.2), the *coordinate vector of x relative to basis B* is

$$[x]_B := (c_1, \dots, c_n) \in \mathbb{F}^n,$$

the *coordinate vector mapping*

$$X \ni x = \sum_{k=1}^{n} c_k x_k \mapsto Tx := [x]_B := (c_1, \dots, c_n) \in \mathbb{F}^n$$

being an *isomorphism* between the vector spaces X and $\mathbb{F}^n$ (see Section 1.4).

Theorem 5.10 (Isomorphism Theorem for Linear Operators). *Let $(X, \|\cdot\|)$ be an n-dimensional ($n \in \mathbb{N}$) Banach space over $\mathbb{F}$ and B be an (ordered) basis for X. Then*

$$L(X) \ni A \mapsto T(A) = [A]_B \in M_{n \times n}$$

is an isomorphism between the vector space L(X) of all linear operators on X and the vector space $M_{n \times n}$ of all $n \times n$ matrices with entries from $\mathbb{F}$, i. e., T is linear,

$$\forall A, C \in L(X) \; \forall \lambda, \mu \in \mathbb{F} : \; [\lambda A + \mu C]_B = [A]_B + \mu [C]_B,$$

and bijective.

Furthermore, T is also multiplicative:

$$\forall A, C \in L(X) : \; [AC]_B = [A]_B [C]_B,$$

i. e., T is an isomorphism between L(X) and $M_{n \times n}$ as algebras.

Exercise 5.29. Prove.

Remarks 5.16.

– Thus, in a finite-dimensional setting, matrix representation of linear operators allows us to reduce linear operator problems to matrix problems.
– The presence of norm in the space in the prior theorem is superfluous.

Furthermore, we have the following:

Corollary 5.4 (Implications of Isomorphism Theorem for Linear Operators). *Let $(X, \|\cdot\|)$ be an n-dimensional ($n \in \mathbb{N}$) Banach space over $\mathbb{F}$ and B be an (ordered) basis for X. Then a (bounded) linear operator A on X (i. e., $A \in L(X)$)*
1. *is idempotent (i. e., $A^2 = A$) iff the matrix $[A]_B$ is idempotent;*
2. *is nilpotent (i. e., $\exists n \in \mathbb{N} : \; A^n = A$) iff the matrix $[A]_B$ is nilpotent;*
3. *has an inverse $A^{-1} \in L(X)$ iff the matrix $[A]_B$ has an inverse $[A]_B^{-1}$ (i. e., $\det([A]_B) \neq 0$), in which case*

$$[A^{-1}]_B = [A]_B^{-1}.$$

Exercise 5.30. Prove.

Based on the prior corollary, we obtain the following statement:

Theorem 5.11 (Spectrum Theorem). *Let $(X, \|\cdot\|)$ be a complex n-dimensional ($n \in \mathbb{N}$) Banach space and B be an (ordered) basis for X. Then, for an arbitrary (bounded) linear operator A on X (i. e., $A \in L(X)$),*

$$\sigma(A) = \sigma_p(A) = \{\lambda \in \mathbb{C} \mid \lambda \text{ is an eigenvalue of the matrix representation } [A]_B\}.$$

Exercise 5.31. Prove.

Remarks 5.17.
- Thus, the spectrum of a linear operator A on a complex n-dimensional ($n \in \mathbb{N}$) Banach space is a *nonempty finite set*, which is entirely a *point spectrum* consisting of the at most n distinct eigenvalues of its matrix representation $[A]_B$ relative to an *arbitrary* basis B for X.
- The fact that all matrix representations of A have the same set of eigenvalues, regardless of the choice of basis for X, is entirely consistent with the following corollary of the *Change of Basis for Linear Operators Theorem* (Theorem 1.8), which implies that all matrix representations of an arbitrary linear operator in a finite-dimensional vector space are *similar matrices* (see Remark 1.9):

Corollary 5.5 (Implications of Change of Basis for Linear Operators Theorem). *Let X be a complex n-dimensional ($n \in \mathbb{N}$) vector space, and let $A : X \to X$ be a linear operator on X. The matrix representations of A relative to all possible bases of X are similar matrices and hence have the same*
1. *determinant,*
2. *characteristic polynomial,*
3. *characteristic equation, and*
4. *the set of eigenvalues with identical algebraic and geometric multiplicities.*

The prior statement allows us to define the following notions inherent only to linear operators on finite-dimensional vector spaces.

Definition 5.11 (Characteristic Polynomial and Equation). For a linear operator A on a complex finite-dimensional vector space X,
- the *characteristic polynomial* of A is the characteristic polynomial p of degree n shared by all matrix representations of A;
- the *characteristic equation* of A is the equation

$$p(\lambda) = 0.$$

Definition 5.12 (Algebraic Multiplicity of an Eigenvalue). The *algebraic multiplicity* of an eigenvalue $\lambda \in \mathbb{C}$ of a linear operator A on a complex finite-dimensional vector space X is its multiplicity as a zero of the characteristic polynomial of A, i. e., as a root of the characteristic equation.

Applying the *Isomorphism Theorem for Linear Operators* (Theorem 5.10), we obtain the following generalization of the celebrated *Cayley–Hamilton Theorem* (Theorem 1.9).

Theorem 5.12 (Generalized Cayley–Hamilton Theorem). *For an arbitrary linear operator A on a complex finite-dimensional vector space X,*

$$p(A) = 0,$$

where p is the characteristic polynomial of A.

Exercise 5.32. Prove.

5.6.2 Special Matrix Representations

Linear operators on finite-dimensional vector spaces allowing matrix representations, the principal problem in this context is of finding the simplest possible such representations.

5.6.2.1 Diagonal Matrix Representations
The first effort is, certainly, to be given to *diagonal* matrix representations.

Theorem 5.13 (Diagonal Matrix Representations). *The matrix representation $[A]_B$ of a linear operator A on a (real or complex) finite-dimensional vector space X relative to a basis B for X is diagonal iff B is an eigenbasis of A, i. e., B consists of the eigenvectors of A.*

Exercise 5.33. Prove.

Hint. Apply the *Matrix Representation Theorem* (Theorem 1.5) (see Remarks 1.6).

Remarks 5.18.
- Thus, a linear operator A on an n-dimensional ($n \in \mathbb{N}$) vector space X allows a *diagonal matrix representation* $[A]_B$ *iff* A has precisely n linearly independent eigenvectors, which is the case *iff* all eigenvalues of A are *complete*, i. e., have equal algebraic and geometric multiplicities (see Definition 5.4). In particular, this occurs when all eigenvalues of A are *distinct*.
- Generally, the geometric multiplicity of an eigenvalue of a linear operator on a complex finite-dimensional vector space does not exceed its algebraic multiplicity. When the former is less than the latter, the eigenvalue is called *defective* (see, e. g., [34, 49, 54]).

  Hence, any *simple eigenvalue*, i. e., an eigenvalue of algebraic multiplicity 1 (see Definition 5.12), is *complete* and only a *repeated eigenvalue*, i. e., an eigenvalue of algebraic multiplicity greater than 1, can be defective.

Examples 5.11.

1. On $\mathbb{F}^2$ ($\mathbb{F} = \mathbb{R}$ or $\mathbb{F} = \mathbb{C}$), the operator A of multiplication by the *real symmetric* 2×2 matrix

$$\begin{bmatrix} 1 & 1 \\ 1 & 1 \end{bmatrix}$$

has *two distinct* eigenvalues $\lambda_1 = 0$ and $\lambda_2 = 2$ with the corresponding linearly independent eigenvectors

$$\begin{bmatrix} 1 \\ -1 \end{bmatrix} \quad \text{and} \quad \begin{bmatrix} 1 \\ 1 \end{bmatrix}$$

(see the *Linear Independence of Eigenvectors Proposition* (Proposition 5.8, Section 5.8, Problem 5)) forming an (ordered) *eigenbasis*

$$B := \left\{ \begin{bmatrix} 1 \\ -1 \end{bmatrix}, \begin{bmatrix} 1 \\ 1 \end{bmatrix} \right\}$$

for $\mathbb{F}^2$, relative to which the matrix representation

$$[A]_B = \begin{bmatrix} 0 & 0 \\ 0 & 2 \end{bmatrix}$$

of A is *diagonal*.

2. On $\mathbb{C}^2$, the operator A of multiplication by the 2×2 matrix

$$\begin{bmatrix} 0 & 1 \\ -1 & 0 \end{bmatrix}$$

has the pair of *distinct* complex conjugate eigenvalues $\lambda_1 = i$ and $\lambda_2 = -i$ with the corresponding linearly independent eigenvectors

$$\begin{bmatrix} -i \\ 1 \end{bmatrix} \quad \text{and} \quad \begin{bmatrix} i \\ 1 \end{bmatrix}$$

(see the *Linear Independence of Eigenvectors Proposition* (Proposition 5.8, Section 5.8, Problem 5)) forming an (ordered) *eigenbasis*

$$B := \left\{ \begin{bmatrix} -i \\ 1 \end{bmatrix}, \begin{bmatrix} i \\ 1 \end{bmatrix} \right\}$$

for $\mathbb{C}^2$ relative to which the matrix representation

$$[A]_B = \begin{bmatrix} i & 0 \\ 0 & -i \end{bmatrix}$$

of A is *diagonal*.

Remark 5.19. It is noteworthy that the same linear operator on the real space $\mathbb{R}^2$ has no eigenvalues, and hence, no eigenvectors whatsoever.

3. On $\mathbb{F}^2$ ($\mathbb{F} = \mathbb{R}$ or $\mathbb{F} = \mathbb{C}$), the operator A of multiplication by the 2×2 matrix

$$\begin{bmatrix} 0 & 1 \\ 0 & 0 \end{bmatrix},$$

having one *repeated* $\lambda = 0$ of algebraic multiplicity $k = 2$ and geometric multiplicity $p = 1$, i. e., a *defective eigenvalue*, does not allow a diagonal matrix representation.

4. The *differentiation operator*

$$P_3 \ni p \mapsto Dp := p' \in P_3,$$

where P_3 is the four-dimensional space of polynomials with coefficients from $\mathbb{F}$ ($\mathbb{F} = \mathbb{R}$ or $\mathbb{F} = \mathbb{C}$) of degree at most 3, relative to the *standard basis*

$$B := \{1, x, x^2, x^3\},$$

has the matrix representation

$$[D]_B = \begin{bmatrix} 0 & 1 & 0 & 0 \\ 0 & 0 & 2 & 0 \\ 0 & 0 & 0 & 3 \\ 0 & 0 & 0 & 0 \end{bmatrix}$$

(see Examples 1.4). However, having one *repeated* $\lambda = 0$ of algebraic multiplicity $k = 4$ and geometric multiplicity $p = 1$, i. e., a *defective eigenvalue*, it does not allow a diagonal matrix representation.

Exercise 5.34. Verify.

5.6.2.2 Jordan Canonical Matrix Representations

Although a linear operator A on a *complex* n-dimensional ($n \in \mathbb{N}$) vector space X may not have precisely n linearly independent eigenvectors to form an eigenbasis, one can still choose a special basis B, called a *Jordan*[7] *basis* comprised of m linearly independent eigenvectors of A ($1 \le m \le n$) and $n - m$ *generalized eigenvectors* relative to which the matrix representation $[A]_B$ of A is a *Jordan canonical matrix* (see, e. g., [34, 49]), which, when $m = n$, is *diagonal*, the corresponding Jordan basis being merely an eigenbasis.

---

[7] Camille Jordan (1838–1922).

The existence of a *Jordan canonical matrix representation* for an arbitrary linear operator on a finite-dimensional complex vector space has a number of profound implications; in particular, it underlies the *Cayley–Hamilton Theorem* (Theorem 1.9).

An $n \times n$ ($n \in \mathbb{N}$) Jordan canonical matrix representation of a linear operator A on a complex n-dimensional vector space X is an upper triangular *block diagonal* matrix, which has the following structure (see, e. g., [34, 49]).

- It consists of *Jordan blocks* corresponding to each eigenvalue of A aligned along the main diagonal and 0's elsewhere.
- A *Jordan block* of size $1 \le s \le n$ corresponding to an eigenvalue $\lambda \in \mathbb{C}$ of A, if $s = 1$, is the 1×1 matrix

$$[\lambda]$$

or, if $2 \le s \le n$, is the $s \times s$ upper triangular matrix

$$\begin{bmatrix} \lambda & 1 & 0 & 0 & \ldots & 0 & 0 \\ 0 & \lambda & 1 & 0 & \ldots & 0 & 0 \\ & & & \ddots & & & \\ 0 & 0 & 0 & 0 & \ldots & \lambda & 1 \\ 0 & 0 & 0 & 0 & \ldots & 0 & \lambda \end{bmatrix}$$

with λ's on the main diagonal, 1's on the *superdiagonal* (right above the main diagonal), and 0's elsewhere.

- The number $1 \le p \le n$ of Jordan blocks corresponding to an eigenvalue $\lambda \in \mathbb{C}$ of A is equal to the maximum number of linearly independent eigenvectors associated with λ, i. e., the dimension of its eigenspace:

$$p = \dim \ker(A - \lambda I),$$

which is the *geometric multiplicity* of λ (see Definition 5.4).

This number does not exceed the *algebraic multiplicity* $1 \le k \le n$ of λ (see Definition 5.12 and Remarks 5.18), the sizes of all Jordan blocks corresponding to λ adding up to the latter.

Thus, if λ is a *complete eigenvalue* (see Remarks 5.18), i. e., $p = k$, all Jordan blocks corresponding to λ are of size 1, i. e., of the form

$$[\lambda] .$$

- The size of the largest Jordan block corresponding to an eigenvalue $\lambda \in \mathbb{C}$ of A is the multiplicity of λ as a zero of the *minimal polynomial* of A, i. e., a polynomial q of the lowest degree such that

$$q(A) = 0$$

(see the *Generalized Cayley–Hamilton Theorem*, Theorem 5.12).

- To each Jordan block of size s associated with an eigenvalue $\lambda \in \mathbb{C}$ of A, there correspond precisely *one* eigenvector $x_1 \in X$ and $s - 1$ *generalized eigenvectors* $x_2, \ldots, x_s \in X$, i. e., such vectors that

$$(A - \lambda I)x_1 = 0 \quad \text{and} \quad (A - \lambda I)x_{i+1} = x_i, \ i = 1, \ldots, s - 1.$$

Exercise 5.35. Explain.

Examples 5.12.

1. On $\mathbb{C}^2$, the operator A of multiplication by the 2×2 matrix

$$\begin{bmatrix} 0 & 1 \\ 0 & 0 \end{bmatrix}$$

does not allow a diagonal matrix representation (see Examples 5.11). It has one linearly independent eigenvector

$$e_1 := \begin{bmatrix} 1 \\ 0 \end{bmatrix}$$

and one generalized eigenvector

$$e_2 := \begin{bmatrix} 0 \\ 1 \end{bmatrix}$$

$(Ae_2 = e_1)$ corresponding to the unique *defective eigenvalue* $\lambda_1 = 0$. They form the *standard basis B* for $\mathbb{C}^2$, which is the *Jordan basis of A* relative to which, the above matrix itself is precisely the unique *Jordan canonical matrix representation* of A, the Jordan block of size 2 correspond to $\lambda_1 = 0$.

2. On $\mathbb{C}^3$, the operator A of multiplication by the 3×3 matrix

$$\begin{bmatrix} 1 & 1 & -2 \\ 0 & 1 & -1 \\ 0 & 0 & 0 \end{bmatrix}$$

has two eigenvalues: *simple* $\lambda_1 = 0$ and *repeated* $\lambda_2 = 1$ of algebraic multiplicity $k_2 = 2$.

The simple eigenvalue $\lambda_1 = 0$ is *complete* with one corresponding linearly independent eigenvector

$$x_1 := \begin{bmatrix} 1 \\ 1 \\ 1 \end{bmatrix}.$$

The repeated eigenvalue $\lambda_2 = 1$ is *defective* with one corresponding linearly independent eigenvector

$$x_2 := \begin{bmatrix} 1 \\ 0 \\ 0 \end{bmatrix}.$$

Solving the linear system

$$(A - I)x = x_2,$$

we find a generalized eigenvector

$$x_3 := \begin{bmatrix} 0 \\ 1 \\ 0 \end{bmatrix}.$$

The linearly independent vectors x_1, x_2, and x_3 form an (ordered) *Jordan basis*
$B := \{x_1, x_2, x_3\}$ (not an eigenbasis!) relative to which a *Jordan canonical matrix
representation* of A is

$$[D]_B = \begin{bmatrix} 0 & 0 & 0 \\ 0 & 1 & 1 \\ 0 & 0 & 1 \end{bmatrix}.$$

It has two Jordan blocks:

$$[0]$$

of size 1 corresponding to the *simple eigenvalue* $\lambda_1 = 0$ and

$$\begin{bmatrix} 1 & 1 \\ 0 & 1 \end{bmatrix}$$

of size 2 corresponding to the *defective eigenvalue* $\lambda_2 = 1$.
Changing the order of the basis vectors, we obtain another (ordered) *Jordan basis*
$B' := \{x_2, x_3, x_1\}$ relative to which a *Jordan canonical matrix representation* of A is

$$[D]_{B'} = \begin{bmatrix} 1 & 1 & 0 \\ 0 & 1 & 0 \\ 0 & 0 & 0 \end{bmatrix}.$$

Thus, such a rearrangement of the basis vectors leads to the corresponding rearrangement of Jordan blocks.

Remark 5.20. To preserve the structure of a Jordan canonical matrix representation, the vectors x_2, x_3 should remain precisely in this order. Thus, in this case, B and B' are the only two possible *Jordan bases*.

3. The *differentiation operator*

$$P_3 \ni p \mapsto Dp := p' \in P_3,$$

where P_3 is the four-dimensional space of polynomials with complex coefficients of degree at most 3, does not allow a diagonal matrix representation (see Examples 5.11). However, relative to the *Jordan basis*

$$B := \{p_1(x) := 1, p_2(x) := x, p_3(x) := x^2/2, p_4(x) := x^3/6\}$$

of D, where p_1 is one linearly independent eigenpolynomial and three generalized eigenpolynomials p_2, p_3, and p_4 $(Dp_{i+1} = p_i, i = 1, 2, 3)$ corresponding to the unique defective eigenvalue $\lambda_1 = 0$ relative to which the *Jordan canonical matrix representation* of A is

$$[D]_B = \begin{bmatrix} 0 & 1 & 0 & 0 \\ 0 & 0 & 1 & 0 \\ 0 & 0 & 0 & 1 \\ 0 & 0 & 0 & 0 \end{bmatrix},$$

the Jordan block of size 4 corresponding to $\lambda_1 = 0$.

5.6.3 Spectral Decomposition

By the *Isomorphism Theorem for Linear Operators* (Theorem 5.10), the existence of a *Jordan canonical matrix representation* for an arbitrary linear operator A in a complex n-dimensional $(n \in \mathbb{N})$ Banach space implies that $(X, \| \cdot \|)$, i.e., any such operator admits the following *spectral decomposition*:

$$A = \sum_{j=1}^{m}(\lambda_j P_j + Q_j), \tag{5.8}$$

where

- $\lambda_j, j = 1, \dots, m$, are all distinct *eigenvalues* of A forming its *spectrum* $\sigma(A)$ $(m \in \mathbb{N}, 1 \le m \le n)$,
- $P_j, j = 1, \dots, m$, are *projection operators*, and
- $Q_j := (A - \lambda_j I)P_j = P_j(A - \lambda_j I), j = 1, \dots, m$, are *nilpotent operators*

(see, e. g., [16, 39, 24, 12]).

A few important observations concerning the structure of the *spectral decomposition* are in order.

- The projection $P_j = P(\lambda_j, A), j = 1, \ldots, m$, called the *spectral projection*, or the *Riesz projection*, of the operator A at λ_j, is the projection operator, whose matrix representation relative to a Jordan basis of A is the $n \times n$ matrix obtained from the associated Jordan canonical matrix representation of A via replacing λ_j's in all Jordan blocks corresponding to λ_j with 1's and all nonzero entries, if any, with 0's. The range $R(P(\lambda_j, A))$ of the spectral projection $P(\lambda_j, A), j = 1, \ldots, m$, is a subspace of X, which is not to be confused with the *eigenspace* of λ_j $\ker(A - \lambda_j I)$. In fact, the former contains the latter:

$$\ker(A - \lambda_j I) \subseteq R(P(\lambda_j, A)),$$

dim $R(P(\lambda, A))$ being the *algebraic multiplicity* k_j of λ_j $(1 \le k_j \le n)$, i.e., the sum of the sizes of all Jordan blocks corresponding to λ_j, and dim $\ker(A - \lambda I)$ being the *geometric multiplicity* p_j of λ_j $(1 \le p_j \le k_j)$, i.e., the number of Jordan blocks corresponding to λ_j.
In fact,

$$R(P(\lambda_j, A)) = \ker(A - \lambda_j I)$$

iff λ is a *complete eigenvalue* of A, i.e., $p_j = k_j$, in which case all Jordan blocks corresponding to λ_j are of size 1.

Remark 5.21. For any *regular point* $\lambda \in \rho(A)$, we can naturally define

$$P(\lambda, A) := 0.$$

- The operators $Q_j, j = 1, \ldots, m$, are linear operators whose matrix representation relative to a Jordan basis of A is the $n \times n$ matrix obtained from the associated Jordan canonical matrix representation of A via replacing all nonzero entries, if any, except 1's in all Jordan blocks corresponding to λ_j, with 0's, is *nilpotent*, the *index* of λ_j

$$i_j := \min\{i \in \mathbb{N} \mid Q_j^i = 0\} \tag{5.9}$$

being the multiplicity of λ_j as a zero of the *minimal polynomial* of A, i.e., the size of the largest Jordan block of λ_j (see, e. g.., [24, 34, 39]).

Remark 5.22. Observe that

$$R(P(\lambda_j, A)) = \ker(A - \lambda_j I)^{i_j}, j = 1, \ldots, m.$$

- The operators P_j and $Q_j, j = 1, \ldots, m$, satisfy the following relations:

$$\sum_{j=1}^{m} P_j = I,$$

$$P_i P_j = \delta_{ij} P_i, \ i, j = 1, \ldots, m,$$

$$P_i Q_j = Q_j P_i = \delta_{ij} Q_j, \ i, j = 1, \ldots, m,$$

$$Q_i Q_j = \delta_{ij} Q_i^2, \ i, j = 1, \ldots, m, \tag{5.10}$$

where δ_{ij} is the *Kronecker delta*, the first two of which allow saying that the spectral projections P_j, $j = 1, \ldots, m$, form a *resolution of the identity* with the following direct sum decomposition of X in place:

$$X = \bigoplus_{j=1}^{m} R(P_j).$$

Remark 5.23. In particular, if A allows a *diagonal matrix representation*, which is the case *iff* A furnishes an *eigenbasis* B for X, i. e., all Jordan blocks in a Jordan canonical (diagonal) matrix representation of A are of size 1, and hence,

$$Q_j = 0, \ j = 1, \ldots, m.$$

In this case, *spectral decomposition* (5.8) for A acquires the following form:

$$A = \sum_{j=1}^{m} \lambda_j P_j, \tag{5.11}$$

each spectral projection P_j, $j = 1, \ldots, m$, being a projection onto the *eigenspace* of λ_j.

Examples 5.13.
1. For the *differentiation operator*

$$P_3 \ni p \mapsto Dp := p' \in P,$$

where P_3 is the four-dimensional space of polynomials with complex coefficients of degree at most 3, whose a *Jordan canonical matrix representation*

$$[D]_B = \begin{bmatrix} 0 & 1 & 0 & 0 \\ 0 & 0 & 1 & 0 \\ 0 & 0 & 0 & 1 \\ 0 & 0 & 0 & 0 \end{bmatrix}$$

relative to the *Jordan basis*

$$B := \{p_1(x) := 1, p_2(x) := x, p_3(x) := x^2/2, p_4(x) := x^3/6\}$$

is the Jordan block of size 4 corresponding to $\lambda_1 = 0$ (see Examples 5.12), the *spectral decomposition* is

$$A = 0P_1 + Q_1 = Q_1,$$

where the *spectral projection* at the defective eigenvalue $\lambda = 0$ is $P = I$, and the *nilpotent operator* is $Q_1 = D$ with $D^4 = 0$.

2. A linear operator A on a complex four-dimensional vector space X with a *Jordan canonical matrix representation*

$$[A]_B = \begin{bmatrix} -1 & 0 & 0 & 0 \\ 0 & 2 & 0 & 0 \\ 0 & 0 & 2 & 1 \\ 0 & 0 & 0 & 2 \end{bmatrix}$$

relative to a Jordan basis B has two eigenvalues: *simple* $\lambda_1 = -1$ of algebraic multiplicity $k_1 = 1$ and *repeated* $\lambda_2 = 2$ of algebraic multiplicity $k_2 = 3$, which is *defective* of geometric multiplicity $p_2 = 2$.

Exercise 5.36. Explain.

The *spectral decomposition* of A is

$$A = ((-1)P_1 + Q_1) + (2P_2 + Q_2),$$

where the *spectral projections* P_j, $j = 1, 2$, have the following matrix representations relative to the Jordan basis B:

$$[P_1]_B = \begin{bmatrix} 1 & 0 & 0 & 0 \\ 0 & 0 & 0 & 0 \\ 0 & 0 & 0 & 0 \\ 0 & 0 & 0 & 0 \end{bmatrix} \quad \text{and} \quad [P_2]_B = \begin{bmatrix} 0 & 0 & 0 & 0 \\ 0 & 1 & 0 & 0 \\ 0 & 0 & 1 & 0 \\ 0 & 0 & 0 & 1 \end{bmatrix},$$

respectively, and the *nilpotent operators* Q_j, $j = 1, 2$, are as follows:

$$Q_1 = 0$$

and Q_2 with $Q_2^2 = 0$ has the matrix representation

$$[Q_2]_B = \begin{bmatrix} 0 & 0 & 0 & 0 \\ 0 & 0 & 0 & 0 \\ 0 & 0 & 0 & 1 \\ 0 & 0 & 0 & 0 \end{bmatrix}$$

relative to the Jordan basis B.

5.6.4 Application: Abstract Evolution Equations

In this section, we tackle certain applications of the operator exponentials in a finite-dimensional setting, in particular, to prove the celebrated *Lyapunov*[8] *Stability Theorem*.

---

8 Aleksandr Lyapunov (1857–1918).

5.6.4.1 Operator Exponentials in a Finite-Dimensional Setting

As is known (see Section 5.5.4), the *general solution* of the abstract evolution equation

$$y'(t) = Ay(t), \ t \geq 0, \tag{5.12}$$

with a bounded linear operator on a Banach space $(X, \| \cdot \|)$ over $\mathbb{F}$ is given by the exponential formula familiar from ordinary differential equations

$$y(t) = e^{tA}x, \ t \geq 0, x \in X, \tag{5.13}$$

where the operator series

$$e^{tA} = \sum_{k=0}^{\infty} \frac{t^k A^k}{k!}$$

converges uniformly for any $t \in \mathbb{F}$.

In a finite-dimensional setting, based on the *spectral decomposition* for A (see Section 5.6.3), the above operator series representation of the operator exponentials e^{tA}, $t \geq 0$, can be rewritten as the following finite sum:

$$e^{tA} = \sum_{j=1}^{m} e^{\lambda_j t} P_j \sum_{k=0}^{i_j-1} \frac{t^k}{k!} Q_j^k, \ t \in \mathbb{F} \tag{5.14}$$

(see Section 5.6.3 and, e. g., [24, 12, 39, 21], cf. [21, Proposition 2.6]).

Exercise 5.37. Verify using (5.8)–(5.10).

Remark 5.24. In particular, if A allows a *diagonal matrix representation*, which is the case *iff* A furnishes an *eigenbasis* B for X (see the *Diagonal Matrix Representations Theorem* (Theorem 5.13)), i. e., all eigenvalues of A are *complete* (see Remarks 5.18), its *spectral decomposition* is given by (5.11) (see Remark 5.23), and hence, its exponential representation (5.14) acquires a still simpler form:

$$e^{tA} = \sum_{j=1}^{m} e^{t\lambda_j} P_j, \ t \in \mathbb{F}, \tag{5.15}$$

i. e., the matrix representation $[e^{tA}]_B$ of the operator exponential e^{tA} relative to B is a diagonal matrix with the entries $e^{t\lambda_j}$ on the main diagonal, each occurring as many times as the algebraic multiplicity (same as geometric multiplicity) of λ_j, $j = 1, \ldots, m$.

Examples 5.14.
1. On the complex two-dimensional Banach space $l_p^{(2)}$ ($1 \leq p \leq \infty$), for the operator of multiplication by the 2×2 real symmetric matrix

$$\begin{bmatrix} 1 & 1 \\ 1 & 1 \end{bmatrix},$$

there exists an *eigenbasis*

$$B := \left\{ \begin{bmatrix} 1 \\ -1 \end{bmatrix}, \begin{bmatrix} 1 \\ 1 \end{bmatrix} \right\}$$

relative to which the matrix representation of A is the diagonal matrix

$$[A]_B = \begin{bmatrix} 0 & 0 \\ 0 & 2 \end{bmatrix}$$

(see Examples 5.11).

Hence, by representation (5.15) and the *Isomorphism Theorem for Linear Operators* (Theorem 5.10),

$$[e^{tA}]_B = [e^{0t}P_1 + e^{2t}P_2]_B = e^{0t}\begin{bmatrix} 1 & 0 \\ 0 & 0 \end{bmatrix} + e^{2t}\begin{bmatrix} 0 & 0 \\ 0 & 1 \end{bmatrix} = \begin{bmatrix} 1 & 0 \\ 0 & e^{2t} \end{bmatrix}, \ t \in \mathbb{C}.$$

2. On the complex two-dimensional Banach space $l_p^{(2)}$ $(1 \le p \le \infty)$, for the operator of multiplication by the 2×2 matrix

$$\begin{bmatrix} 0 & 1 \\ -1 & 0 \end{bmatrix},$$

there exists an *eigenbasis*

$$B := \left\{ \begin{bmatrix} -i \\ 1 \end{bmatrix}, \begin{bmatrix} i \\ 1 \end{bmatrix} \right\}$$

relative to which the matrix representation of A is the diagonal matrix

$$[A]_B = \begin{bmatrix} i & 0 \\ 0 & -i \end{bmatrix}$$

(see Examples 5.11).

Hence, by representation (5.15) and the *Isomorphism Theorem for Linear Operators* (Theorem 5.10),

$$[e^{tA}]_B = [e^{it}P_1 + e^{-it}P_2]_B = e^{it}\begin{bmatrix} 1 & 0 \\ 0 & 0 \end{bmatrix} + e^{-it}\begin{bmatrix} 0 & 0 \\ 0 & 1 \end{bmatrix} = \begin{bmatrix} e^{it} & 0 \\ 0 & e^{-it} \end{bmatrix}, \ t \in \mathbb{C}.$$

3. On the complex two-dimensional Banach space $l_p^{(2)}$ $(1 \le p \le \infty)$, for the operator of multiplication by the 2×2 matrix

$$\begin{bmatrix} 0 & 1 \\ 0 & 0 \end{bmatrix},$$

the standard basis B is a *Jordan Basis*, relative to which the matrix representation of A is the Jordan block of size 2

$$[A]_B = \begin{bmatrix} 0 & 1 \\ 0 & 0 \end{bmatrix}$$

(see Examples 5.12).

Hence, by representation (5.14) and the *Isomorphism Theorem for Linear Operators* (Theorem 5.10), in view of $P_1 = I$,

$$[e^{tA}]_B = [e^{0t}(I + tQ_1)]_B = e^{0t}\left(\begin{bmatrix} 1 & 0 \\ 0 & 1 \end{bmatrix} + t \begin{bmatrix} 0 & 1 \\ 0 & 0 \end{bmatrix} \right) = \begin{bmatrix} 1 & t \\ 0 & 1 \end{bmatrix}, \; t \in \mathbb{C}.$$

4. For the nilpotent *differentiation operator*

$$P_3 \ni p \mapsto Dp := p' \in P_3,$$

where P_3 is the four-dimensional space of polynomials with complex coefficients of degree at most 3, with a *Jordan canonical matrix representation*

$$[D]_B = \begin{bmatrix} 0 & 1 & 0 & 0 \\ 0 & 0 & 1 & 0 \\ 0 & 0 & 0 & 1 \\ 0 & 0 & 0 & 0 \end{bmatrix}$$

relative to the *Jordan basis*

$$B := \{p_1(x) := 1, p_2(x) := x, p_3(x) := x^2/2, p_4(x) := x^3/6\}$$

(see Examples 5.12 and 5.13), by representation (5.14) and the *Isomorphism Theorem for Linear Operators* (Theorem 5.10), in view of $P_1 = I$,

$$[e^{tD}]_B = \left[e^{0t}\left(I + tQ_1 + \frac{t^2}{2!}Q_1^2 + \frac{t^3}{3!}Q_1^3 \right) \right]_B$$

$$= e^{0t}\left(\begin{bmatrix} 1 & 0 & 0 & 0 \\ 0 & 1 & 0 & 0 \\ 0 & 0 & 1 & 0 \\ 0 & 0 & 0 & 1 \end{bmatrix} + t \begin{bmatrix} 0 & 1 & 0 & 0 \\ 0 & 0 & 1 & 0 \\ 0 & 0 & 0 & 1 \\ 0 & 0 & 0 & 0 \end{bmatrix} + \frac{t^2}{2!} \begin{bmatrix} 0 & 0 & 1 & 0 \\ 0 & 0 & 0 & 1 \\ 0 & 0 & 0 & 0 \\ 0 & 0 & 0 & 0 \end{bmatrix} \right.$$

$$\left. + \frac{t^3}{3!} \begin{bmatrix} 0 & 0 & 0 & 1 \\ 0 & 0 & 0 & 0 \\ 0 & 0 & 0 & 0 \\ 0 & 0 & 0 & 0 \end{bmatrix} \right) = \begin{bmatrix} 1 & t & \frac{t^2}{2!} & \frac{t^3}{3!} \\ 0 & 1 & t & \frac{t^2}{2!} \\ 0 & 0 & 1 & t \\ 0 & 0 & 0 & 1 \end{bmatrix}, \; t \in \mathbb{C}.$$

5. For a linear operator A on a complex four-dimensional Banach space $(X, \|\cdot\|)$ with a *Jordan canonical matrix representation*

$$[A]_B = \begin{bmatrix} -1 & 0 & 0 & 0 \\ 0 & 2 & 0 & 0 \\ 0 & 0 & 2 & 1 \\ 0 & 0 & 0 & 2 \end{bmatrix}$$

relative to a Jordan basis B (see Examples 5.13), by representation (5.14) and the *Isomorphism Theorem for Linear Operators* (Theorem 5.10), in view of (5.10),

$$[e^{tA}]_B = [e^{(-1)t}P_1 + e^{2t}(P_2 + tQ_2)]_B$$

$$= e^{(-1)t} \begin{bmatrix} 1 & 0 & 0 & 0 \\ 0 & 0 & 0 & 0 \\ 0 & 0 & 0 & 0 \\ 0 & 0 & 0 & 0 \end{bmatrix} + e^{2t} \left(\begin{bmatrix} 0 & 0 & 0 & 0 \\ 0 & 1 & 0 & 0 \\ 0 & 0 & 1 & 0 \\ 0 & 0 & 0 & 1 \end{bmatrix} + t \begin{bmatrix} 0 & 0 & 0 & 0 \\ 0 & 0 & 0 & 0 \\ 0 & 0 & 0 & 1 \\ 0 & 0 & 0 & 0 \end{bmatrix} \right)$$

$$= \begin{bmatrix} e^{-t} & 0 & 0 & 0 \\ 0 & e^{2t} & 0 & 0 \\ 0 & 0 & e^{2t} & te^{2t} \\ 0 & 0 & 0 & e^{2t} \end{bmatrix}, \; t \in \mathbb{C}.$$

5.6.4.2 Asymptotic Stability and Lyapunov Stability Theorem

Definition 5.13 (Asymptotic Stability). Let A be a bounded linear operator on a Banach space $(X, \|\cdot\|)$ (i. e., $A \in L(X)$). Evolution equation (5.12) is called *asymptotically stable* if, for each solution $y(\cdot)$ of the equation

$$\lim_{t \to \infty} y(t) = 0,$$

i. e., in view of (5.13),

$$\forall x \in X: \lim_{t \to \infty} e^{tA}x = 0.$$

Operator exponential representation (5.14) is instrumental for proving the following celebrated theorem ([42], cf. [20, Theorem 2.10]).

Theorem 5.14 (Lyapunov Stability Theorem). *Let A be a linear operator on a complex finite-dimensional Banach $(X, \|\cdot\|)$. Evolution equation (5.12) is asymptotically stable iff all eigenvalues of A have negative real parts, i. e.,*

$$\sigma(A) \subseteq \{\lambda \in \mathbb{C} \mid \mathrm{Re}\,\lambda < 0\}.$$

Proof. "*Only if*" part. Let us prove this part *by contrapositive*, assuming that

$$\sigma(A) \nsubseteq \{\lambda \in \mathbb{C} \mid \mathrm{Re}\,\lambda < 0\},$$

i. e., there is an eigenvalue λ of A with $\mathrm{Re}\,\lambda \geq 0$.

Then, for the corresponding eigenvalue solution of (5.12),

$$y(t) = e^{\lambda t}x, \ t \geq 0,$$

where $x \neq 0$ is an eigenvector associated with λ,

$$y(t) = e^{\lambda t}x \not\to 0, \ t \to \infty,$$

since

$$\|y(t)\| = \|e^{\lambda t}x\| = |e^{\lambda t}|\|x\| = e^{\mathrm{Re}\,\lambda t}\|x\| \geq e^{0t}\|x\| = \|x\| > 0, \ t \geq 0.$$

Hence, we infer that equation (5.12) is not asymptotically stable, which completes the proof of the *"only if"* part by contrapositive.

"*If*" part. Suppose that

$$\sigma(A) \subseteq \{\lambda \in \mathbb{C} \mid \mathrm{Re}\,\lambda < 0\},$$

and hence, the *spectral bound* of A

$$\omega := \max_{\lambda \in \sigma(A)} \mathrm{Re}\,\lambda < 0.$$

Let $y(\cdot)$ be an arbitrary solution of Equation (5.12). Then, by (5.13) and (5.14),

$$y(t) = e^{tA}x = \left[\sum_{j=1}^{m} e^{\lambda_j t} P_j \sum_{k=0}^{i_j-1} \frac{t^k}{k!} Q_j^k \right] x, \ t \geq 0, \tag{5.16}$$

with some $y(0) = x \in X$.

By *subadditivity*, *absolute scalability*, and *submultiplicativity* of operator norm (see Remark 4.9), also designated by $\| \cdot \|$, for any $t \geq 0$,

$$\|y(t)\| = \|e^{tA}x\| = \left\| \left[\sum_{j=1}^{m} e^{\lambda_j t} P_j \sum_{k=0}^{i_j-1} \frac{t^k}{k!} Q_j^k \right] x \right\| \leq \left\| \sum_{j=1}^{m} e^{\lambda_j t} P_j \sum_{k=0}^{i_j-1} \frac{t^k}{k!} Q_j^k \right\| \|x\|$$

$$\leq \left[\sum_{j=1}^{m} |e^{\lambda_j t}| \|P_j\| \sum_{k=0}^{i_j-1} \frac{t^k}{k!} \|Q_j\|^k \right] \|x\| \leq \left[\sum_{j=1}^{m} \|P_j\| \sum_{k=0}^{i_j-1} \frac{t^k}{k!} \|Q_j\|^k \right] e^{\omega t} \|x\|.$$

Whence, considering that $\omega < 0$, we conclude that

$$\|y(t)\| = \|e^{tA}x\| \to 0, \ t \to \infty.$$

Exercise 5.38. Explain.

This completes the proof of the "*if*" part and the entire statement. □

Remarks 5.25.

- The *Lyapunov Stability Theorem* is a purely *qualitative* statement allowing to pre-
  dict the *asymptotic behavior* (i. e., behavior as $t \to \infty$) of the solutions of evolution
  equation (5.12) without actually solving it, based on the location of the spectrum
  of the operator A in the complex plane $\mathbb{C}$.
- As a particular case, when A is the operator of multiplication by an $n \times n$ matrix $[a_{ij}]$
  with complex entries on the complex n-dimensional $l_p^{(n)}(\mathbb{C})$ ($n \in \mathbb{N}$, $1 \le p \le \infty$)
  (see Examples 4.2), we obtain the familiar version of the *Lyapunov Stability Theo-
  rem* for constant-coefficient homogeneous linear systems of differential equations
  (see the *Global Existence and Uniqueness Theorem for Constant-Coefficient Homogeneous Linear Systems of Differential Equations* (Theorem 2.42)).

Example 5.15. On the complex two-dimensional space $l_p^{(2)}$ ($1 \le p \le \infty$), for the operators A, C, and D of multiplication by the 2×2 matrices

$$\begin{bmatrix} 1 & 0 \\ 0 & 2 \end{bmatrix}, \begin{bmatrix} 0 & 1 \\ 0 & 0 \end{bmatrix}, \text{ and } \begin{bmatrix} 0 & 1 \\ -1 & 0 \end{bmatrix},$$

respectively (see Examples 5.2), by the *Lyapunov Stability Theorem* (Theorem 5.14),
none of the corresponding evolution equations

$$y'(t) = Ay(t), \ y'(t) = Cy(t), \text{ and } y'(t) = Dy(t), \ t \ge 0,$$

is asymptotically stable, whereas for the operator E of multiplication by the 2×2 matrix

$$\begin{bmatrix} 1 & -1 \\ 3 & -2 \end{bmatrix},$$

the corresponding evolution equation

$$y'(t) = Ey(t), \ t \ge 0,$$

is asymptotically stable.

Exercise 5.39. Verify.

5.6.4.3 Mean Ergodicity of Bounded Solutions

Operator exponential representation (5.14) is also pivotal for proving the following
statement on the existence of the limit at infinity of the *Cesàro*[9] *means*:

$$\frac{1}{t} \int_0^t y(s)\,ds,$$

---

[9] Ernesto Cesàro (1859–1906).

with integration understood in the classical *Riemann sense* (see, e. g., [16, 47]), for every *bounded solution* $y(\cdot)$ (i. e., $\sup_{t \geq 0} \|y(t)\| < \infty$) of evolution equation (5.12) with a linear operator A on a complex finite-dimensional Banach space $(X, \| \cdot \|)$.

The existence of such limits is referred to as *mean ergodicity*.

Remark 5.26. For a continuous function $y : [0, \infty) \to X$, the notion of the *Cesàro limit*

$$\lim_{t \to \infty} \frac{1}{t} \int_0^t y(s) \, ds$$

extends that of the regular one

$$\lim_{t \to \infty} y(t),$$

i. e., the existence of the latter implies the existence of the former and its coincidence with the latter.

Exercise 5.40. Verify.

The converse, however, is not true. For instance, in $X = \mathbb{C}$ with the absolute-value norm, all solutions

$$y(t) = e^{it} x, \; t \geq 0, \; x \in X,$$

of Equation (5.12), with A being the multiplication operator by the *imaginary unit i*, are *bounded*, and

$$\lim_{t \to \infty} \frac{1}{t} \int_0^t e^{is} x \, ds = \lim_{t \to \infty} \frac{e^{it} x - x}{it} = 0,$$

whereas

$$\lim_{t \to \infty} e^{it} x$$

exists only for the trivial one (with $x = 0$).

Thus, for a bounded solution of evolution equation (5.12), the Cesàro limit may exist when the regular one does not.

Theorem 5.15 (Mean Ergodicity Theorem, M. V. Markin [48, Theorem 4.1]). *Let A be a linear operator on a complex finite-dimensional Banach space $(X, \| \cdot \|)$. For every bounded solution $y(\cdot)$ of evolution equation (5.12),*

$$\lim_{t \to \infty} \frac{1}{t} \int_0^t y(s) \, ds = P(0, A) y(0),$$

where $P(0, A)$ is the spectral projection of A at 0.

If the spectrum of the operator A contains no pure imaginary values, then

$$\lim_{t \to \infty} y(t) = P(0, A) y(0).$$

Proof. Let $y(\cdot)$ be an arbitrary bounded solution of Equation (5.12). Then, by (5.13) and (5.14),

$$y(t) = e^{tA}x = \sum_{j=1}^{m} e^{\lambda_j t} P_j \sum_{k=0}^{i_j-1} \frac{t^k}{k!} Q_j^k x, \; t \geq 0, \tag{5.17}$$

with some $y(0) = x \in X$.

Since, by (5.10), the *spectral projections* P_j, $j = 1, \ldots, m$, form a *resolution of the identity*, we can introduce a new norm on X as follows:

$$X \ni x \mapsto \|x\|_1 := \sum_{j=1}^{m} \|P_j x\|.$$

Exercise 5.41. Verify the norm axioms for $\| \cdot \|_1$.

Since X is *finite-dimensional*, by the *Norm Equivalence Theorem* (Theorem 3.9), the new norm is *equivalent* to the original one (see, e. g., [67]).

This implies that the boundedness of $y(\cdot)$ is equivalent to the boundedness of each summand

$$e^{\lambda_j t} P_j \sum_{k=0}^{i_j-1} \frac{t^k}{k!} Q_j^k x, \; t \geq 0, \; j = 1, \ldots, m, \tag{5.18}$$

in representation (5.17).

For each $j = 1, \ldots, m$, we have the following cases:

(1) If $\operatorname{Re}\lambda_j < 0$, summand (5.18) is *bounded* and converges to 0 as $t \to \infty$.

Exercise 5.42. Verify by reasoning as in the proof of the *Lyapunov Stability Theorem* (Theorem 5.14).

(2) If $\operatorname{Re}\lambda_j \geq 0$, in view of (5.10), by (3.2) and *subadditivity* and *absolute scalability* of norm, for any $t \geq 0$,

$$\left\| e^{\lambda_j t} P_j \sum_{k=0}^{i_j-1} \frac{t^k}{k!} Q_j^k x \right\| \geq e^{\operatorname{Re}\lambda_j t} \left[\left\| \frac{t^{i_j-1}}{(i_j-1)!} Q_j^{i_j-1} P_j x \right\| - \left\| \sum_{k=0}^{i_j-2} \frac{t^k}{k!} Q_j^k P_j x \right\| \right]$$

$$\geq e^{\operatorname{Re}\lambda_j t} \left[\frac{t^{i_j-1}}{(i_j-1)!} \|Q_j^{i_j-1} P_j x\| - \sum_{k=0}^{i_j-2} \frac{t^k}{k!} \|Q_j^k P_j x\| \right],$$

and hence, the *boundedness* of summand (5.18) necessarily implies that

$$Q_j^{i_j-1} P_j x = 0.$$

Continuing in this fashion, we arrive at the following conclusion:

- if $\operatorname{Re} \lambda_j > 0$,

$$Q_j^k P_j x = 0, \ k = 0, \ldots, i_j - 1,$$

i. e., $P_j x = 0$, and hence,

$$e^{\lambda_j t} P_j \sum_{k=0}^{i_j-1} \frac{t^k}{k!} Q_j^k x = 0, \ t \geq 0;$$

- if $\operatorname{Re} \lambda_j = 0$,

$$Q_j^k P_j x = 0, \ k = 1, \ldots, i_j - 1,$$

i. e., in view of (5.10), $Q_j x = Q_j P_j x = 0$, and hence,

$$e^{\lambda_j t} P_j \sum_{k=0}^{i_j-1} \frac{t^k}{k!} Q_j^k x = e^{\lambda_j t} P_j x, \ t \geq 0.$$

Thus, for a bounded solution $y(\cdot)$ of Equation (5.12), representation (5.17) acquires the form

$$y(t) = \sum_{j:\, \operatorname{Re} \lambda_j < 0} e^{\lambda_j t} P_j \sum_{k=0}^{i_j-1} \frac{t^k}{k!} Q_j^k x + \sum_{j:\, \operatorname{Re} \lambda_j = 0} e^{\lambda_j t} P_j x, \ t \geq 0, \tag{5.19}$$

in which
- the sum corresponding to the eigenvalues of A with negative real parts vanishes at infinity:

$$\lim_{t \to \infty} \sum_{j:\, \operatorname{Re} \lambda_j < 0} e^{\lambda_j t} P_j \sum_{k=0}^{i_j-1} \frac{t^k}{k!} Q_j^k y = 0;$$

- the sum corresponding to the *pure imaginary* eigenvalues of A vanishes at infinity in the *Cesáro sense* since, for each $\lambda_j \in i\mathbb{R} \setminus \{0\}$, where $i\mathbb{R}$ stands for the *imaginary axis*,

$$\lim_{t \to \infty} \frac{1}{t} \int_0^t e^{\lambda_j s} P_j y \, ds = \lim_{t \to \infty} \frac{e^{\lambda_j t} - 1}{t \lambda_j} P_j y = 0;$$

- provided, for some $j = 1, \ldots, m$, $\lambda_j = 0$, the corresponding constant term in (5.19) is $P_j y$, where $P_j = P(0, A)$ is the *spectral projection* at 0.
  If $0 \in \rho(A)$, $P(0, A) := 0$ (see Remark 5.21).

Whence, the conclusion of the statement follows immediately. □

Example 5.16. A linear operator A on a complex five-dimensional Banach space $(X, \| \cdot \|)$ with a *Jordan canonical matrix representation*

$$[A]_B = \begin{bmatrix} 1 & 0 & 0 & 0 & 0 \\ 0 & 0 & 0 & 0 & 0 \\ 0 & 0 & i & 1 & 0 \\ 0 & 0 & 0 & i & 1 \\ 0 & 0 & 0 & 0 & i \end{bmatrix}$$

relative to a Jordan basis B has three eigenvalues: *simple* $\lambda_1 = 1$ and $\lambda_2 = 0$ and *defective* $\lambda_3 = i$ of algebraic multiplicity $k_3 = 3$ and geometric multiplicity $p_3 = 1$ (cf. Examples 5.13).

Exercise 5.43. Explain.

Hence, the *spectral decomposition* of A is

$$A = (1P_1 + Q_1) + (0P_2 + Q_2) + (iP_3 + Q_3),$$

where the *spectral projections* $P_j, j = 1, 2, 3$, have the following matrix representations relative to the Jordan basis B:

$$[P_1]_B = \begin{bmatrix} 1 & 0 & 0 & 0 & 0 \\ 0 & 0 & 0 & 0 & 0 \\ 0 & 0 & 0 & 0 & 0 \\ 0 & 0 & 0 & 0 & 0 \\ 0 & 0 & 0 & 0 & 0 \end{bmatrix}, \quad [P_2]_B = \begin{bmatrix} 0 & 0 & 0 & 0 & 0 \\ 0 & 1 & 0 & 0 & 0 \\ 0 & 0 & 0 & 0 & 0 \\ 0 & 0 & 0 & 0 & 0 \\ 0 & 0 & 0 & 0 & 0 \end{bmatrix},$$

$$[P_3]_B = \begin{bmatrix} 0 & 0 & 0 & 0 & 0 \\ 0 & 0 & 0 & 0 & 0 \\ 0 & 0 & 1 & 0 & 0 \\ 0 & 0 & 0 & 1 & 0 \\ 0 & 0 & 0 & 0 & 1 \end{bmatrix},$$

respectively, and the *nilpotent operators* $Q_j, j = 1, 2, 3$, are as follows:

$$Q_1 = Q_2 = 0,$$

and Q_3 with $Q_3^3 = 0$ has the matrix representation

$$[Q_3]_B = \begin{bmatrix} 0 & 0 & 0 & 0 & 0 \\ 0 & 0 & 0 & 0 & 0 \\ 0 & 0 & 0 & 1 & 0 \\ 0 & 0 & 0 & 0 & 1 \\ 0 & 0 & 0 & 0 & 0 \end{bmatrix}$$

relative to the Jordan basis B (see Examples 5.13).

By representation (5.14),

$$e^{tA} = e^{1t}P_1 + e^{0t}P_2 + e^{it}P_3\left(I + tQ_3 + \frac{t^2}{2!}Q_3^2\right), \ t \in \mathbb{C}.$$

A solution

$$y(t) = e^{tA}x = e^t P_1 x + P_2 x + e^{it}P_3\left(I + tQ_3 + \frac{t^2}{2!}Q_3^2\right)x, \ t \geq 0, x \in X,$$

of evolution equation (5.12) (see Section 5.6.4.1) is bounded *iff*

$$P_1 x = 0 \quad \text{and} \quad Q_3 x = 0 \tag{5.20}$$

(see the proof of the prior theorem), which is equivalent to

$$\exists c_2, c_3 \in \mathbb{C} : [y(0)]_B = [x]_B = (0, c_2, c_3, 0, 0) \in \mathbb{C}^5$$

(see Section 1.4).

Exercise 5.44. Explain.

Therefore, any bounded solution $y(\cdot)$ of evolution equation (5.12) is of the form

$$y(t) = P_2 x + e^{it}P_3 x, \ t \geq 0,$$

where $x \in X$ satisfying (5.20) is arbitrary.

Thus, consistently with the prior theorem, for any bounded solution $y(\cdot)$ of (5.12),

$$\lim_{t \to \infty} \frac{1}{t} \int_0^t y(s) \, ds = P(0, A)y(0) = P_2 y(0) \in \ker A,$$

where

$$\mathbb{C}^5 \ni (0, c_2, c_3, 0, 0) = [y(0)]_B \mapsto [P_2 y(0)]_B = (0, c_2, 0, 0, 0) \in \mathbb{C}^5$$

(see the *Cesàro Limit Proposition* (Proposition 5.13, Section 5.8, Problem 19)).

Remark 5.27. Observe that, in this case, the *spectral projection* $P(0, A) = P_2$ of the operator A at $\lambda_2 = 0$ is the projection onto the corresponding eigenspace $\ker A$ along the range $R(A)$ (see Section 5.6.3 and [48, Section 5.3]).

5.7 Compact Linear Operators

Compact linear operators constitute a very important class of bounded linear operators arising in the study of integral equations. In this chapter, we consider some examples and properties of such operators, and develop the spectral theory for them.

5.7.1 Definition, Characterization, Examples

Recall that $\mathbb{F}$ stands for the *scalar field* of $\mathbb{R}$ or $\mathbb{C}$.

Definition 5.14 (Compact Linear Operator). Let $(X, \|\cdot\|_X)$ and $(Y, \|\cdot\|_Y)$ be normed vector spaces over $\mathbb{F}$.

A linear operator $A : X \to Y$ is called *compact* (or *completely continuous*) if it maps *bounded sets* of $(X, \|\cdot\|_X)$ to *precompact sets* of $(Y, \|\cdot\|_Y)$.

The set of all such operators is designated by $K(X, Y)$. Provided $Y = X$, the notation $K(X)$ is used.

Remarks 5.28.
- In view of the linearity of A,

$$A \in K(X, Y) \;\Leftrightarrow\; A\overline{B}_X(0,1) \text{ is a precompact set in } (Y, \|\cdot\|_Y).$$

- Bounded linear operators mapping bounded sets to bounded sets (see *Characterizations of Bounded Linear Operators* (Theorem 4.4)) and precompact sets being bounded (see the *Precompactness and Total Boundedness Proposition* (Proposition 2.8) and the *Properties of Totally Bounded Sets* (Proposition 2.43)), every compact linear operator is necessarily bounded, and hence, the inclusion

$$K(X, Y) \subseteq L(X, Y)$$

holds. However, as we see below, generally, a bounded linear operator need not be compact.
- The image of any precompact set in $(X, \|\cdot\|_X)$ under an arbitrary bounded linear operator $A \in L(X, Y)$ is a precompact set in $(Y, \|\cdot\|_Y)$.

Exercise 5.45. Verify.

Hint. For the last statement, apply the *Sequential Characterization of Precompactness* (Theorem 2.51).

Proposition 5.2 (Sequential Characterization of Compact Operators). *Let* $(X, \|\cdot\|_X)$ *and* $(Y, \|\cdot\|_Y)$ *be normed vector spaces over* $\mathbb{F}$. *A linear operator* $A : X \to Y$ *is compact (i. e., $A \in K(X, Y)$) iff, for any bounded sequence* $(x_n)_{n \in \mathbb{N}}$ *in* $(X, \|\cdot\|_X)$, *there exists a subsequence* $(x_{n(k)})_{k \in \mathbb{N}}$ *such that the sequence* $(Ax_{n(k)})_{k \in \mathbb{N}}$ *converges in* $(Y, \|\cdot\|_Y)$, *i. e., the image sequence* $(Ax_n)_{n \in \mathbb{N}}$ *contains a convergent subsequence.*

Exercise 5.46. Prove.

Hint. Apply the *Sequential Characterization of Precompactness* (Theorem 2.51).

Examples 5.17.

1. By the *Generalized Heine–Borel Theorem* (Theorem 3.10), for normed vector spaces $(X, \| \cdot \|_X)$ and $(Y, \| \cdot \|_Y)$ over $\mathbb{F}$, each operator $A \in L(X, Y)$ of *finite rank*, i.e., such that $\dim R(A) = n$ with some $n \in \mathbb{Z}_+$, is compact.

    In particular, the *zero operator* on X is compact.

    If $\dim R(A) = n$ $(n \in \mathbb{N})$, then relative to an arbitrary basis $B = \{y_1, \dots, y_n\}$ for $R(A)$, the operator A allows the following representation:

    $$Ax = \sum_{k=1}^{n} c_k(Ax)y_k = \sum_{k=1}^{n} f_k(x)y_k,$$

    where $f_k := c_k \circ A \in X^*$, $k = 1, \dots, n$, with $c_1, \dots, c_n \in R(A)^*$ being the *(Hamel) coordinate functionals* relative to B (see Examples 4.1).

    Exercise 5.47. Verify.

2. The target space Y being *finite-dimensional*, all linear operators $A : X \to Y$ are of *finite rank*, and hence, *compact*, i.e., in this case,

    $$K(X, Y) = L(X, Y).$$

    In particular all linear functionals in the *dual space* X^* are *compact linear operators* on X.

3. On the (real or complex) space l_p $(1 \le p < \infty)$ or $(c_0, \| \cdot \|_\infty)$,
    (a) in view of the characterizations of *precompactness* in these spaces (see Theorems 2.48, 2.49 and Remark 2.69), for an arbitrary *vanishing sequence* $a := (a_n)_{n \in \mathbb{N}} \in c_0$, the multiplication operator

    $$x := (x_n)_{n \in \mathbb{N}} \mapsto Ax := (a_n x_n)_{n \in \mathbb{N}}$$

    is *compact*; in particular, when $a := (a_n)_{n \in \mathbb{N}} \in c_{00}$, the operator A is of *finite rank* (see the *Compact Multiplication Operators on Sequence Spaces Corollary* (Corollary 5.7));
    (b) the *right shift operator*

    $$x := (x_1, x_2, \dots) \mapsto Ax := (0, x_1, x_2, \dots)$$

    and the *left shift operator*

    $$x := (x_1, x_2, \dots) \mapsto Bx := (x_2, x_3, x_4, \dots)$$

    are bounded, but *not compact*.

    Exercise 5.48. Verify.

4. On the *incomplete space* $(c_{00}, \|\cdot\|_\infty)$, for the *vanishing sequence* $(1/n)_{n\in\mathbb{N}} \in c_0$, the operator of multiplication

$$x := (x_n)_{n\in\mathbb{N}} \mapsto Ax := (x_n/n)_{n\in\mathbb{N}}$$

is *not* compact by the *Sequential Characterization of Compact Operators* (Proposition 5.2) since, for the sequence $(x_n := (\underbrace{1,\ldots,1}_{n \text{ terms}},0,0,\ldots))_{n\in\mathbb{N}} \subseteq \overline{B}(0,1)$,

$$Ax_n = (1,1/2,\ldots,1/n,0,0,\ldots) \to (1/n)_{n\in\mathbb{N}} \in c_0 \setminus c_{00}, \ n \to \infty, \ \text{in } (c_0, \|\cdot\|_\infty),$$

and hence, the image sequence $(Ax_n)_{n\in\mathbb{N}}$ has no subsequence convergent in the space $(c_{00}, \|\cdot\|_\infty)$.

5. All the subsequent examples are in the (real or complex) space $(C[a,b], \|\cdot\|_\infty)$.

(a) The bounded linear operator of multiplication by a function $m \in C[a,b]$,

$$C[a,b] \ni f \to [Ax](t) := m(t)f(t) \in C[a,b]$$

(see Examples 4.2), is compact *iff* $m(t) \equiv 0$ on $[a,b]$.

Exercise 5.49. Verify.

(b) **Theorem 5.16** (Compactness of Integral Operators). *Any integral operator of the form*

$$C[a,b] \ni x \to [Ax](t) := \int_a^b K(t,s)x(s)\, ds \in C[a,b]$$

with an arbitrary $K \in C([a,b] \times [a,b])$ *is compact on the space* $(C[a,b], \|\cdot\|_\infty)$ *(i.e., $A \in K(C[a,b])$).*

Proof. Suppose that $K \in C([a,b] \times [a,b])$.
Let us first show that

$$A : C[a,b] \to C[a,b].$$

Indeed, by the *Heine–Cantor Uniform Continuity Theorem* (Theorem 2.56) applied to the function $K(\cdot,\cdot)$ on the *compact* in $l_2^{(2)}$ set $[a,b] \times [a,b]$ (see *Heine–Borel Theorem* (Theorem 2.47)),

$$\forall \varepsilon > 0 \ \exists \delta > 0 \ \forall (t_1,s), (t_2,s) \in [a,b] \times [a,b] \text{ with } |t_1 - t_2| < \delta :$$
$$|K(t_1,s) - K(t_2,s)| < \frac{\varepsilon}{b-a}.$$

Hence, for an arbitrary $x \in C[a,b]$,

$$\left| [Ax](t_1) - [Ax](t_2) \right| = \left| \int_a^b K(t_1,s)x(s)\,ds - \int_a^b K(t_2,s)x(s)\,ds \right|$$

$$\le \int_a^b \left| K(t_1,s) - K(t_2,s) \right| \left| x(s) \right| ds \le \frac{\varepsilon}{b-a}(b-a)\|x\|_\infty = \varepsilon\|x\|_\infty, \qquad (5.21)$$

which implies that the function Ax is *uniformly continuous* on $[a,b]$, and hence, $Ax \in C[a,b]$.

Also, by the *Weierstrass Extreme Value Theorem* (Theorem 2.55),

$$M := \max_{(t,s)\in[a,b]^2} \left| K(t,s) \right| \in [0,\infty),$$

and hence, for an arbitrary $x \in C[a,b]$,

$$\|Ax\|_\infty = \max_{a\le t\le b} \left| \int_a^b K(t,s)x(s)\,ds \right| \le \max_{a\le t\le b} \int_a^b \left| K(t,s) \right| \left| x(s) \right| ds$$

$$\le M(b-a)\|x\|_\infty, \qquad (5.22)$$

which shows that

$$A \in L\big(C[a,b]\big) \quad \text{and} \quad \|A\| \le M(b-a).$$

Estimates (5.22) and (5.21) also show that, for any bounded set S in the space $(C[a,b], \|\cdot\|_\infty)$ ($\sup_{x\in S}\|x\|_\infty < \infty$), the image $A(S)$ is *uniformly bounded* and *equicontinuous* on $[a,b]$.

Exercise 5.50. Explain.

Hence, for any bounded set S in $(C[a,b], \|\cdot\|_\infty)$, by the *Arzelà–Ascoli Theorem* (Theorem 2.59), the image $A(S)$ is *precompact* in $(C[a,b], \|\cdot\|_\infty)$, which implies that the operator A is *compact* (i. e., $A \in K(C[a,b])$). □

Remark 5.29. The *Existence and Uniqueness Theorem for Fredholm Integral Equations of the Second Kind* (Theorem 2.37), stating that such an equation

$$x(t) - \lambda \int_a^b K(t,s)x(s)\,ds = y(t)$$

with an arbitrary $K \in C([a,b] \times [a,b])$ has a unique solution in $C[a,b]$ for each $y \in C[a,b]$ whenever $|\lambda| < \frac{1}{M(b-a)}$, implies that

$$\{\lambda \in \mathbb{C} \mid |\lambda| > M(b-a)\} \subseteq \rho(A).$$

Exercise 5.51. Explain.

By *Gelfand's Spectral Radius Theorem* (Theorem 5.8), this is consistent with the fact that

$$\|A\| \le M(b-a).$$

(see the proof of the prior theorem).

(c) Based on the *Arzelà–Ascoli Theorem* (Theorem 2.59), one can similarly prove

Theorem 5.17 (Compactness of Volterra Integral Operators). *Any Volterra integral operator*

$$C[a,b] \ni x \to [Ax](t) := \int_a^t K(t,s)x(s)\, ds \in C[a,b],$$

where $K \in C([a,b] \times [a,b])$, is compact on the space $(C[a,b], \| \cdot \|_\infty)$ (i. e., $A \in K(C[a,b])$).

Exercise 5.52. Prove.

In particular, for $K(t,s) := 1$, $(t,s) \in [a,b] \times [a,b]$, the integration operator

$$C[a,b] \ni x \to [Ax](t) := \int_a^t x(s)\, ds \in C[a,b]$$

(see Examples 4.2, 5.2, and 5.4) is *compact* on the space $(C[a,b], \| \cdot \|_\infty)$ (i. e., $A \in K(C[a,b])$).

5.7.2 Compactness of the Identity Operator

The case of the identity operator on a normed vector space is of particular importance.

Proposition 5.3 (Compactness of the Identity Operator). *The identity operator I on a normed vector space $(X, \| \cdot \|_X)$ is compact iff the space X is finite-dimensional.*

Exercise 5.53. Prove.

Hint. Apply the *Characterization of Finite-Dimensional Normed Vector Spaces* (Corollary 3.6).

The prior statement, when rephrased as in Proposition 5.4, becomes another characterization of finite-dimensional normed vector spaces (see the *Characterization of Finite-Dimensional Banach Spaces* (Proposition 3.17, Section 3.5, Problem 17) and the *Characterization of Finite-Dimensional Banach Spaces* (Proposition 3.19, Section 3.5, Problem 19)).

Proposition 5.4 (Characterization of Finite-Dimensional Normed Vector Spaces).
A normed vector space $(X, \| \cdot \|_X)$ is finite-dimensional iff the identity operator I on X is compact.

Exercise 5.54. Prove.

5.7.3 Properties

Theorem 5.18 (Properties of Compact Linear Operators). *Let $(X, \| \cdot \|_X)$, $(Y, \| \cdot \|_Y)$, and $(Z, \| \cdot \|_Z)$ be normed vector spaces over $\mathbb{F}$.*
(1) *The set $K(X, Y)$ is a subspace of $L(X, Y)$.*
(2) *If $A \in K(X, Y)$ and $B \in L(Y, Z)$, then $BA \in K(X, Z)$.*
   If $A \in L(X, Y)$ and $B \in K(Y, Z)$, then $BA \in K(X, Z)$.
(3) *The set $K(X)$ is closed under operator multiplication, i.e., is a subalgebra of the operator algebra $L(X)$. Furthermore, $K(X)$ is an ideal of $L(X)$, i.e.,*

$$\forall B \in L(X) : K(X) \cdot B := \{AB \mid A \in K(X)\} \subseteq K(X) \text{ and}$$

$$B \cdot K(X) := \{BA \mid A \in K(X)\} \subseteq K(X).$$

(4) *Provided $(Y, \| \cdot \|_Y)$ is a Banach space, for an arbitrary operator sequence $(A_n)_{n\in\mathbb{N}}$ in $K(X, Y)$ uniformly convergent to an operator $A \in L(X, Y)$, i.e.,*

$$A_n \to A \in L(X, Y) \text{ in } (L(X, Y), \| \cdot \|),$$

   where $\| \cdot \|$ is the operator norm, $A \in K(X, Y)$.
   In particular, if $(X, \| \cdot \|_X)$ is a Banach space, for an arbitrary operator sequence $(A_n)_{n\in\mathbb{N}}$ in $K(X)$ uniformly convergent to an operator $A \in L(X)$, $A \in K(X)$.
(5) *Provided $(Y, \| \cdot \|_Y)$ is a Banach space, $(K(X, Y), \| \cdot \|)$ is a Banach space.*
   In particular, if $(X, \| \cdot \|_X)$ is a Banach space, $(K(X), \| \cdot \|)$ is a Banach space.

Proof. We prove only parts (1), (4), and (5) here, proving parts (2) and (3) being left as an exercise.
(1) For arbitrary $A, B \in K(X, Y)$ and $\lambda, \mu \in \mathbb{F}$, and any bounded sequence $(x_n)_{n\in\mathbb{N}}$ in $(X, \| \cdot \|_X)$, since the operator A is compact, by the *Sequential Characterization of Compact Operators* (Proposition 5.2), there exists a subsequence $(x_{n(k)})_{K\in\mathbb{N}}$ such that the sequence $(Ax_{n(k)})_{K\in\mathbb{N}}$ converges in $(Y, \| \cdot \|_Y)$.
   Similarly, since the operator B is also compact, the subsequence $(x_{n(k)})_{K\in\mathbb{N}}$ in its turn contains a subsequence $(x_{n(k(j))})_{j\in\mathbb{N}}$ such that the sequence $(Bx_{n(k(j))})_{j\in\mathbb{N}}$ converges in $(Y, \| \cdot \|_Y)$.
   Therefore, for the subsequence $(x_{n(k(j))})_{j\in\mathbb{N}}$ of the initial sequence $(x_n)_{n\in\mathbb{N}}$, the sequence $((\lambda A + \mu B)x_{n(k(j))})_{j\in\mathbb{N}}$ converges in $(Y, \| \cdot \|_Y)$.

Exercise 5.55. Explain (see Remarks 3.18).

By the *Sequential Characterization of Compact Operators* (Proposition 5.2), this proves that $\lambda A + \mu B \in K(X, Y)$, and hence, $K(X, Y)$ is a *subspace* of $L(X, Y)$ (see Remarks 3.4).

(4) Suppose that an operator sequence $(A_n)_{n\in\mathbb{N}}$ in $K(X, Y)$ *uniformly converges* to an operator $A \in L(X, Y)$, i. e.,

$$A_n \to A \in L(X, Y) \text{ in } (L(X, Y), \|\cdot\|),$$

and let a *bounded* sequence $(x_n)_{n\in\mathbb{N}}$ in $(X, \|\cdot\|_X)$ (i. e., $\sup_{k\in\mathbb{N}} \|x_k\| < \infty$) be arbitrary.

Then, by the *Sequential Characterization of Compact Operators* (Proposition 5.2), since the operator A_1 is compact, the sequence $(x_n)_{n\in\mathbb{N}}$ contains a subsequence $(x_{1,n})_{n\in\mathbb{N}}$ such that the sequence $(A_1 x_{1,n})_{n\in\mathbb{N}}$ converges in $(Y, \|\cdot\|_Y)$. Similarly, since the operator A_2 is compact, the subsequence $(x_{1,n})_{n\in\mathbb{N}}$ in its turn contains a subsequence $(x_{2,n})_{n\in\mathbb{N}}$ such that the sequence $(A_2 x_{2,n})_{n\in\mathbb{N}}$ converges in $(Y, \|\cdot\|_Y)$. Continuing inductively, we obtain a countable collection of sequences

$$\{(x_{m,n})_{n\in\mathbb{N}} \mid m \in \mathbb{Z}_+\},$$

such that

$$(x_{0,n})_{n\in\mathbb{N}} := (x_n)_{n\in\mathbb{N}},$$

for each $m \in \mathbb{N}$, $(x_{m,n})_{n\in\mathbb{N}}$ is a subsequence of $(x_{(m-1),n})_{n\in\mathbb{N}}$, and $(A_m x_{m,n})_{n\in\mathbb{N}}$ converges in $(Y, \|\cdot\|_Y)$.

Let us show now that, for the *"diagonal subsequence"* $(x_{n,n})_{n\in\mathbb{N}}$ of the sequence $(x_n)_{n\in\mathbb{N}}$, the sequence $(A x_{nn})_{n\in\mathbb{N}}$ converges in $(Y, \|\cdot\|_Y)$.

Indeed, since the operator sequence $(A_n)_{n\in\mathbb{N}}$ converges in $(L(X, Y), \|\cdot\|)$,

$$\forall \varepsilon > 0 \,\exists M \in \mathbb{N}: \|A - A_M\| < \varepsilon / \left(4\left(\sup_{k\in\mathbb{N}} \|x_k\| + 1\right)\right). \tag{5.23}$$

Furthermore, the sequence $(A_M x_{n,n})_{n\in\mathbb{N}}$ *converges* in $(Y, \|\cdot\|_Y)$ being a subsequence of the convergent sequence $(A_M x_{M,n})_{n\in\mathbb{N}}$ for $n \geq M$, and hence, is *fundamental* in $(Y, \|\cdot\|_Y)$, i. e.,

$$\exists N \in \mathbb{N} \,\forall m, n \geq N: \|A_M x_{n,n} - A_M x_{m,m}\|_Y < \varepsilon/2. \tag{5.24}$$

In view of (5.23) and (5.24), by *subadditivity* of norm,

$$\forall m, n \geq N: \|A x_{n,n} - A x_{m,m}\|_Y$$
$$= \|A x_{n,n} - A_M x_{n,n} + A_M x_{n,n} - A_M x_{m,m} + A_M x_{m,m} - A x_{m,m}\|_Y$$
$$\leq \|A x_{n,n} - A_M x_{n,n}\|_Y + \|A_M x_{n,n} - A_M x_{m,m}\|_Y + \|A_M x_{m,m} - A x_{m,m}\|_Y$$

$$\leq \|A - A_M\|\|x_{n,n}\|_X + \|A_M x_{n,n} - A_M x_{m,m}\|_Y + \|A_M - A\|\|x_{m,m}\|_X$$

$$\leq 2\|A - A_M\| \sup_{k\in\mathbb{N}} \|x_k\| + \|A_N x_{n,n} - A_N x_{m,m}\|_Y < 2(\varepsilon/4) + \varepsilon/2 = \varepsilon,$$

which implies that the sequence $(Ax_{n,n})_{n\in\mathbb{N}}$ is *fundamental* in $(Y, \|\cdot\|_Y)$, and hence, since the space $(Y, \|\cdot\|_Y)$ is Banach, converges in it.

By the *Sequential Characterization of Compact Operators* (Proposition 5.2), we conclude that $A \in K(X, Y)$.

(5) As follows from part (4) by the *Sequential Characterization of a Closed Set* (Theorem 2.19), $K(X, Y)$ is a closed subspace of the Banach space $(L(X, Y), \|\cdot\|)$ (see the *Space of Bounded Linear Operators Theorem* (Theorem 4.5)). Whence, by the *Characterization of Completeness* (Theorem 2.27), we conclude that $(K(X, Y), \|\cdot\|)$ is a Banach space. $\qquad\square$

Exercise 5.56. Prove parts (2) and (3).

Remarks 5.30.

– In part (4) of the prior theorem, the requirement of *completeness* for the target space $(Y, \|\cdot\|_Y)$ is essential and cannot be dropped. Indeed, on the *incomplete* normed vector space $(c_{00}, \|\cdot\|_\infty)$, for the *vanishing sequence* $(1/n)_{n\in\mathbb{N}} \in c_0$, the multiplication operator

$$x := (x_n)_{n\in\mathbb{N}} \mapsto Ax := (x_n/n)_{n\in\mathbb{N}}$$

is *not compact* (see Examples 5.17), although A is the *uniform limit* of the sequence of *finite-rank*, and hence compact, operators

$$x := (x_k)_{k\in\mathbb{N}} \mapsto A_n x := P_n Ax = (x_1, x_2/2, \dots, x_n/n, 0, 0, \dots),$$

where

$$c_{00} \ni (x_k)_{k\in\mathbb{N}} \mapsto P_n x := (x_1, x_2, \dots, x_n, 0, 0, \dots) \in c_{00}, \ n \in \mathbb{N},$$

is the *projection operator* onto the n-dimensional subspace

$$Y_n := \{(x_k)_{k\in\mathbb{N}} \in c_{00} \mid x_k = 0, \ k \geq n+1\} \in c_{00}$$

along the closed infinite-dimensional subspace

$$Z_n := \{(x_k)_{k\in\mathbb{N}} \in c_{00} \mid x_k = 0, \ k = 1, \dots, n\}$$

in $(c_{00}, \|\cdot\|_\infty)$ (also $(c_0, \|\cdot\|_\infty)$) (see Examples 6.10).

Exercise 5.57. Verify.

– In part (4) of the prior theorem, the condition of the *uniform convergence* of the operator sequence is essential and cannot be tempered to strong convergence. Indeed, on the Banach space $(c_0, \| \cdot \|_\infty)$, the sequence of *finite-rank*, and hence compact, projection operators

$$c_0 \ni x := (x_k)_{k \in \mathbb{N}} \mapsto P_n x := (x_1, x_2, \ldots, x_n, 0, 0, \ldots) \in c_0, \; n \in \mathbb{N},$$

strongly converges to the *identity operator*, since

$$\forall x := (x_k)_{k \in \mathbb{N}} \in c_0 : \; P_n x \to x =: Ix, \; n \to \infty.$$

Exercise 5.58. Verify.

However, since the space c_0 is *infinite-dimensional*, by the *Compactness of the Identity Operator Proposition* (Proposition 5.3), the identity operator on $(c_0, \| \cdot \|_\infty)$ is *not compact*.

5.7.4 Approximation by Finite-Rank Operators

Here, we see that, under a certain condition, compact operators can be uniformly approximated by finite-rank operators. We start with an example showing that, generally, the uniform limit of a sequence of finite-rank operators need not be of finite rank.

Example 5.18. On the Banach space l_p $(1 \le p < \infty)$ or $(c_0, \| \cdot \|_\infty)$, for the *vanishing sequence* $(1/n)_{n \in \mathbb{N}} \in c_0$, the operator of multiplication

$$x := (x_n)_{n \in \mathbb{N}} \mapsto Ax := (x_n/n)_{n \in \mathbb{N}}$$

is *compact* (see Examples 5.17), but *not of finite rank*.

Exercise 5.59. Explain why A is *not of finite rank*.

However, A is the *uniform limit* of the sequence of finite-rank operators

$$(x_k)_{k \in \mathbb{N}} \mapsto A_n x := P_n A x = (x_1, x_2/2, \ldots, x_n/n, 0, 0, \ldots),$$

where

$$(x_k)_{k \in \mathbb{N}} \mapsto P_n x := (x_1, x_2, \ldots, x_n, 0, 0, \ldots), \; n \in \mathbb{N}.$$

Exercise 5.60. Verify (see Remarks 5.30).

Theorem 5.19 (Approximation by Finite-Rank Operators). *Let $(X, \| \cdot \|_X)$ be a Banach space and $(Y, \| \cdot \|_Y)$ be a normed vector space over $\mathbb{F}$, for which there exists a sequence*

$(P_n)_{n \in \mathbb{N}} \subset K(Y)$ *of finite-rank operators strongly convergent to the identity operator on* Y, *i. e.,*

$$\forall y \in Y : P_n y \to y, \ n \to \infty, \ in \ (Y, \| \cdot \|_Y). \tag{5.25}$$

Then every compact operator $A \in K(X, Y)$ *is the uniform limit, i. e., the limit in* $(L(X, Y), \| \cdot \|)$, *where* $\| \cdot \|$ *is the operator norm, of the sequence* $(P_n A)_{n \in \mathbb{N}} \subset K(X, Y)$ *of finite-rank operators.*

Proof. Let $A \in K(X, Y)$ be arbitrary.

The fact that $(P_n A)_{n \in \mathbb{N}} \subset K(X, Y)$ is a sequence of finite-rank operators is quite clear.

Exercise 5.61. Explain.

Let us prove that $(P_n A)_{n \in \mathbb{N}}$ uniformly converges to A *by contradiction,* assuming that

$$P_n A \not\to A, \ n \to \infty, \ in \ (L(X, Y), \| \cdot \|).$$

Then there exist an $\varepsilon > 0$ and a subsequence $(P_{n(k)} A)_{k \in \mathbb{N}}$ such that

$$\| P_{n(k)} A - A \| \geq \varepsilon, \ k \in \mathbb{N},$$

which, by the definition of the operator norm (see Definition 4.4), implies that there exists a sequence $(x_k)_{k \in \mathbb{N}}$ of vectors in X with

$$\| x_k \|_X = 1, \ k \in \mathbb{N},$$

such that

$$\| P_{n(k)} A x_k - A x_k \|_Y = \| (P_{n(k)} A - A) x_k \|_Y \geq \varepsilon/2, \ k \in \mathbb{N}. \tag{5.26}$$

Since the operator A is *compact,* by the *Sequential Characterization of Compact Operators* (Proposition 5.2), there exists a subsequence $(x_{k(j)})_{j \in \mathbb{N}}$ such that

$$A x_{k(j)} \to y \in Y, \ j \to \infty, \ in \ (Y, \| \cdot \|_Y). \tag{5.27}$$

Considering (5.25) and the fact that every convergent sequence is *bounded* (see the *Properties of Fundamental Sequences* (Theorem 2.22)), by the *Uniform Boundedness Principle* (Theorem 4.6), we infer that

$$\sup_{n \in \mathbb{N}} \| P_n \| < \infty.$$

Therefore, by *subadditivity* and *absolute scalability* of norm and *submultiplicativity* of operator norm, and in view of (5.27) and (5.25), we have the following:

$$\|P_{n(k(j))}Ax_{k(j)} - Ax_{k(j)}\|_Y$$
$$= \|P_{n(k(j))}Ax_{k(j)} - P_{n(k(j))}y + P_{n(k(j))}y - y + y - Ax_{k(j)}\|_Y$$
$$\leq \|P_{n(k(j))}Ax_{k(j)}y - P_{n(k(j))}y\|_Y + \|P_{n(k(j))}y - y\|_Y + \|y - Ax_{k(j)}\|_Y$$
$$\leq \|P_{n(k(j))}\|\|Ax_{k(j)} - y\|_Y + \|Ax_{k(j)} - y\|_Y + \|P_{n(k(j))}y - y\|_Y$$
$$= (\|P_{n(k(j))}\| + 1)\|Ax_{k(j)} - y\|_Y + \|P_{n(k(j))}y - y\|_Y$$
$$\leq \left(\sup_{n \in \mathbb{N}} \|P_n\| + 1\right)\|Ax_{k(j)} - y\|_Y + \|P_{n(k(j))}y - y\|_Y \to 0, \ j \to \infty,$$

which *contradicts* (5.26) and hence completes the proof. □

We immediately obtain

Corollary 5.6 (Approximation by Finite-Rank Operators). *Let $(X, \| \cdot \|_X)$ be a Banach space and $(Y, \| \cdot \|_Y)$ be a Banach space with a Schauder basis $E := \{e_k\}_{k \in \mathbb{N}}$ over $\mathbb{F}$. Then every compact operator $A \in K(X, Y)$ is the uniform limit, i. e., the limit in $(L(X, Y), \| \cdot \|)$, where $\| \cdot \|$ is the operator norm, of the sequence $(P_n A)_{n \in \mathbb{N}}$ of finite-rank operators, where*

$$Y \ni y := \sum_{k=1}^{\infty} c_k e_k \mapsto P_n y := \sum_{k=1}^{n} c_k e_k, \ n \in \mathbb{N}.$$

Exercise 5.62. Prove.

Remarks 5.31.
- For each $n \in \mathbb{N}$, P_n is the *projection operator* onto the n-dimensional subspace $\mathrm{span}(\{e_1, \dots, e_n\})$ along $\overline{\mathrm{span}(\{e_{n+1}, e_{n+2}, \dots\})}$.
- In particular, the statement of the prior corollary holds if Y is a *separable Hilbert space*, which necessarily has an *orthonormal* Schauder basis (see Chapter 6).

Now, we arrive at the following description of compact multiplication operators on the spaces l_p $(1 \leq p < \infty)$ and $(c_0, \| \cdot \|_\infty)$ (see Examples 5.17).

Corollary 5.7 (Compact Multiplication Operators on Sequence Spaces). *On the (real or complex) space l_p $(1 \leq p < \infty)$ or $(c_0, \| \cdot \|_\infty)$, the operator of multiplication*

$$x := (x_k)_{k \in \mathbb{N}} \mapsto Ax := (a_k x_k)_{k \in \mathbb{N}}$$

by a bounded sequence $a := (a_k)_{k \in \mathbb{N}} \in l_\infty$ is compact iff $a \in c_0$, i. e.,

$$\lim_{k \to \infty} a_k = 0,$$

the operator being of finite rank iff $a \in c_{00}$, i. e.,

$$\exists N \in \mathbb{N} \ \forall k \geq N : a_k = 0.$$

Proof. Here, we prove the *compactness part* only, the *finite-rank part* being left as an exercise.

"*If*" part is stated and given as an exercise in Examples 5.17.

"*Only if*" part. Suppose that, on the space on l_p $(1 \le p < \infty)$ or $(c_0, \| \cdot \|_\infty)$, for a bounded sequence $(a_k)_{k\in\mathbb{N}} \in l_\infty$, the multiplication operator

$$x := (x_k)_{k\in\mathbb{N}} \mapsto Ax := (a_k x_k)_{k\in\mathbb{N}}$$

is *compact*.

Recall that the set $E := \{e_k := (\delta_{km})_{m\in\mathbb{N}}\}_{k\in\mathbb{N}}$, where δ_{km} is the *Kronecker delta*, is a *Schauder basis* for the spaces l_p $(1 \le p < \infty)$ and $(c_0, \| \cdot \|_\infty)$, any sequence $x := (x_k)_{k\in\mathbb{N}}$ being represented as

$$x = \sum_{k=0}^{\infty} x_k e_k$$

(see Examples 3.17 and 3.52).

By the prior corollary, with the spaces X and Y being, respectively, l_p $(1 \le p < \infty)$ or $(c_0, \| \cdot \|_\infty)$, we conclude that, for the sequence $(P_n)_{n\in\mathbb{N}}$ of the *finite-rank projections*

$$(x_k)_{k\in\mathbb{N}} \mapsto P_n x := (x_1, x_2, \dots, x_n, 0, 0, \dots), \ n \in \mathbb{N},$$

the sequence $(P_n A)_{n\in\mathbb{N}}$ of finite-rank operators uniformly converges to A.

Hence,

$$\sup_{k \ge n+1} |a_k| = \sup_{\|x\|=1} \| (\underbrace{0, \dots, 0}_{n \text{ terms}}, a_{n+1} x_{n+1}, a_{n+2} x_{n+2}, \dots) \|$$
$$= \sup_{\|x\|=1} \| (A - P_n A)x \| = \| A - P_n A \| \to 0, \ n \to \infty, \tag{5.28}$$

where the same notation $\| \cdot \|$ is used to designate the norm on l_p or c_0 and the operator norm whenever appropriate.

This implies that

$$\lim_{k\to\infty} a_k = 0.$$

Exercise 5.63.

(a) Explain the first equality in (5.28) (see Examples 4.2).

(b) Prove the *finite-rank part*. □

5.7.5 Fredholm Alternative

By the *Rank-Nullity Theorem* (Theorem 4.1) and the *Codimension of a Subspace Corollary* (Corollary 3.4) (see Remark 4.3), for arbitrary linear operator on a finite-dimensional vector space X,

$$\dim \ker A = \operatorname{codim} R(A).$$

Here, we stretch this equality to the operators of the form $I - A$, where A is a compact linear operator on an infinite-dimensional Banach space.

Theorem 5.20 (Fredholm Alternative). *Let A be a compact linear operator on a Banach space $(X, \|\cdot\|)$ (i. e., $A \in K(X)$). Then*

$$\operatorname{codim} R(I - A) = \dim \ker(I - A) = n$$

for some $n \in \mathbb{Z}_+$.
    In particular,

$$R(I - A) = X \iff \ker(I - A) = \{0\},$$

i. e., either the equation

$$(I - A)x = y$$

has a unique solution for each $y \in X$ or the equation

$$(I - A)x = 0$$

has a nontrivial solution $x \neq 0$.

A proof for the general case can be found in [60]. A proof under the additional condition that there exists a *finite-rank operator* $B \in K(X)$ such that

$$\|A - B\| < 1,$$

which, by the *Approximation by Finite-Rank Operators Corollary* (Corollary 5.6) holds if the Banach space $(X, \|\cdot\|)$ has a Schauder basis, e. g., for such spaces as l_p ($1 \leq p < \infty$), $(c_0, \|\cdot\|_\infty)$, $(c, \|\cdot\|_\infty)$, and $(C[a, b], \|\cdot\|_\infty)$ (see Section 3.2.5), is given in [26, Chapter XI, Theorem 4.1]. In Section 6.11.1, we are to deliver a proof in a Hilbert space setting.

Remark 5.32. To be precise, the *Fredholm alternative* is the concluding statement of the prior theorem.

5.7.6 Spectrum of a Compact Operator, Riesz–Schauder Theorem

By the *Spectrum Theorem* (Theorem 5.11), in a complex n-dimensional Banach space ($n \in \mathbb{N}$) $(X, \|\cdot\|)$, the spectrum of an arbitrary linear operator $A : X \to X$, which automatically is a *finite-rank*, and hence, *compact operator* (see Examples 5.17), consists of at most n eigenvalues of a finite number of eigenvalues with finite geometric multiplicities.

Here, we prove the *Riesz–Schauder Theorem* generalizing the aforementioned *Spectrum Theorem* (Theorem 5.11) to compact linear operators.
    Let us first prove the following:

Lemma 5.2 (Classification of Nonzero Points). *Let A be a compact linear operator on a complex Banach space* $(X, \|\cdot\|)$ *(i. e., $A \in K(X)$). Then each $\lambda \in \mathbb{C} \setminus \{0\}$ is either a regular point of A or its eigenvalue of finite geometric multiplicity, i. e.,*

$$\dim \ker(A - \lambda I) = n$$

with some $n \in \mathbb{N}$.

Proof. The statement immediately follows from the *Fredholm Alternative* (Theorem 5.20) in view of the fact that

$$\forall \lambda \in \mathbb{C} \setminus \{0\} : \ A - \lambda I = -\lambda\left(I - \frac{A}{\lambda}\right).$$

Exercise 5.64. Fill in the details. ☐

Theorem 5.21 (Riesz–Schauder Theorem). *Let A be a compact linear operator on a complex infinite-dimensional Banach space* $(X, \|\cdot\|)$ *(i. e., $A \in K(X)$). The spectrum $\sigma(A)$ of A consists of 0 and a countable set of nonzero eigenvalues with finite geometric multiplicities.*

If the set $\sigma(A) \setminus \{0\}$ is countably infinite, for its arbitrary countable arrangement $\{\lambda_n\}_{n=1}^{\infty}$,

$$\lambda_n \to 0, \ n \to \infty.$$

The operator A being of finite rank, its spectrum $\sigma(A)$ consists of 0 and a finite set of nonzero eigenvalues with finite geometric multiplicities.

Proof. Let a compact linear operator $A \in K(X)$ be arbitrary.

Since the space X is *infinite-dimensional*, applying the *Properties of Compact Operators* (Theorem 5.18) and the *Compactness of the Identity Operator Proposition* (Proposition 5.3), one can easily show that $0 \in \sigma(A)$ (see Section 5.8, Problem 23).

By the *Classification of Nonzero Points Lemma* (Lemma 5.2), the set $\sigma(A) \setminus \{0\}$, if nonempty, consists of nonzero eigenvalues of A with finite geometric multiplicities.

We have

$$\sigma(A) \setminus \{0\} = \bigcup_{n=1}^{\infty} \sigma_n,$$

where

$$\sigma_n := \{\lambda \in \sigma(A) \mid |\lambda| \geq 1/n\}, \ n \in \mathbb{N}.$$

Exercise 5.65. Explain.

Thus, to prove the fact that the set $\sigma(A) \setminus \{0\}$ is *countable*, by the *Properties of Countable Sets* (Proposition 1.3), it suffices to show that, for each $n \in \mathbb{N}$, the set σ_n is finite.

Let us reason *by contradiction*, assuming that there exists an $N \in \mathbb{N}$ such that the set σ_N is *infinite*. Then, by the *Properties of Countable Sets* (Theorem 1.3), σ_N contains a countably infinite subset $\{\lambda_i\}_{i \in \mathbb{N}}$ of nonzero eigenvalues of A.

Let $x_i \in X$ be an eigenvector associated with the eigenvalue λ_i, $i \in \mathbb{N}$, and

$$Y_n := \mathrm{span}(\{x_1, \ldots, x_n\}), \ n \in \mathbb{N}.$$

Since the eigenvectors x_i, $i \in \mathbb{N}$, are *linearly independent* by the *Linear Independence of Eigenvectors Proposition* (Proposition 5.8, Section 5.8, Problem 5), the *proper inclusions*

$$Y_{n-1} \subset Y_n, \ n = 2, 3, \ldots,$$

hold.

Also,

$$AY_n \subseteq Y_n, \ n \in \mathbb{N}.$$

Exercise 5.66. Explain.

Since, for each $n = 2, 3, \ldots$, Y_{n-1} is a *proper subspace* of Y_n, by *Riesz's Lemma in Finite-Dimensional Spaces* (Proposition 3.18, Section 3.5, Problem 18), there exists $e_n \in Y_n$ such that

$$\|e_n\| = 1 = \rho(e_n, Y_{n-1}).$$

Then, for any $n = 2, 3, \ldots$ and $m = 1, \ldots, n - 1$, by *absolute scalability* of norm,

$$\|Ae_n - Ae_m\| = \|\lambda_n e_n - (\lambda_n e_n - Ae_n + Ae_m)\| = |\lambda_n| \left\| e_n - \frac{1}{\lambda_n}(\lambda_n e_n - Ae_n + Ae_m) \right\|.$$

As is easily seen, for any $n = 2, 3, \ldots$ and $m = 1, \ldots, n - 1$,

$$Ae_m \in Y_{n-1} \quad \text{and} \quad \lambda_n e_n - Ae_n \in Y_{n-1}.$$

Exercise 5.67. Explain.

Hence, for any $n = 2, 3, \ldots$ and $m = 1, \ldots, n - 1$,

$$\|Ae_n - Ae_m\| = |\lambda_n| \left\| e_n - \frac{1}{\lambda_n}(\lambda_n e_n - Ae_n + Ae_m) \right\| \geq |\lambda_n| \rho(e_n, Y_{n-1}) = |\lambda_n| \geq 1/N,$$

which, by the *Sequential Characterization of Compact Operators* (Proposition 5.2), contradicts the *compactness* of the operator A.

Exercise 5.68. Explain.

The obtained contradiction proves the fact that the set σ_n is *finite* for each $n \in \mathbb{N}$, which implies that the set $\sigma(A)\backslash\{0\}$ is *countable* and also that, provided the set $\sigma(A)\backslash\{0\}$ is *countably infinite*, its only *limit point* is 0.

Exercise 5.69. Explain.

Therefore, if the set $\sigma(A) \setminus \{0\}$ is countably infinite, for its arbitrary countable arrangement $\{\lambda_n\}_{n=1}^\infty$,

$$\lambda_n \to 0, \ n \to \infty.$$

Exercise 5.70. Explain.

If the compact operator A is of *finite rank*, then, since all eigenvectors corresponding to nonzero eigenvalues belong to its range $R(A)$, which is *finite-dimensional*, by the *Linear Independence of Eigenvectors Proposition* (Proposition 5.8, Section 5.8, Problem 5), there can be only a *finite set* of such eigenvalues.

Exercise 5.71. Explain. □

Examples 5.19.

1. For the *finite-rank* zero operator 0 on a complex infinite-dimensional Banach space $(X, \|\cdot\|)$,

$$\sigma(0) = \sigma_p(0) = \{0\},$$

the geometric multiplicity of the eigenvalue 0 being equal to $\dim X > \aleph_0$ (see Examples 5.17 and 5.2, the *Basis of a Banach Space Theorem* (Theorem 3.14), and Remark 3.29).

2. On the complex Banach space l_p $(1 \le p < \infty)$,
   (a) for the compact operator A of multiplication by the vanishing sequence $(0, 1/2, 1/3, \dots)$,

   $$\sigma(A) = \sigma_p(A) = \{0, 1/2, 1/3, \dots\},$$

   the geometric multiplicity of each eigenvalue, including 0, being equal to 1 (see Examples 5.17 and 5.2);
   (b) for the compact operator A of multiplication by the vanishing sequence $(1/n)_{n \in \mathbb{N}}$,

   $$\sigma(A) = \{0\} \cup \{1/n\}_{n \in \mathbb{N}} \text{ with } \sigma_c(A) = \{0\} \text{ and } \sigma_p(A) = \{1/n\}_{n \in \mathbb{N}}$$

   (see Examples 5.17 and 5.2).

3. On the complex Banach space $(C[a, b], \|\cdot\|_\infty)$ $(-\infty < a < b < \infty)$, for the compact integration operator

$$C[a, b] \ni x \to [Ax](t) := \int_a^t x(s)\, ds \in C[a, b],$$

$$\sigma(A) = \sigma_r(A) = \{0\}$$

(see Examples 5.17 and 5.2).

Observe that, although, in this case, the spectrum of A is a finite set, A is *not* of finite rank.

Exercise 5.72. Verify.

Remarks 5.33.
- Thus, for a compact operator A on a complex Banach space $(X, \| \cdot \|)$, $\sigma(A)$ is a countable set.
- The point $\lambda = 0$ may belong to $\sigma_p(A)$, i. e., be an *eigenvalue* of A, whose geometric multiplicity need not be finite to $\sigma_c(A)$, or to $\sigma_r(A)$.

5.7.7 Application: Fredholm Integral Equations of the Second Kind

Applying the prior theorem to the compact *integral operator*

$$C[a,b] \ni x \to [Ax](t) := \int_a^b K(t,s)x(s)\,ds \in C[a,b],$$

where $K \in C([a,b] \times [a,b])$, in the complex space $(C[a,b], \| \cdot \|_\infty)$ $(-\infty < a < b < \infty)$ (see the *Compactness of Integral Operators Theorem* (Theorem 5.16)), we obtain the following generalization of the *Existence and Uniqueness Theorem for Fredholm Integral Equations of the Second Kind* (Theorem 2.37) (see also Remark 5.29).

Corollary 5.8 (Fredholm Alternative for Integral Equations). *Let $K \in C([a,b] \times [a,b])$ $(-\infty < a < b < \infty)$. Then, for any $\lambda \neq 0$, the Fredholm integral equation of the second kind*

$$\int_a^b K(t,s)x(s)\,ds - \lambda x(t) = y(t) \tag{5.29}$$

has a unique solution in $C[a,b]$ for each $y \in C[a,b]$ iff the homogeneous equation

$$\int_a^b K(t,s)x(s)\,ds - \lambda x(t) = 0 \tag{5.30}$$

has only the trivial solution in $C[a,b]$.

Except for a countable set of nonzero λ's, which has 0 as the only possible limit point, Equation (5.29) has a unique solution in $C[a,b]$ for each $y \in C[a,b]$.

For any $\lambda \neq 0$, Equation (5.30) has a finite number of linearly independent solutions.

For an example, see Section 5.8, Problems 22 and 28.

5.8 Problems

In the subsequent problems, $\mathbb{F}$ stands for the scalar field of real or complex numbers (i. e., $\mathbb{F} = \mathbb{R}$ or $\mathbb{F} = \mathbb{C}$).

1. Let $(A, D(A))$ be a *closed* linear operator in a normed vector space $(X, \| \cdot \|)$ and $B \in L(X)$.

   (a) Prove that
   - $A+B$ with the domain $D(A+B) = D(A) \cap X = D(A)$ is a *closed* linear operator in $(X, \| \cdot \|)$, and
   - AB with the domain $D(AB) = \{x \in X \mid Bx \in D(A)\}$ is a *closed* linear operator in $(X, \| \cdot \|)$.

   (b) Give an example showing that the product BA with the domain $D(AB) = D(A)$ need not be a closed linear operator.

2. Prove

   Proposition 5.5 (Invertibility of Linear Operators). *Let $(A, D(A))$ and $(B, D(B))$ be linear operators from a vector space X to a vector space Y over $\mathbb{F}$.*

   (1) *If A is invertible, then its inverse A^{-1} is invertible, and*

   $$\left(A^{-1}\right)^{-1} = A.$$

   (2) *If A is invertible and $\lambda \in \mathbb{F} \setminus \{0\}$, then λA is invertible, and*

   $$(\lambda A)^{-1} = \frac{1}{\lambda} A^{-1}.$$

   (3) *If A and B are invertible, then AB is invertible, and*

   $$(AB)^{-1} = B^{-1}A^{-1}.$$

3. Prove

   Proposition 5.6 (Existence of Bounded Inverse). *Let $(X, \| \cdot \|_X)$ and $(Y, \| \cdot \|_Y)$ be normed vector spaces over $\mathbb{F}$ and $(A, D(A))$ be a linear operator from $(X, \| \cdot \|_X)$ to $(Y, \| \cdot \|_Y)$. The inverse $A^{-1} : (R(A), \| \cdot \|_Y) \to (X, \| \cdot \|_X)$ exists, and $A^{-1} \in L(R(A), X)$ iff*

   $$\exists c > 0 : \|Ax\|_Y \geq c\|x\|_X,$$

   in which case $\|A^{-1}\| \leq 1/c$.

   (See the *Sufficient Condition of Boundedness* (Proposition 5.7, Section 4.7, Problem 4)).

4. Prove

   Proposition 5.7 (Sufficient Condition of Boundedness). *Let $(X, \| \cdot \|_X)$ and $(Y, \| \cdot \|_Y)$ be Banach spaces over $\mathbb{F}$ and $A : X \to Y$ be a surjective linear operator from X onto*

Y such that

$$\exists c > 0 \; \forall x \in X : \|Ax\|_Y \geq c\|x\|_X.$$

Then A is bounded, i. e., $A \in L(X, Y)$.

Hint. Apply the result of Problem 3, the *Closedness of Inverse Operator Proposition* (Proposition 4.23, Section 5.8, Problem 20), and the *Closed Graph Theorem* (Theorem 4.10).

5. Prove

Proposition 5.8 (Linear Independence of Eigenvectors). *The eigenvectors of a linear operator A on a vector space X corresponding to distinct eigenvalues, if any, are linearly independent.*

6. Let $(A, D(A))$ be a one-to-one linear operator in a vector space X over $\mathbb{F}$ and $\mu \in \mathbb{F}$. Prove that
   (a) $\lambda \in \mathbb{F}$, λ is an eigenvalue of A iff $\lambda - \mu$ is an eigenvalue of $A - \mu I$, with the corresponding eigenspaces being the same:

$$\ker(A - \lambda I) = \ker(A - \mu I - (\lambda - \mu)I);$$

   (b) for any $\lambda \in \mathbb{F} \setminus \{0\}$, λ is an eigenvalue of A iff $\frac{1}{\lambda}$ is an eigenvalue of its inverse A^{-1}, with the corresponding eigenspaces being the same:

$$\ker(A - \lambda I) = \ker\left(A^{-1} - \frac{1}{\lambda}I\right).$$

7. Prove that, if A is a closed linear operator in a complex Banach space $(X, \|\cdot\|)$ with $\rho(A) \neq \emptyset$, then

$$\forall \lambda \in \rho(A) : \; AR(\lambda, A) \in L(X).$$

8. Prove

Theorem 5.22 (Spectral Mapping Theorem for Inverse Operators). *Let A be an invertible closed linear operator in a complex Banach space $(X, \| \cdot \|)$ with $\sigma(A) \neq \emptyset, \{0\}, \mathbb{C}$. Then*

$$\rho(A^{-1}) \setminus \{0\} = \{\lambda^{-1} \mid \lambda \in \rho(A) \setminus \{0\}\}.$$

Equivalently,

$$\sigma(A^{-1}) \setminus \{0\} = \{\lambda^{-1} \mid \lambda \in \sigma(A) \setminus \{0\}\}.$$

In particular, if $A, A^{-1} \in L(X)$, i. e., $0 \in \rho(A) \cap \rho(A^{-1})$,

$$\sigma(A^{-1}) = \{\lambda^{-1} \mid \lambda \in \sigma(A)\}.$$

Hint. Show that

$$\forall \lambda \in \rho(A) \setminus \{0\} : \lambda^{-1} \in \rho(A^{-1}) \text{ with } R(\lambda^{-1}, A^{-1}) = -\lambda A R(\lambda, A).$$

9. Show that, on the complex space l_p $(1 \le p \le \infty)$,
    (a) for the bounded linear *right shift operator*

$$l_p \ni x := (x_1, x_2, \dots) \mapsto Ax := (0, x_1, x_2, \dots) \in l_p$$

(see Examples 4.2),

$$0 \in \sigma_r(A),$$

and the inverse A^{-1} is the restriction of the bounded linear *left shift operator*

$$l_p \ni x := (x_1, x_2, \dots) \mapsto Bx := (x_2, x_3, x_4, \dots) \in l_p$$

(see Examples 4.2) to the closed hyperplane

$$R(A) = \{(x_n)_{n \in \mathbb{N}} \in l_p \mid x_1 = 0\};$$

(b)

$$\sigma_p(B) = \{\lambda \in \mathbb{C} \mid |\lambda| < 1\}$$

(see Examples 6.18, 3 (b) and Remark 6.26).
10. On the complex space $(C[a, b], \|\cdot\|_\infty)$ $(-\infty < a < b < \infty)$, determine and classify the spectrum of the bounded linear operator of multiplication by the function
    (a) $m_1(t) := t, t \in [a, b],$
    (b) $m_2(t) := \begin{cases} 0, & a \le t \le (a+b)/2, \\ t - (a+b)/2, & (a+b)/2 \le t \le b. \end{cases}$
11. (a) Prove

> **Proposition 5.9** (Left/Right Inverse). *Let* $(X, \|\cdot\|)$ *be a normed vector space over* $\mathbb{F}$. *An operator* $A \in L(X)$ *has an inverse* $A^{-1} \in L(X)$ *iff* A *has a left inverse* $B \in L(X)$, *i. e.,*
>
> $$BA = I,$$
>
> *and a right inverse* $C \in L(X)$, *i. e.,*
>
> $$AC = I,$$
>
> *in which case*
>
> $$B = C = A^{-1}.$$

(b) Give an example showing that a left/right inverse of an operator $A \in L(X)$ need not be its inverse.

Hint. Consider the *right* and *left shift operators* on l_p $(1 \le p \le \infty)$ (see Examples 4.2 and Problem 9).

12. Prove

Proposition 5.10 (Invertibility of Bounded Linear Operators). *Let* $(X, \| \cdot \|)$ *be a normed vector space over* $\mathbb{F}$ *and*

$$\mathcal{I} := \{A \in L(X) \mid \exists A^{-1} \in L(X)\},$$

i. e., $\mathcal{I}$ *is the set of all operators in* $L(X)$ *with an inverse in* $L(X)$.
(1) *If* $A \in \mathcal{I}$, *then* $A^{-1} \in \mathcal{I}$, *and* $(A^{-1})^{-1} = A$.
(2) *The identity operator* $I \in \mathcal{I}$, *and* $I^{-1} = I$.
(3) *If* $A \in \mathcal{I}$ *and* $\lambda \in \mathbb{F} \setminus \{0\}$, *then* $\lambda A \in \mathcal{I}$, *and*

$$(\lambda A)^{-1} = \frac{1}{\lambda} A^{-1}.$$

(4) *If* $A, B \in \mathcal{I}$, *then* $AB \in \mathcal{I}$, *and*

$$(AB)^{-1} = B^{-1} A^{-1}.$$

(5) *If* $A, AB \in \mathcal{I}$, *then* $B \in \mathcal{I}$.
(6) *If* $B, AB \in \mathcal{I}$, *then* $A \in \mathcal{I}$.
(7) *If* $AB, BA \in \mathcal{I}$, *then* $A, B \in \mathcal{I}$.
(8) *The set* $\mathcal{I}$ *is a group relative to the operator multiplication.*

13. Prove

Proposition 5.11 (Sufficient Condition of Invertibility). *Let* $(X, \| \cdot \|)$ *be a Banach space and an operator* $A \in L(X)$ *be such that* $\exists A^{-1} \in L(X)$. *Then, for any* $B \in L(X)$ *with*

$$\|A - B\| < \frac{1}{\|A^{-1}\|},$$

$$\exists B^{-1} = A^{-1} \sum_{n=0}^{\infty} [(A - B)A^{-1}]^n \in L(X),$$

the operator series converging uniformly, and

$$\|B^{-1} - A^{-1}\| \le \frac{\|A^{-1}\| \|A - B\| \|A^{-1}\|}{1 - \|A - B\| \|A^{-1}\|} \to 0, \ B \to A, \ in \ (L(X), \| \cdot \|).$$

Remarks 5.34. In particular, this implies that
– the group $\mathcal{I}$ of all operators in $L(X)$ with an inverse in $L(X)$ (see Problem 12) is an *open set* in $(L(X), \| \cdot \|)$ and

– the inversion operation

$$\mathscr{I} \ni A \mapsto A^{-1} \in \mathscr{I}$$

is *continuous* in the operator norm.

14. Let $(X, \|\cdot\|)$ be a complex Banach space and $A \in L(X)$. Show that the estimate

$$\|\lambda R(\lambda, A) + I\| \le \frac{\|A\|}{|\lambda| - \|A\|}$$

holds whenever $|\lambda| > \|A\|$, which implies that

$$\lambda R(\lambda, A) \to -I, \ |\lambda| \to \infty, \ \text{in } (L(X), \|\cdot\|).$$

Hint. Apply *Gelfand's Spectral Radius Theorem* (Theorem 5.8).

15. In $\mathbb{C}^3$, find the *transition matrix*, and write the *change of coordinates formula* from the standard basis

$$B := \left\{ e_1 := \begin{bmatrix} 1 \\ 0 \\ 0 \end{bmatrix}, e_2 := \begin{bmatrix} 0 \\ 1 \\ 0 \end{bmatrix}, e_3 := \begin{bmatrix} 0 \\ 0 \\ 1 \end{bmatrix} \right\}$$

to the ordered basis

$$B' := \left\{ \begin{bmatrix} 1 \\ 0 \\ 0 \end{bmatrix}, \begin{bmatrix} 1 \\ 1 \\ 0 \end{bmatrix}, \begin{bmatrix} 1 \\ 1 \\ 1 \end{bmatrix} \right\}.$$

16. Prove

Proposition 5.12 (Quasinilpotence in a Finite-Dimensional Setting). *Let* $(X, \|\cdot\|)$ *be a complex finite-dimensional Banach space. An operator* $A \in L(X)$ *is quasinilpotent, i. e.,* $\sigma(A) = \{0\}$, *iff it is nilpotent, i. e.,* $A^m = 0$ *for some* $m \in \mathbb{N}$.

17. On the complex two-dimensional space $l_p^{(2)}$ $(1 \le p \le \infty)$, for the linear operator E of multiplication by the 2×2 matrix

$$\begin{bmatrix} 1 & -1 \\ 3 & -2 \end{bmatrix},$$

find

$$[e^{tE}]_B, \ t \ge 0,$$

relative to an *eigenbasis* B of E for $\mathbb{C}^2$ (see Example 5.15).

18. On the complex three-dimensional space $l_p^{(3)}$ ($1 \le p \le \infty$), for the linear operator A of multiplication by the 3×3 matrix

$$
\begin{bmatrix}
1 & 0 & 3 \\
2 & 1 & 2 \\
0 & 0 & -2
\end{bmatrix},
$$

(a) find a *Jordan canonical matrix representation* $[A]_B$ of A relative to a *Jordan basis B* for $\mathbb{C}^3$ without finding the basis itself;
(b) find $[e^{tA}]_B$, $t \ge 0$, relative to the Jordan basis B;
(c) apply the *Lyapunov Stability Theorem* (Theorem 5.14) to determine whether evolution equation (5.12) is *asymptotically stable*.

19. Prove

Proposition 5.13 (Cesàro Limit). *Let A be a bounded linear operator on a Banach $(X, \| \cdot \|)$ (i. e., $A \in L(X)$). If, for a bounded solution $y(\cdot)$ (i. e., $\sup_{t \ge 0} \|y(t)\| < \infty$) of evolution equation (5.12), there exists the Cesàro limit*

$$
\lim_{t \to \infty} \frac{1}{t} \int_0^t y(s) \, ds =: y_\infty \in X,
$$

then $y_\infty \in \ker A$.

Hint. Use the *total change formula*

$$
y(t) - y(0) = \int_0^t y'(s) \, ds = \int_0^t Ay(s) \, ds = A \int_0^t y(s) \, ds, \ t \ge 0.
$$

20. On the complex three-dimensional space $l_p^{(3)}$ ($1 \le p \le \infty$), for the linear operator A of Problem 18, apply the *Mean Ergodicity Theorem* (Theorem 5.15)
(a) to determine the general form of bounded solutions of evolution equation (5.12);
(b) to find the *Cesàro limit*

$$
\lim_{t \to \infty} \frac{1}{t} \int_0^t y(s) \, ds
$$

for any bounded solution $y(\cdot)$ of (5.12);
(c) to determine whether the regular limit

$$
\lim_{t \to \infty} y(t)
$$

also exists for any bounded solution $y(\cdot)$ of (5.12).
See Example 5.16.

21. For a linear operator A on a complex four-dimensional Banach space $(X, \|\cdot\|)$ with a *Jordan canonical matrix representation*

$$[A]_B = \begin{bmatrix} 0 & 1 & 0 & 0 \\ 0 & 0 & 0 & 0 \\ 0 & 0 & 1 & 1 \\ 0 & 0 & 0 & 1 \end{bmatrix}$$

relative to a Jordan basis B, apply the *Mean Ergodicity Theorem* (Theorem 5.15)
(a) to determine the general form of bounded solutions of evolution equation (5.12);
(b) to find the limit

$$\lim_{t \to \infty} y(t)$$

for any bounded solution $y(\cdot)$ of (5.12).
See Example 5.16.

22. Let

$$K(t,s) := \begin{cases} (1-s)t, & 0 \le t \le s, \\ (1-t)s, & s \le t \le 1, \end{cases} \quad (t,s) \in [0,1] \times [0,1].$$

(a) Prove that the integral operator

$$C[0,1] \ni x \to [Ax](t) := \int_0^1 K(t,s)x(s)\,ds \in C[0,1]$$

is *compact* on the Banach space $(C[0,1], \|\cdot\|_\infty)$ (i. e., $A \in K(C[0,1])$).
(b) Show that for an arbitrary $y \in C[0,1]$,

$$x(t) := [Ay](t) = \int_0^1 K(t,s)y(s)\,ds, \ t \in [0,1],$$

is the *unique* solution on $[0,1]$ of the boundary value problem

$$x'' + y = 0, \ x(0) = x(1) = 0.$$

Remark 5.35. The latter implies that, in the space $(C[0,1], \|\cdot\|_\infty)$, for the unbounded closed linear differential operator

$$Bx := -\frac{d^2 x}{dt^2}$$

with the domain

$$D(B) := \{x \in C^2[a,b] \mid x(0) = x(1) = 0\}$$

(see Section 3.5, Problem 10), is invertible, and $B^{-1} = A \in K(C[0,1])$.

23. Let $(X, \| \cdot \|)$ be an *infinite-dimensional* normed vector space. Prove that
    (a) if $A \in L(X)$ is invertible, then $A^{-1} \notin K(X)$;
    (b) if $A \in K(X)$ is invertible, then $A^{-1} \notin L(X)$.
24. Let $(X, \| \cdot \|)$ be an *infinite-dimensional* normed vector space over $\mathbb{F}$ and $A \in K(X)$. Prove that, if, for $\lambda \in \mathbb{F}$, there exists $(A - \lambda I)^{-1} \in L(X)$, then
    (a) $(A - \lambda I)^{-1} \notin K(X)$;
    (b) provided $\lambda \neq 0$, $I + \lambda(A - \lambda I)^{-1} \in K(X)$.
25. Suppose that $(X, \| \cdot \|)$ is a Banach space, $A \in K(X)$, and the operator $I - A$ is *injective*. Prove that, if $S \subseteq X$ and the set $(I - A)S$ is *bounded*, then the set S is *bounded*.

    Hint. Apply the *Fredholm Alternative* (Theorem 5.20).

26. Prove

    Proposition 5.14 (Operators with Compact Resolvent). *Let A be a closed linear operator on a complex Banach space $(X, \| \cdot \|)$. Prove that, if*

    $$\rho(A) \neq \emptyset \quad and \quad \exists \lambda_0 \in \rho(A) : R(\lambda, A) \in K(X),$$

    (1) *A is unbounded;*
    (2) *$\forall \lambda \in \rho(A) : R(\lambda, A) \in K(X)$;*
    (3) *the spectrum $\sigma(A)$ of A is a countable set of eigenvalues with finite geometric multiplicities and, provided $\sigma(A) = \sigma_p(A) = \{\lambda_n\}_{n \in \mathbb{N}}$ is a countably infinite set,*

    $$|\lambda_n| \to \infty, \; n \to \infty.$$

    Hint. To prove part (2), apply the *Riesz–Schauder Theorem* (Theorem 5.21), and use the results of Problems 8 and 6.

27. Apply the result of Problem 26 to show that, in the complex Banach space $(C[0, 1], \| \cdot \|_\infty)$, for the unbounded closed linear differential operator B introduced in Remark 5.35,
    (a) $\rho(B) \neq \emptyset$ and $\forall \lambda \in \rho(B) : R(\lambda, B) \in K(C[0, 1])$;
    (b) $\sigma(B) = \sigma_p(B) = \{(n\pi)^2\}_{n \in \mathbb{N}}$, each eigenvalue $(n\pi)^2$, $n \in \mathbb{N}$, having geometric multiplicity 1 with the corresponding one-dimensional eigenspace $\mathrm{span}(\{\sin(n\pi t)\})$.

    Remark 5.36. The unbounded closed linear differentiation operator A from Examples 5.2, 4(d) is another example of an *operator with compact resolvent*. In this case, however, $\sigma(A) = \emptyset$ (see the *Compactness of Volterra Integral Operators Theorem* (Theorem 5.17)).

28. Apply the results of Problems 27, 8, and 6 to show that, in the complex Banach space $(C[0,1], \|\cdot\|_\infty)$, for the compact linear integral operator A of Problem 22,

$$\sigma(A) = \{0\} \cup \{1/(n\pi)^2\}_{n\in\mathbb{N}} \text{ with } \sigma_r(A) = \{0\} \text{ and } \sigma_p(A) = \{1/(n\pi)^2\}_{n\in\mathbb{N}},$$

each eigenvalue $1/(n\pi)^2$, $n \in \mathbb{N}$, having geometric multiplicity 1 with the corresponding one-dimensional eigenspace $\mathrm{span}(\{\sin(n\pi t)\})$.

6 Elements of Spectral Theory in a Hilbert Space Setting

In this chapter, we study certain elements of spectral theory of linear operators in Hilbert spaces, whose inner product structure and *self-duality* (see Section 6.4) bring to life the important concepts of *symmetry* and *self-adjointness* for linear operators (see Section 6.10).

6.1 Inner Product and Hilbert Spaces

We start our discourse with the discussion of inner product and Hilbert spaces, whose structure allows us to define the notions of *orthogonality* and the angle between vectors, and hence, even further justifies the use of the conventional geometric terms and intuition.

6.1.1 Definitions and Examples

Recall that $\mathbb{F}$ stands for the *scalar field* $\mathbb{R}$ or $\mathbb{C}$.

Definition 6.1 (Inner Product Space). An *inner product space* (or a *pre-Hilbert space*) over $\mathbb{F}$ is a *vector space X* over $\mathbb{F}$ equipped with an *inner product* (or a *scalar product*), i. e., a mapping

$$(\cdot,\cdot) : X \times X \to \mathbb{F}$$

subject to the following *inner product axioms*.

1. $(x,x) \geq 0,\ x \in X$, and $(x,x) = 0$ iff $x = 0$.   *Positive Definiteness*
2. $(x,y) = \overline{(y,x)},\ x,y \in X$.   *Conjugate Symmetry*
3. $(\lambda x + \mu y, z) = \lambda(x,z) + \mu(y,z),\ \lambda,\mu \in \mathbb{F},\ x,y,z \in X$.   *Linearity in the First Argument*

The space is said to be *real* if $\mathbb{F} = \mathbb{R}$, and *complex* if $\mathbb{F} = \mathbb{C}$.

Notation. $(X,(\cdot,\cdot))$.

Remarks 6.1.
-   For a *real* inner product space, the axiom of *conjugate symmetry* turns into
    2R. $(x,y) = (y,x),\ x,y \in X$ *Symmetry*
-   From *conjugate symmetry* and *linearity in the first argument*, we immediately derive the following property:
    4. $(x,\lambda y + \mu z) = \overline{\lambda}(x,y) + \overline{\mu}(x,z),\ \lambda,\mu \in \mathbb{F},\ x,y,z \in X$ *Conjugate Linearity in the Second Argument*

https://doi.org/10.1515/9783110600988-006

also called *antilinearity*, or *semilinearity*.

For a *real* inner product space, the latter turns into

4R. $(x, \lambda y + \mu z) = \lambda(x, y) + \mu(x, z),\ \lambda, \mu \in \mathbb{R},\ x, y, z \in X$ *Linearity in the Second Argument*

–   From *linearity* and *conjugate symmetry* of inner product, we immediately infer that

$$\forall x \in X: \ (0, x) = (x, 0) = 0.$$

Exercise 6.1. Prove in two different ways.

Examples 6.1.

1.   On the space $l_2^{(n)}$ ($n \in \mathbb{N}$), the mapping

$$l_2^{(n)} \ni x := (x_1, \ldots, x_n), y := (y_1, \ldots, y_n) \mapsto (x, y) := \sum_{k=1}^{n} x_k \overline{y}_k \in \mathbb{F}$$

is an *inner product*, the conjugation being superfluous when the space is real.

2.   On the space l_2, the mapping

$$l_2 \ni x := (x_k)_{k \in \mathbb{N}}, y := (y_k)_{k \in \mathbb{N}} \mapsto (x, y) := \sum_{k=1}^{\infty} x_k \overline{y}_k \in \mathbb{F}$$

is an *inner product*, the conjugation being superfluous when the space is real.

Remark 6.2. The inner product is well defined due to the *Cauchy–Schwarz inequality* for sequences (see (2.2)).

3.   On the space $C[a, b]$ ($-\infty < a < b < \infty$), the mapping

$$C[a, b] \ni f, g \mapsto (f, g) := \int_a^b f(t) \overline{g(t)}\, dt \in \mathbb{F}$$

is an *inner product*, the conjugation being superfluous when the space is real.

4.   When integration is understood in the *Lebesgue sense* (see, e. g., [30, 25, 59, 62, 46]), the mapping

$$f, g \mapsto (f, g) := \int_a^b f(t) \overline{g(t)}\, dt \in \mathbb{F}$$

is an inner product on the space $L_2(a, b)$ ($-\infty \le a < b \le \infty$) of all equivalence classes of equal almost everywhere relative to the *Lebesgue measure* square integrable on (a, b) functions:

$$\int_a^b |f(t)|^2\, dt < \infty,$$

$f(\cdot)$ and $g(\cdot)$ being arbitrary representatives of the equivalence classes f and g, respectively (see, e. g., [25, 62, 46, 70]).

Exercise 6.2. Verify 1–3.

Definition 6.2 (Subspace of an Inner Product Space). If $(X, (\cdot, \cdot))$ is an *inner product space* and $Y \subseteq X$ is a *linear subspace* of X, then the restriction of the inner product $(\cdot, \cdot)$ to $Y \times Y$ is an inner product on Y and the inner product space $(Y, (\cdot, \cdot))$ is called a *subspace* of $(X, (\cdot, \cdot))$.

Examples 6.2.
1. The space c_{00} can be considered to be a subspace of l_2, but c_0 cannot.
2. The space $C^1[a, b]$ ($-\infty < a < b < \infty$) can be considered to be a subspace of $(C[a, b], (\cdot, \cdot))$, but $R[a, b]$ cannot.

Exercise 6.3. Explain.

6.1.2 Inner Product Norm, Cauchy–Schwarz Inequality

Inner product generates a norm with certain distinctive properties.

Theorem 6.1 (Inner Product Norm). *On an inner product space* $(X, (\cdot, \cdot))$, *the mapping*

$$X \ni x \mapsto \|x\| := (x, x)^{1/2} = \sqrt{(x, x)} \in \mathbb{R}$$

is a norm, called the inner product norm.

One can easily verify that the norm axioms of *nonnegativity*, *separation*, and *absolute scalability* (see Definition 3.17) hold.

Exercise 6.4. Verify.

To show that *subadditivity* holds as well, we are to prove the following important inequality generalizing the *Cauchy–Schwarz inequalities* for n-tuples and sequences (see (2.2)).

Theorem 6.2 (Cauchy–Schwarz Inequality). *In an inner product space* $(X, (\cdot, \cdot))$,

$$\forall x, y \in X : |(x, y)| \le \|x\| \|y\|,$$

with

$$|(x, y)| = \|x\| \|y\| \iff x \text{ and } y \text{ are linearly dependent.}$$

Proof. If $y = 0$, we have

$$|(x, y)| = 0 = \|x\| \|y\| \tag{6.1}$$

(see Remarks 6.1), and hence, the inequality is trivially true.

Suppose that $y \neq 0$. Then, for any $\lambda \in \mathbb{F}$, by *linearity* and *conjugate linearity* of inner product and the definition of $\|\cdot\|$,

$$0 \le (x + \lambda y, x + \lambda y) = (x, x + \lambda y) + \lambda(y, x + \lambda y) = (x, x) + \bar{\lambda}(x, y) + \lambda(y, x) + \lambda\bar{\lambda}(y, y)$$
$$= \|x\|^2 + \bar{\lambda}(x, y) + \lambda(y, x) + |\lambda|^2\|y\|^2.$$

Setting $\lambda = -\frac{(x,y)}{\|y\|^2}$, in view of *conjugate symmetry*, we arrive at

$$0 \le \|x\|^2 - \frac{|(x, y)|^2}{\|y\|^2} - \frac{|(x, y)|^2}{\|y\|^2} + \frac{|(x, y)|^2}{\|y\|^2}.$$

Whence, the *Cauchy–Schwarz inequality* follows immediately.

If the vectors x and y are *linearly dependent*, then (see Examples 3.5) either at least one of them is *zero*, in which case equality (6.1) (see Remarks 6.1) trivially holds, or

$$x, y \neq 0 \text{ and } y = \lambda x$$

with some $\lambda \in \mathbb{F} \setminus \{0\}$, in which case, by *conjugate linearity* and the definition of $\|\cdot\|$,

$$|(x, y)| = |(x, \lambda x)| = |\bar{\lambda}|\|x\|^2 = |\lambda|\|x\|^2 = \|x\|\|\lambda x\| = \|x\|\|y\|.$$

Conversely, if for $x, y \in X$,

$$|(x, y)| = \|x\|\|y\|,$$

then, by the definition of $\|\cdot\|$ and *linearity* and *conjugate linearity* of inner product,

$$\begin{aligned}
\|\|y\|^2 x - (x, y)y\|^2 &= (\|y\|^2 x - (x, y)y, \|y\|^2 x - (x, y)y) \\
&= \|y\|^2(x, \|y\|^2 x - (x, y)y) - (x, y)(y, \|y\|^2 x - (x, y)y) \\
&= \|y\|^4\|x\|^2 - \|y\|^2|(x, y)|^2 - \|y\|^2|(x, y)|^2 + |(x, y)|^2\|y\|^2 \\
&= 2\|y\|^4\|x\|^2 - 2\|y\|^4\|x\|^2 = 0.
\end{aligned}$$

Whence, by *positive definiteness*,

$$\|y\|^2 x - (x, y)y = 0,$$

which proves the linear dependence of x and y.

Exercise 6.5. Explain. □

Remark 6.3. The *Cauchy–Schwarz Inequality* (Theorem 6.2) in the spaces $l_2^{(n)}$ ($n \in \mathbb{N}$) and l_2 yields the *Cauchy–Schwarz inequalities* for n-tuples and sequences (see (2.2)) as particular cases.

Now, let us prove the *subadditivity* of $\|\cdot\|$ in Theorem 6.1.

Proof. For every $x, y \in X$, by *linearity* and *conjugate linearity* of inner product and the *Cauchy–Schwarz Inequality* (Theorem 6.2),

$$\|x + y\|^2 = (x + y, x + y) = (x, x + y) + (y, x + y) = (x, x) + (x, y) + (y, x) + (y, y)$$
$$= \|x\|^2 + (x, y) + \overline{(x, y)} + \|y\|^2 = \|x\|^2 + 2\operatorname{Re}(x, y) + \|y\|^2 \le \|x\|^2 + 2|(x, y)| + \|y\|^2$$
$$\le \|x\|^2 + 2\|x\|\|y\| + \|y\|^2 = (\|x\| + \|y\|)^2.$$

Whence the *subadditivity* of $\| \cdot \|$ follows immediately, and hence, the latter is a norm on X. $\qquad\square$

Remarks 6.4.

- While proving the *Inner Product Norm Theorem* (Theorem 6.1), we have come across the following identity:

$$\|x \pm y\|^2 = \|x\|^2 \pm 2\operatorname{Re}(x, y) + \|y\|^2, \ x, y \in X. \tag{6.2}$$

For the case of "$-$", using the similarity of the above to the geometric *Law of Cosines*, for any vectors $x, y \in X$ with $(x, y) \ne 0$, which, in particular, implies that $x, y \in X \setminus \{0\}$, one can define the *angle* between them as follows:

$$\theta := \arccos \frac{\operatorname{Re}(x, y)}{\|x\|\|y\|} \in [0, \pi].$$

The notion is well defined due to the *Cauchy–Schwarz Inequality* (Theorem 6.2) since

$$-1 \le \frac{\operatorname{Re}(x, y)}{\|x\|\|y\|} \le 1.$$

For vectors $x, y \in X$ with $(x, y) = 0$, i. e., *"orthogonal"* (see Section 6.1.4.3), $\theta := \pi/2$.

- The triple $(X, (\cdot, \cdot), \| \cdot \|)$ is used to designate an inner product space $(X, (\cdot, \cdot))$ with the *inner product norm* $\| \cdot \|$.

The *Cauchy–Schwarz Inequality* (Theorem 6.2) has the following immediate important implication:

Proposition 6.1 (Joint Continuity of Inner Product). *On an inner product space $(X, (\cdot, \cdot), \| \cdot \|)$, inner product is jointly continuous, i. e., if*

$$X \ni x_n \to x \in X \text{ and } X \ni y_n \to y \in X, \ n \to \infty, \text{ in } (X, (\cdot, \cdot), \| \cdot \|),$$

then

$$(x_n, y_n) \to (x, y), \ n \to \infty.$$

Exercise 6.6. Prove.

6.1.3 Hilbert Spaces

Definition 6.3 (Hilbert Space). A *Hilbert space* is an inner product space $(X, (\cdot, \cdot), \|\cdot\|)$, which is a Banach space relative to the *inner product norm* $\|\cdot\|$.

Examples 6.3.

1. On the space $l_2^{(n)}$ ($n \in \mathbb{N}$), the inner product

$$l_2^{(n)} \ni x := (x_1, \ldots, x_n), y := (y_1, \ldots, y_n) \mapsto (x, y) := \sum_{k=1}^{n} x_k \bar{y}_k \in \mathbb{F},$$

generates the *2-norm*

$$l_2^{(n)} \ni x := (x_1, \ldots, x_n) \mapsto \|x\|_2 := \left[\sum_{k=1}^{n} |x_k|^2 \right]^{1/2},$$

which is the *Euclidean norm*.

Hence, $l_2^{(n)}$ is a *Hilbert space* (see Examples 3.13).

The real $l_2^{(n)}$ is called the *Euclidean n-space*, and the complex $l_2^{(n)}$ is called the *unitary n-space*.

2. On the space l_2 ($n \in \mathbb{N}$), the inner product

$$l_2 \ni x := (x_k)_{k \in \mathbb{N}}, y := (y_k)_{k \in \mathbb{N}} \mapsto (x, y) := \sum_{k=1}^{\infty} x_k \bar{y}_k \in \mathbb{F},$$

generates the *2-norm*

$$l_2 \ni x := (x_k)_{k \in \mathbb{N}} \mapsto \|x\|_2 := \left[\sum_{k=1}^{\infty} |x_k|^2 \right]^{1/2},$$

and hence, l_2 is a *Hilbert space* (see Examples 3.13).

3. The space $(c_{00}, (\cdot, \cdot), \|\cdot\|_2)$, considered as a subspace of l_2, is an *incomplete inner product space*.

Exercise 6.7. Verify.

4. The space $C[a, b]$ ($-\infty < a < b < \infty$) is an *incomplete inner product space* relative to the *integral inner product*

$$C[a, b] \ni f, g \mapsto (f, g) := \int_a^b f(t)\overline{g(t)} \, dt \in \mathbb{F}$$

generating the *2-norm*

$$C[a, b] \ni f \mapsto \|f\|_2 := \left[\int_a^b |f(t)|^2 dt \right]^{1/2}.$$

(see Examples 3.13).

5. The space $L_2(a, b)$ $(-\infty \le a < b \le \infty)$ is a *Hilbert space* relative to the *integral inner product*

$$f, g \mapsto (f, g) := \int_a^b f(t)\overline{g(t)}\, dt \in \mathbb{F}$$

generating the 2-*norm*

$$L_2(a, b) \ni f \mapsto \|f\|_2 := \left[\int_a^b |f(t)|^2 dt\right]^{1/2}$$

(see Examples 6.1).

6.1.4 Geometric Properties of Inner Product Norm

The following are geometric properties inherent to inner product norm resonating with familiar ones from the two- and three-dimensional cases.

6.1.4.1 Polarization Identities
The following *polarization identities* relate inner product with the norm generated by it.

Proposition 6.2 (Polarization Identities).
(1) *In a real inner product space* $(X, (\cdot, \cdot), \|\cdot\|)$,

$$\forall x, y \in X : \ (x, y) = \frac{1}{4}\big[\|x + y\|^2 - \|x - y\|^2\big].$$

(2) *In a complex inner product space* $(X, (\cdot, \cdot), \|\cdot\|)$,

$$\forall x, y \in X : \ (x, y) = \frac{1}{4}\big[\|x + y\|^2 - \|x - y\|^2\big] + \frac{i}{4}\big[\|x + iy\|^2 - \|x - iy\|^2\big].$$

Exercise 6.8. Prove.

Hint. Apply identity (6.2).

6.1.4.2 Parallelogram Law
The following geometric property is characteristic for an inner product norm.

Theorem 6.3 (Parallelogram Law). *In an inner product space* $(X, (\cdot, \cdot), \|\cdot\|)$, *the following parallelogram law holds:*

$$\forall x, y \in X : \ \|x + y\|^2 + \|x - y\|^2 = 2\big[\|x\|^2 + \|y\|^2\big].$$

Conversely, if $(X, \| \cdot \|)$ is a normed vector space, whose norm satisfies the parallelogram law, then X is an inner product space, i. e., the norm is generated by an inner product $(\cdot, \cdot)$ on X:

$$\|x\| := (x, x)^{1/2}, \ x \in X.$$

Proof. The *parallelogram law* is derived directly from identity (6.2).

Exercise 6.9. Derive.

Suppose now that $(X, \| \cdot \|)$ is a normed vector space, whose norm satisfies the *parallelogram law*.

Let us first consider the *real case* and show that

$$(x, y) := \frac{1}{4}[\|x + y\|^2 - \|x - y\|^2], \ x, y \in X,$$

is an *inner product* on X generating the norm $\| \cdot \|$.

The generation follows from the fact that

$$(x, x) := \|x\|^2, \ x \in X,$$

which also immediately implies *positive definiteness* for $(\cdot, \cdot)$.

The *symmetry* property also directly follows from the definition via *absolute scalability* of norm.

Exercise 6.10. Verify.

It remains to prove the *linearity* in the first argument.

To prove the *additivity* in the first argument, consider the function

$$F(x, y, z) := 4[(x + y, z) - (x, z) - (y, z)], \ x, y, z \in X.$$

By the definition of $(\cdot, \cdot)$,

$$F(x, y, z) = \|x + y + z\|^2 - \|x + y - z\|^2 - \|x + z\|^2 + \|x - z\|^2 - \|y + z\|^2 + \|y - z\|^2. \quad (6.3)$$

Whence, by the *parallelogram law*,

$$\|x + y \pm z\|^2 = 2\|x \pm z\|^2 + 2\|y\|^2 - \|x - y \pm z\|^2,$$

we arrive at

$$F(x, y, z) = -\|x - y + z\|^2 + \|x - y - z\|^2 + \|x + z\|^2 - \|x - z\|^2 - \|y + z\|^2 + \|y - z\|^2. \quad (6.4)$$

Using *absolute scalability* of norm, we have the following:

$$F(x, y, z) = \frac{1}{2}(6.3) + \frac{1}{2}(6.4)$$

$$= \frac{1}{2}\left[\|y + z + x\|^2 + \|y + z - x\|^2\right] - \frac{1}{2}\left[\|y - z + x\|^2 + \|y - z - x\|^2\right]$$

$$- \|y + z\|^2 + \|y - z\|^2 \qquad\qquad \text{by the } \textit{parallelogram law};$$

$$= \|y + z\|^2 + \|x\|^2 - \|y - z\|^2 - \|x\|^2 - \|y + z\|^2 + \|y - z\|^2 = 0, \ x, y, z \in X,$$

which proves that

$$(x + y, z) = (x, z) + (y, z), \ x, y, z \in X,$$

i. e., the *additivity* in the first argument.

To prove the *homogeneity* in the first argument, consider the function

$$f(\lambda, x, y) := (\lambda x, y) - \lambda(x, y), \ \lambda \in \mathbb{R}, x, y \in X.$$

By the definition and *absolute scalability* of norm,

$$f(0, x, y) = (0, y) = \frac{1}{4}\left[\|y\|^2 - \| - y\|^2\right] = 0, \ x, y \in X,$$

i. e.,

$$(0, y) = 0(x, y), \ x, y \in X,$$

and

$$f(-1, x, y) = (-x, y) + (x, y) = \frac{1}{4}\left[\| - x + y\|^2 - \| - x - y\|^2\right]$$

$$+ \frac{1}{4}\left[\|x + y\|^2 - \|x - y\|^2\right] = 0, \ x, y \in X,$$

i. e.,

$$(-x, y) = -(x, y), \ x, y \in X.$$

Considering this and the demonstrated *additivity in the first argument*, for any $m \in \mathbb{Z} \setminus \{0\}$, we have the following:

$$f(m, x, y) = (mx, y) - m(x, y) = (\text{sgn}(m)|m|x, y) - m(x, y)$$

$$= (\text{sgn}(m)(\underbrace{x + \cdots + x}_{|m| \text{ times}}), y) - m(x, y) = \text{sgn}(m)\Big[\underbrace{(x, y) + \cdots + (x, y)}_{|m| \text{ times}}\Big] - m(x, y)$$

$$= \text{sgn}(m) \cdot |m|(x, y) - m(x, y) = m(x, y) - m(x, y) = 0, \ x, y \in X.$$

Thus,

$$(mx, y) = m(x, y), \ m \in \mathbb{Z}, x, y \in X.$$

For any $m \in \mathbb{Z}$ and $n \in \mathbb{N}$,

$$f(m/n, x, y) = ((m/n)x, y) - (m/n)(x, y) = m((1/n)x, y) - (m/n)(x, y)$$

$$= (m/n)n((1/n)x, y) - (m/n)(x, y) = (m/n)(x, y) - (m/n)(x, y) = 0.$$

Therefore,

$$f(\lambda, x, y) = 0, \ \lambda \in \mathbb{Q}, x, y \in X.$$

Considering the *continuity* of the function

$$f(\lambda x, y) = (\lambda x, y) - \lambda(x, y)$$
$$= \frac{1}{4}[\|\lambda x + y\|^2 - \|\lambda x - y\|^2] - \frac{\lambda}{4}[\|x + y\|^2 - \|x - y\|^2]$$

in λ for arbitrary fixed $x, y \in X$, we conclude that

$$f(\lambda, x, y) = 0, \ \lambda \in \mathbb{R}, x, y \in X,$$

i. e.,

$$(\lambda x, y) = \lambda(x, y), \ \lambda \in \mathbb{R}, x, y \in X,$$

which completes the proof for the *real case*.

The *complex case* is proved by defining an *inner product* on X as follows:

$$(x, y) := \frac{1}{4}[\|x + y\|^2 - \|x - y\|^2] + \frac{i}{4}[\|x + iy\|^2 - \|x - iy\|^2], \ x, y \in X,$$

and separating the real and imaginary parts.

The *additivity* and *homogeneity* in the first argument for *real scalars* follow from the *real case*.

Since, by *absolute scalability* of norm,

$$(ix, y) = \frac{1}{4}[\|ix + y\|^2 - \|ix - y\|^2] + \frac{i}{4}[\|ix + iy\|^2 - \|ix - iy\|^2]$$
$$= \frac{1}{4}[\|x - iy\|^2 - \|x + iy\|^2] + \frac{i}{4}[\|x + y\|^2 - \|x - y\|^2] = i(x, y), \ x, y \in X,$$

the *homogeneity* in the first argument holds for *complex scalars* as well, which completes the proof for the *complex case*. □

Remark 6.5. Observe that, in both real and complex cases, the only way to define the potential inner product is necessarily via the *Polarization Identities* (Proposition 6.2).

Corollary 6.1 (When Banach Space is Hilbert). *A Banach space* $(X, \| \cdot \|)$ *is a Hilbert space iff the parallelogram law holds.*

Exercise 6.11. Show that
(a) the Banach spaces $l_p^{(n)}$ and l_p $(1 \le p \le \infty, n \in \mathbb{N})$ are Hilbert spaces *iff* $p = 2$;
(b) the Banach space $(C[a, b], \| \cdot \|_\infty)$ $(-\infty < a < b < \infty)$ is *not* a Hilbert space.

6.1.4.3 Orthogonality and Pythagorean Theorem

The importance of the *orthogonality* concept, inherent to inner product spaces, for their theory and applications cannot be overestimated.

Definition 6.4 (Orthogonality). Vectors x and y in an inner product space $(X, (\cdot, \cdot))$ are said to be *orthogonal* if

$$(x, y) = 0.$$

Notation. $x \perp y$.

Remark 6.6. Due to *conjugate symmetry* of inner product, the relationship of orthogonality of vectors is *symmetric*, i. e., $x \perp y \Leftrightarrow y \perp x$.

Examples 6.4.

1. In the (real or complex) space $l_2^{(n)}$ ($n \in \mathbb{N}$), the vectors of the *standard unit basis*

    $$e_1 := (1, 0, 0, \ldots, 0), e_2 := (0, 1, 0, \ldots, 0), \ldots, e_n := (0, 0, 0, \ldots, 1)$$

    are *pairwise orthogonal*, i. e., any two distinct vectors in it are orthogonal:

    $$e_i \perp e_j, \ i, j = 1, \ldots, n, i \neq j.$$

2. In the (real or complex) space l_2, the vectors of the *standard Schauder basis* $\{e_n := (\delta_{nk})_{k \in \mathbb{N}}\}_{n \in \mathbb{N}}$, where δ_{nk} is the Kronecker delta, are *pairwise orthogonal*.
3. In the complex space $C[0, 2\pi]$ with the integral inner product

    $$C[0, 2\pi] \ni f, g \mapsto (f, g) := \int_0^{2\pi} f(t)\overline{g(t)} \, dt,$$

    the functions $x_k(t) := e^{ikt}$, $k \in \mathbb{Z}$, $t \in [0, 2\pi]$, where i is the *imaginary unit*, are *pairwise orthogonal* as well as their real parts $\cos nt$, $n \in \mathbb{Z}_+$, and imaginary parts $\sin nt$, $n \in \mathbb{N}$, in the real space $(C[0, 2\pi], (\cdot, \cdot))$.
4. The functions of the prior example represent pairwise orthogonal equivalence classes in the space $L_2(0, 2\pi)$, complex and real, respectively.

Exercise 6.12. Verify.

Proposition 6.3 (Orthogonal Characterization of Zero Vector). *In an inner product space $(X, (\cdot, \cdot))$,*

$$x = 0 \Leftrightarrow \forall y \in X : x \perp y.$$

Exercise 6.13. Prove (see Remarks 6.1).

The following is a natural generalization of the classical *Pythagorean Theorem*:

Theorem 6.4 (Pythagorean Theorem). *For any pair of orthogonal vectors x, y, in an inner product space $(X, (\cdot, \cdot), \| \cdot \|)$,*

$$\|x \pm y\|^2 = \|x\|^2 + \|y\|^2.$$

Exercise 6.14.

(a) Prove.

(b) Show that, in a *real* inner product space, the converse is also true, i. e.,

$$x \perp y \iff \|x + y\|^2 = \|x\|^2 + \|y\|^2.$$

(c) Is converse true in a *complex* inner product space?

Hint. For parts (b) and (c), using identity (6.2), show that

$$\|x + y\|^2 = \|x\|^2 + \|y\|^2 \iff \operatorname{Re}(x, y) = 0,$$

then infer and find a *counterexample*, respectively.

6.2 Convexity and Nearest Point Property

The *Nearest Point Property* (Theorem 6.6) is fundamental for the theory of Hilbert spaces and, having far-reaching implications, is instrumental for proving the *Projection Theorem* (Theorem 6.7), which, in its turn, underlies the proof of the *Riesz Representation Theorem* (Theorem 6.8). To proceed, we need to first introduce and discuss the notion of *convexity*.

6.2.1 Convexity

Definition 6.5 (Convex Set). A nonempty set C in a vector space X is called *convex* if

$$\forall x, y \in C, \ \forall 0 \le \lambda \le 1 : \ \lambda x + (1 - \lambda)y \in C,$$

i. e., for each pair x, y of its points, C contains the *line segment*

$$\{\lambda x + (1 - \lambda)y \mid 0 \le \lambda \le 1\}$$

connecting them.

Examples 6.5.

1. In a vector space X,

    (a) every *singleton* $\{x_0\}$ is a *convex set*,

    (b) every *subspace* Y is a *convex set*.

2. In $\mathbb{R}^2$, the sets $\{(x,y) \in \mathbb{R}^2 \mid x \geq 0\}$ and $\{(x,y) \in \mathbb{R}^2 \mid y < 0\}$ are *convex*, and the set $\{(x,y) \in \mathbb{R}^2 \mid xy \geq 0\}$ is *not*.

3. In a normed vector space $(X, \|\cdot\|)$, every (open or closed) *nontrivial ball* is a *convex set*.

This simple fact, in particular, precludes the mapping

$$\mathbb{F}^n \ni x = (x_1, \ldots, x_m) \mapsto \|x\|_p := \left[\sum_{k=1}^n |x_k|^p \right]^{1/p}$$

from being a norm on $\mathbb{F}^n$ ($n = 2, 3, \ldots$) for $0 < p < 1$.

Exercise 6.15.
(a) Verify.
(b) Show that the only convex sets in $\mathbb{R}$ are the *intervals*.
(c) Give more examples of convex sets and of sets, which are not convex, in $\mathbb{R}^2$.

Theorem 6.5 (Properties of Convex Sets). *In a vector space X over $\mathbb{F}$,*
(1) *if C is a convex set, then, for each $x \in X$, the translation $x + C$ is a convex set;*
(2) *if C is a convex set, then, for each $\mu \in \mathbb{F}$, the product μC is a convex set;*
(3) *if $\{C_i\}_{i \in I}$ is a nonempty collection of convex sets, then the intersection $\bigcap_{i \in I} C_i$, if nonempty, is a convex set;*
(4) *provided $(X, \|\cdot\|)$ is a normed vector space, if C is a convex set, then the closure $\overline{C}$ is a convex set.*

Exercise 6.16.
(a) Prove.
(b) Give an example showing that the union of convex sets need not be convex.
(c) Give an example showing that the converse to (4) is *not* true.

By properties (3) and (4) in the prior theorem, the following notions are well defined:

Definition 6.6 (Convex Hull and Closed Convex Hull). Let S be a nonempty subset of a vector space X, then

$$\mathrm{conv}(S) := \bigcap_{C \text{ is convex}, S \subseteq C} C$$

is the *smallest convex set of X containing S* called the *convex hull* of S.

In a normed vector space $(X, \|\cdot\|)$, the *closed convex hull* of a nonempty set S is the smallest closed convex set containing S, i. e., the closure $\overline{\mathrm{conv}(S)}$ of its convex hull.

Remark 6.7. A nonempty set S in a vector space X is convex *iff* $S = \mathrm{conv}(S)$.

Exercise 6.17.

(a) Verify.

(b) Describe the convex hull of a two-point set $\{x, y\}$ in a vector space X.

(c) Describe the convex hull of a finite set in $\mathbb{R}^2$ (see the *Convex Hull's Structure Proposition* (Proposition 6.20), Section 6.12, Problem 7).

6.2.2 Nearest Point Property

Theorem 6.6 (Nearest Point Property). *Let C be a closed convex set in a Hilbert space $(X, (\cdot, \cdot), \| \cdot \|)$. Then, for each $x \in X$, there exists a unique nearest point to x in C, i. e.,*

$$\forall x \in X \, \exists! \, y \in C : \|x - y\| = \rho(x, C) := \inf_{u \in C} \|x - u\|.$$

Proof. If $x \in C$, the statement is trivially true.

Exercise 6.18. Explain.

Suppose $x \notin C$. Then, by the *closedness* of C, $\rho(x, C) > 0$ (see Section 2.18, Problem 17).

Choosing a sequence $(y_n)_{n \in \mathbb{N}}$ in C such that

$$\|x - y_n\| \to \rho(x, C), \, n \to \infty, \tag{6.5}$$

by the *Parallelogram Law* (Theorem 6.3), we have

$$\|y_n - y_m\|^2 = \|y_n - x + x - y_m\|^2 = 2\|y_n - x\|^2 + 2\|x - y_m\|^2 - \|y_n + y_m - 2x\|^2$$

$$= 2\|y_n - x\|^2 + 2\|x - y_m\|^2 - 4\left\|\frac{y_n + y_m}{2} - x\right\|^2, \, m, n \in \mathbb{N}.$$

Since, by the *convexity* of C, $\frac{y_n + y_m}{2} \in C$, $m, n \in \mathbb{N}$, and hence

$$\left\|\frac{y_n + y_m}{2} - x\right\| \geq \rho(x, C), \, m, n \in \mathbb{N},$$

in view of (6.5), we infer

$$0 \leq \|y_n - y_m\|^2 \leq 2\|y_n - x\|^2 + 2\|x - y_m\|^2 - 4\rho^2(x, C) \to 0, \, m, n \to \infty,$$

which, by the *Squeeze Theorem*, implies that the sequence $(y_n)_{n \in \mathbb{N}}$ is *fundamental*. Therefore, by the *completeness* of $(X, (\cdot, \cdot), \| \cdot \|)$ and the *closedness* of C,

$$\exists y \in C : y_n \to y, \, n \to \infty,$$

which, in view of (6.5), by *continuity* of norm, implies that

$$\|x - y\| = \lim_{n \to \infty} \|x - y_n\| = \rho(x, C),$$

i. e., y is a *nearest point* to x in C, which completes the proof of the *existence part*.

To prove the *uniqueness*, suppose

$$\|x - z\| = \rho(x, C)$$

for some $z \in C$. Then, by the *Parallelogram Law* (Theorem 6.3),

$$\|y - z\|^2 = \|y - x + x - z\|^2 = 2\|y - x\|^2 + 2\|x - z\|^2 - \|y + z - 2x\|^2$$

$$= 2\|y - x\|^2 + 2\|x - z\|^2 - 4\left\|\frac{y + z}{2} - x\right\|^2 = 4\rho^2(x, C) - 4\left\|\frac{y + z}{2} - x\right\|^2.$$

Since, by the *convexity* of C, $\frac{y+z}{2} \in C$, and hence,

$$\left\|\frac{y + z}{2} - x\right\| \geq \rho(x, C),$$

we infer

$$0 \leq \|y - z\|^2 \leq 4\rho^2(x, C) - 4\rho^2(x, C) = 0,$$

which, implies that

$$\|y - z\| = 0,$$

and hence, by the *separation* norm axiom, that $z = y$, and completes the proof of the *uniqueness part* and of the entire statement. □

Since every subspace is a convex set (see Examples 6.5), we immediately obtain the following corollary.

Corollary 6.2 (Nearest Point Property Relative to Closed Subspaces). *Let Y be a closed subspace in a Hilbert space $(X, (\cdot, \cdot), \| \cdot \|)$. Then, for each $x \in X$, there exists a unique nearest point to x in Y, i. e.,*

$$\forall x \in X \, \exists! \, y \in Y : \|x - y\| = \rho(x, Y) := \inf_{u \in Y} \|x - u\|.$$

See the *Nearest Point Property Relative to Finite-Dimensional Subspaces* (Theorem 3.13) (see also Remark 3.27).

Remarks 6.8.
- The prior theorem is also known as the *Closest Point Property* (see, e. g., [70]).
- The *closedness* condition on the convex set is essential for the existence of the closest points and cannot be dropped.

   Exercise 6.19. Give a corresponding example.

- The *completeness* condition on the space is also essential for the existence of the nearest points and cannot be dropped.

Example 6.6. In $(c_{00}, (\cdot, \cdot), \| \cdot \|_2)$, which is an incomplete inner product space, when treated as a subspace of the Hilbert space l_2 (see Examples 6.3), consider the nonempty set

$$Y := \left\{ x := (x_n)_{n\in\mathbb{N}} \in c_{00} \,\middle|\, \sum_{n=1}^{\infty} \frac{1}{n} x_n = 0 \right\} = \{x \in c_{00} \mid x \perp y = 0 \text{ in } l_2\}$$

where $y := (1/n)_{n\in\mathbb{N}} \in l_2$.

As is easily seen, Y is a *proper subspace* of c_{00}, and hence, in particular, is a *convex set*.

Exercise 6.20. Verify.

Furthermore, Y is *closed* by the *Sequential Characterization of a Closed Set* (Theorem 2.19), since, if

$$Y \ni x_n \to x \in c_{00}, \; n \to \infty, \text{ in } l_2,$$

in view of $(x_n, y) = 0$, $n \in \mathbb{N}$, by *continuity* of inner product (Proposition 6.1),

$$(x, y) = \lim_{n\to\infty} (x_n, y) = 0,$$

which implies that $x \in Y$.

However, as is shown in Example 6.9 of the following section, in the space $(c_{00}, (\cdot, \cdot), \| \cdot \|_2)$, to no point $x \in Y^c$ is there a nearest point in Y.

- Since the proof of the *uniqueness part* of the prior theorem is entirely based on the *Parallelogram Law* (Theorem 6.3) and uses the *convexity* only, we conclude that, in any inner product space, the *nearest point* in a convex set to a point of the space, if exists, is *unique*.

Exercise 6.21. Give an example of a Banach space, in which the analogue of the *Nearest Point Property* relative to closed convex sets does not hold.

6.3 Projection Theorem

To prove the *Projection Theorem* (Theorem 6.7) underlying the proof of the *Riesz Representation Theorem* (Theorem 6.8), let us first discuss the notion of *orthogonal complement*.

6.3.1 Orthogonal Complements

Definition 6.7 (Orthogonal Complement). In an inner product space $(X, (\cdot, \cdot), \| \cdot \|)$, the *orthogonal complement* $M^\perp$ of a nonempty set M is the set of all elements orthogonal to M, i. e.,

$$M^\perp := \{z \in X \mid z \perp M, \text{ i. e., } \forall y \in M : z \perp y\}.$$

Examples 6.7.
1. In an inner product space $(X, (\cdot, \cdot), \|\cdot\|)$, $\{0\}^\perp = X$ and $X^\perp = \{0\}$.
2. In the Euclidean 2-space $l_2^{(2)}(\mathbb{R})$,

$$\{(x, 0) \mid x \in \mathbb{R}\}^\perp = \{(0, y) \mid y \in \mathbb{R}\}.$$

3. In the Euclidean 3-space $l_2^{(3)}(\mathbb{R})$,

$$\{(x, y, z) \in \mathbb{R}^3 \mid z = 0\}^\perp = \{(x, y, z) \in \mathbb{R}^3 \mid x = y = 0\}.$$

4. For an arbitrary $n \in \mathbb{N}$, in l_2,

$$\{(x_k)_{k \in \mathbb{N}} \in l_2 \mid x_k = 0, \ k \geq n + 1\}^\perp = \{(x_k)_{k \in \mathbb{N}} \in l_2 \mid x_k = 0, \ k = 1, \ldots, n\}.$$

Exercise 6.22. Verify.

Proposition 6.4 (Orthogonal Complement). *In an inner product space $(X, (\cdot, \cdot), \|\cdot\|)$, the orthogonal complement $M^\perp$ of an arbitrary nonempty set $M \subseteq X$ is a closed subspace, and*

$$M \cap M^\perp \subseteq \{0\}.$$

Exercise 6.23. Prove.

Proposition 6.5 (Characterization of Orthogonal Complement of Subspace). *Let Y be a subspace in an inner product space $(X, (\cdot, \cdot), \|\cdot\|)$. Then, for any $x \in X$,*

$$x \in Y^\perp \iff \forall y \in Y : \|x - y\| \geq \|x\|,$$

i. e., $x \in Y^\perp$ iff 0 is the nearest point to x in Y.

Proof. "Only if" part. If $x \in Y^\perp$, then

$$\forall y \in Y : x \perp y$$

and hence, by the *Pythagorean Theorem* (Theorem 6.4),

$$\forall y \in Y : \|x - y\|^2 = \|x\|^2 + \|y\|^2 \geq \|x\|^2.$$

Thus,

$$\forall y \in Y : \|x - y\| \geq \|x\|.$$

"If" part. Let us prove this part *by contradiction* assuming the existence of an $x \in X$ such that

$$\forall y \in Y : \|x - y\| \ge \|x\|, \text{ but } x \not\perp Y.$$

Then, for each $y \in Y$ and any $\lambda \in \mathbb{F}$, since Y is a *subspace*, $\lambda y \in Y$ and, by identity (6.2),

$$\|x\|^2 - 2\operatorname{Re}\bar{\lambda}(x,y) + |\lambda|^2\|y\|^2 = \|x - \lambda y\|^2 \ge \|x\|^2.$$

Whence,

$$-2\operatorname{Re}\bar{\lambda}(x,y) + |\lambda|^2\|y\|^2 \ge 0, \ \lambda \in \mathbb{F}, y \in Y.$$

Since, by the assumption,

$$\exists y \in Y : (x,y) \ne 0,$$

setting $\lambda := t\frac{(x,y)}{|(x,y)|}$ with $t > 0$, we have:

$$\forall t > 0 : -2t|(x,y)| + t^2\|y\|^2 \ge 0.$$

Whence,

$$\forall t > 0 : |(x,y)| \le \frac{t}{2}\|y\|^2.$$

Letting $t \to 0+$, we infer that $(x,y) = 0$, which is a *contradiction* proving the *"if"* part. □

6.3.2 Projection Theorem

Let us now prove the following key theorem.

Theorem 6.7 (Projection Theorem). *If Y is a closed subspace in a Hilbert space $(X, (\cdot, \cdot), \|\cdot\|)$, then every $x \in X$ has a unique decomposition*

$$x = y + z$$

with some $y \in Y$ and $z \in Y^\perp$, where the vector y, called the orthogonal projection of x onto Y, is the nearest point to x in Y, and hence, the space X is decomposed into the orthogonal sum of the subspaces Y and $Y^\perp$:

$$X = Y \oplus Y^\perp.$$

Proof. Let $x \in X$ be arbitrary. By the *Nearest Point Property Relative to Closed Subspaces* (Corollary 6.2), we can choose $y \in Y$ to be the *unique nearest point* to x in Y. Then

$$x = y + z$$

with $z := x - y$.

Since Y is a subspace, for any $y' \in Y$, $y + y' \in Y$ and we have

$$\|z - y'\| = \|x - (y + y')\| \geq \rho(x, Y) = \|x - y\| = \|z\|,$$

which, by the *Characterization of Orthogonal Complement of Subspace* (Proposition 6.5), implies that $z \in Y^{\perp}$.

The uniqueness of the decomposition immediately follows from the fact that the subspaces Y and $Y^{\perp}$ are *disjoint*, i. e.,

$$Y \cap Y^{\perp} = \{0\}$$

(see the *Orthogonal Complement Proposition* (Proposition 6.4)).

Hence, the subspaces Y and $Y^{\perp}$ are *complementary*, and

$$X = Y \oplus Y^{\perp},$$

the direct sum decomposition naturally called *orthogonal*. $\qquad\qquad\square$

Remark 6.9. As follows from the argument based on the *Characterization of Orthogonal Complement of Subspace* (Proposition 6.5) used in the foregoing proof, if, in an inner product space $(X, (\cdot, \cdot), \|\cdot\|)$, a y is the *nearest point* to an $x \in X$ in a subspace Y, which need not be closed, then $x - y \in Y^{\perp}$.

Examples 6.8.

1. $l_2^{(2)}(\mathbb{R}) = \{(x, 0) \mid x \in \mathbb{R}\} \oplus \{(0, y) \mid y \in \mathbb{R}\}$ (see Examples 6.7),

$$\forall\, (x, y) \in \mathbb{R}^2 : \ (x, y) = (x, 0) + (0, y).$$

2. $l_2^{(3)}(\mathbb{R}) = \{(x, y, z) \in \mathbb{R}^3 \mid z = 0\} \oplus \{(x, y, z) \in \mathbb{R}^3 \mid x = y = 0\}$ (see Examples 6.7),

$$\forall\, (x, y, z) \in \mathbb{R}^3 : \ (x, y, z) = (x, y, 0) + (0, 0, z).$$

3. For an arbitrary $n \in \mathbb{N}$,

$$l_2 = \{(x_k)_{k \in \mathbb{N}} \in l_2 \mid x_k = 0, \ k \geq n + 1\} \oplus \{(x_k)_{k \in \mathbb{N}} \in l_2 \mid x_k = 0, \ k = 1, \ldots, n\}$$

(see Examples 6.7).

Proposition 6.6 (Second Orthogonal Complement of a Closed Subspace). *In a Hilbert space* $(X, (\cdot, \cdot), \|\cdot\|)$, *for every closed subspace* Y,

$$(Y^{\perp})^{\perp} = Y.$$

Proof. It is obvious that $Y \subseteq (Y^{\perp})^{\perp}$.

To prove the inverse inclusion, consider an arbitrary $x \in (Y^{\perp})^{\perp}$. By the *Projection Theorem* (Theorem 6.7),

$$x = y + z$$

with some $y \in Y$ and $z \in Y^{\perp}$.

Since $x \in (Y^{\perp})^{\perp}$,

$$0 = (x, z) = (y + z, z) = (y, z) + \|z\|^2 = 0 + \|z\|^2 = \|z\|^2,$$

which, by the *separation* norm axiom, implies that $z = 0$, and hence, $x = y \in Y$, i.e., $(Y^{\perp})^{\perp} \subseteq Y$.

Thus, we conclude that $(Y^{\perp})^{\perp} = Y$. □

Remark 6.10. As the following example shows, in the *Projection Theorem* (Theorem 6.7), the requirement of the *completeness* of the space is essential and cannot be dropped.

Example 6.9. As is shown in Example 6.6,

$$Y := \{x \in c_{00} \mid x \perp y \text{ in } l_2\},$$

where $y := (1/n)_{n \in \mathbb{N}} \in l_2$, is a *proper closed subspace* in the incomplete inner product space $(c_{00}, (\cdot, \cdot), \|\cdot\|_2)$, the latter being treated as a subspace in l_2. Observe that the closedness of Y also immediately follows by the *Orthogonal Complement Proposition* (Proposition 6.4) from the fact that

$$Y = c_{00} \cap \{y\}^{\perp},$$

where $\{y\}^{\perp}$ is the orthogonal complement of $\{y\}$ in l_2.

Then, for the orthogonal complement $Y^{\perp}$ of Y in $(c_{00}, (\cdot, \cdot), \|\cdot\|_2)$, we have

$$Y^{\perp} = c_{00} \cap (\{y\}^{\perp})^{\perp},$$

where $(\{y\}^{\perp})^{\perp}$ is the second orthogonal complement of $\{y\}$ in l_2. By the *Second Orthogonal Complement of a Set Proposition* (Proposition 6.23, Section 6.12, Problem 13), in l_2,

$$(\{y\}^{\perp})^{\perp} = \mathrm{span}(\{y\}) = \{\lambda y \mid \lambda \in \mathbb{F}\},$$

and hence, in $(c_{00}, (\cdot, \cdot), \|\cdot\|_2)$,

$$Y^{\perp} = c_{00} \cap \mathrm{span}(\{y\}) = \{0\}.$$

Exercise 6.24. Explain.

Therefore,

$$c_{00} \neq Y \oplus Y^\perp = Y.$$

Furthermore, for any $x \in c_{00} \setminus Y$, assume that $y \in Y$ is the nearest point to x in Y. Then $x - y \in Y^\perp$ (see Remark 6.9), which, since $Y^\perp = \{0\}$, implies that $x = y \in Y$, which is a *contradiction*, showing that, in $(c_{00}, (\cdot, \cdot), \|\cdot\|_2)$, to no point $x \in Y^c$ is there a nearest point in Y.

Remark 6.11. Example 6.9 also shows that the requirement of the *completeness* of the space is essential and cannot be dropped in the *Second Orthogonal Complement of a Closed Subspace Proposition* (Proposition 6.6) and the *Second Orthogonal Complement of a Set Proposition* (Proposition 6.23, Section 6.12, Problem 13). Indeed, in $(c_{00}, (\cdot, \cdot), \|\cdot\|_2)$,

$$(Y^\perp)^\perp = \{0\}^\perp = c_{00} \neq Y.$$

Proposition 6.7 (Characterization of Denseness of Subspace). *For a subspace Y in a Hilbert space $(X, (\cdot, \cdot), \|\cdot\|)$,*

$$\overline{Y} = X \iff Y^\perp = \{0\}.$$

Proof. Follows immediately from the fact that

$$Y^\perp = \overline{Y}^\perp$$

(see Section 6.12, Problem 9), the *Closure of a Subspace Proposition* (Proposition 3.13, Section 3.5, Problem 6), and the *Projection Theorem* (Theorem 6.7).

Exercise 6.25. Fill in the details. □

Remark 6.12. Example 6.9 also shows that, in the prior proposition, the requirement of the *completeness* of the space is essential and cannot be dropped. Indeed, in it, $\overline{Y} = Y \neq X$, but $Y^\perp = \{0\}$.

6.3.3 Orthogonal Projections

Thus, in a Hilbert space $(X, (\cdot, \cdot), \|\cdot\|)$, with every closed subspace Y associated are the decomposition of X into the *orthogonal sum* of Y and $Y^\perp$,

$$X = Y \oplus Y^\perp,$$

and hence, the projection operator P onto Y along $Y^\perp$, called the *orthogonal projection* onto Y:

$$X \ni x = y + z \; (y \in Y, z \in Y^\perp) \mapsto Px := y \in Y.$$

Then $I - P$ is the *orthogonal projection* onto $Y^\perp$:

$$X \ni x = y + z \, (y \in Y, z \in Y^\perp) \mapsto (I - P)x := z \in Y^\perp,$$

and

$$\forall x, y \in X : \ Px \perp (I - P)y.$$

Examples 6.10.

1. In the Euclidean 2-space $l_2^{(2)}(\mathbb{R})$,

$$P(x, y) = (x, 0), \ (x, y) \in \mathbb{R}^2,$$

is the *orthogonal projection* onto the subspace $\{(x, 0) \mid x \in \mathbb{R}\}$, i. e., the x-axis, and

$$(I - P)(x, y) = (0, y), \ (x, y) \in \mathbb{R}^2,$$

is the *orthogonal projection* onto the subspace $\{(0, y) \mid y \in \mathbb{R}\}$, i. e., the y-axis (see Examples 6.8).

2. In the Euclidean 3-space $l_2^{(3)}(\mathbb{R})$,

$$P(x, y, z) = (x, y, 0), \ (x, y, z) \in \mathbb{R}^3,$$

is the *orthogonal projection* onto the subspace $\{(x, y, z) \in \mathbb{R}^3 \mid z = 0\}$, i. e., the xy-plane, and

$$(I - P)(x, y, z) = (0, 0, z), \ (x, y, z) \in \mathbb{R}^3,$$

is the *orthogonal projection* onto the subspace $\{(x, y, z) \in \mathbb{R}^3 \mid x = y = 0\}$, i. e., the z-axis (see Examples 6.8).

3. For an arbitrary $n \in \mathbb{N}$, in l_2,

$$l_2 \ni x := (x_k)_{k \in \mathbb{N}}^\infty \mapsto Px = (x_1, \dots, x_n, 0, 0, \dots)$$

is the *orthogonal projection* onto the n-dimensional subspace

$$Y_n := \mathrm{span}(\{e_1, \dots, e_n\}) = \{(x_k)_{k \in \mathbb{N}} \in l_2 \mid x_k = 0, \ k \geq n + 1\},$$

where $e_i = (\delta_{ik})_{k=1}^\infty, \ i = 1, 2, \dots, \ (\delta_{ik}$ is the Kronecker delta), and

$$(I - P)x = \{\underbrace{0, \dots, 0}_{n \text{ terms}}, x_{n+1}, x_{n+2}, \dots\}$$

is the *orthogonal projection* onto

$$Y_n^\perp = \overline{\mathrm{span}(\{e_{n+1}, e_{n+2}, \dots\})} = \{(x_k)_{k \in \mathbb{N}} \mid x_k = 0, \ k = 1, \dots, n\}$$

(see Examples 6.8).

6.4 Riesz Representation Theorem

The following celebrated result describes all bounded linear functionals on Hilbert spaces and shows that such spaces are *self-dual*:

Theorem 6.8 (Riesz Representation Theorem). *Let* $(X, (\cdot, \cdot), \| \cdot \|)$ *be a Hilbert space. Then*

$$\forall f \in X^* \; \exists! \, y_f \in X \; \forall x \in X : f(x) = (x, y_f).$$

The mapping

$$X^* \ni f \mapsto y_f \in X$$

is an isometric linear (if the space is real) or conjugate linear (if the space is complex) isomorphism between X^* *and* X, *in which sense we can write* $X^* = X$, *and thus, regard* X *to be self-dual.*

Proof. Consider an arbitrary $f \in X^*$.

If $f = 0$, $y_f = 0$. Indeed,

$$\forall x \in X : f(x) = 0 = (x, 0) \tag{6.6}$$

and, by the *Orthogonal Characterization of the Zero Vector* (Proposition 6.3), $y_f = 0$ is the only vector in X satisfying (6.6). Observe that, in this case, $\|f\| = \|y_f\| = 0$.

If $f \neq 0$,

$$Y := \ker f := \{x \in X \mid f(x) = 0\}$$

is a *closed proper subspace* (*hyperplane*) in X (see the *Kernel of a Linear Functional* (Proposition 4.16, Section 4.7, Problem 10) and the *Null Space of a Linear Functional Proposition* (Proposition 4.1)) and, by the *Projection Theorem* (Theorem 6.7),

$$X = Y \oplus Y^\perp$$

with $Y^\perp \neq \{0\}$.

Choosing a *nonzero* vector $z \in Y^\perp$, without loss of generality, we can assume that $f(z) = 1$.

Exercise 6.26. Explain.

Then, for any $x \in X$,

$$x = [x - f(x)z] + f(x)z$$

and, since

$$f(x - f(x)z) = f(x) - f(x)f(z) = f(x) - f(x) = 0,$$

i. e., $x - f(x)z \in Y$, we infer that

$$\forall x \in X : x - f(x)z \perp z, \text{ i. e., } (x - f(x)z, z) = 0,$$

which, by *linearity* of inner product in the first argument, implies that

$$\forall x \in X : (x, z) = (f(x)z, z) = f(x)\|z\|^2.$$

Let

$$y_f := \frac{1}{\|z\|^2} z \in X \setminus \{0\},$$

then, by *linearity/conjugate linearity* of inner product in the second argument,

$$\forall x \in X : (x, y_f) = \frac{1}{\|z\|^2}(x, z) = f(x),$$

which proves the *existence part*.

The *uniqueness* immediately follows by the *Inner Product Separation Property* (Proposition 6.17, Section 6.12, Problem 1).

Exercise 6.27. Explain.

By the *Cauchy–Schwarz Inequality* (Theorem 6.2), for any $f \in X^* \setminus \{0\}$ with the corresponding $y_f \in X \setminus \{0\}$,

$$\forall x \in X : |f(x)| = |(x, y_f)| \le \|y_f\|\|x\|,$$

which implies that

$$\|f\| \le \|y_f\|.$$

Since furthermore, for the *unit vector* $x := \frac{1}{\|y_f\|} y_f$ ($\|x\| = 1$),

$$\|f\| \ge |f(x)| = \frac{1}{\|y_f\|}|f(y_f)| = \frac{1}{\|y_f\|}(y_f, y_f) = \frac{1}{\|y_f\|}\|y_f\|^2 = \|y_f\|,$$

we conclude that

$$\forall f \in X^* \setminus \{0\} : \|f\| = \|y_f\|.$$

As follows from the *linearity/conjugate linearity* of inner product in the second argument, the mapping

$$X^* \ni f \mapsto y_f \in X \tag{6.7}$$

is *linear* when X is real, and *conjugate linear* when X is complex.

Exercise 6.28. Verify.

It remains to show that the mapping given by (6.7) is *surjective* (i. e., *onto*). Indeed, for each $y \in X$, by *linearity* of inner product in the first argument and the *Cauchy–Schwarz Inequality* (Theorem 6.2),

$$f(x) := (x, y), \ x \in X,$$

is a *bounded linear functional* on X, and hence, $y = y_f$.

Thus, the mapping given by (6.7) is an isometric linear (if the space is real) or conjugate linear (if the space is complex) isomorphism between X^* and X, in which sense, we can write $X^* = X$, and thus, regard X to be *self-dual*. ☐

Remarks 6.13.
- The *Riesz Representation Theorem* is also known as the *Riesz–Fréchet*[1] *Theorem* (see, e. g., [70]).
- By the *Riesz Representation Theorem*, the dual X^* of a Hilbert space $(X, (\cdot, \cdot), \|\cdot\|)$ is also a Hilbert space relative to the inner product

$$X^* \ni f, g \mapsto (f, g)_{X^*} := \overline{(y_f, y_g)} = (y_g, y_f), \ f, g \in X^*,$$

  where

$$f(x) = (x, y_f) \text{ and } g(x) = (x, y_g), \ x \in X.$$

- By the *Riesz Representation Theorem*, each bounded linear (if the space is real) or conjugate linear (if the space is complex) functional g on a Hilbert space $(X, (\cdot, \cdot), \|\cdot\|)$ is of the form:

$$g(x) = (y_g, x)$$

  with some *unique* $y_g \in X$ and the mapping

$$g \mapsto y_g$$

  is an isometric linear isomorphism between the space of bounded linear/conjugate linear functionals on X and X.
- The condition of the *completeness* of the space in the *Riesz Representation Theorem* is essential and cannot be dropped.
  Indeed, on $(c_{00}, (\cdot, \cdot), \|\cdot\|_2)$, which is an incomplete inner product space, when treated as a subspace of the Hilbert space l_2 (see Examples 6.3),

$$c_{00} \ni x := (x_n)_{n\in\mathbb{N}} \mapsto f(x) := \sum_{n=1}^{\infty} \frac{x_n}{n}$$

---

[1] Maurice Fréchet (1878–1973).

is a bounded linear functional, being a restriction of such on l_2 (see the *Linear Bounded Functionals on Certain Hilbert Spaces Corollary* (Corollary 6.3)), but

$$\nexists y_f \in c_{00} : f(x) = (x, y_f)$$

(see Example 6.6).

Exercise 6.29. Verify.

The following description is an immediate corollary of the *Riesz Representation Theorem* (Theorem 6.8) (see Examples 6.3).

Corollary 6.3 (Linear Bounded Functionals on Certain Hilbert Spaces).
1. *For each $f \in l_2^{(n)*}$ ($n \in \mathbb{N}$), there exists a unique element $y_f := (y_1, \dots, y_n) \in l_2^{(n)}$ such that*

$$f(x) = \sum_{k=1}^{n} x_k y_k, \ x := (x_1, \dots, x_n) \in l_2^{(n)},$$

and $\|f\| = \|y_f\|_2 = [\sum_{k=1}^{n} |y_k|^2]^{1/2}$.
2. *For each $f \in l_2^*$, there exists a unique element $y_f := (y_n)_{n \in \mathbb{N}} \in l_2$ such that*

$$f(x) = \sum_{k=1}^{\infty} x_k y_k, \ x := (x_n)_{n \in \mathbb{N}} \in l_2,$$

and $\|f\| = \|y_f\|_2 = [\sum_{k=1}^{\infty} |y_k|^2]^{1/2}$.
3. *For each $f \in L_2^*(a, b)$ ($-\infty \le a < b \le \infty$) (see Examples 6.3), there exists a unique element $y_f \in L_2(a, b)$ such that*

$$f(x) = \int_a^b x(t) y_f(t) \, dt, \ x \in L_2(a, b),$$

and $\|f\| = \|y_f\|_2 = [\int_a^b |y_f(t)|^2 dt]^{1/2}$.

6.5 Orthogonal and Orthonormal Sets

Definition 6.8 (Orthogonal and Orthonormal Sets). A set $\{x_i\}_{i \in I}$ in an inner product space $(X, (\cdot, \cdot), \| \cdot \|)$ is called *orthogonal* if its elements are pairwise orthogonal, i. e.,

$$x_i \perp x_j, \ i, j \in I, i \ne j.$$

If further $\|x_i\| = 1$, for each $i \in I$, the orthogonal set is called *orthonormal*.

Remarks 6.14.

- Often, the terms *"orthogonal/orthonormal system"* and *"orthogonal/orthonormal sequence"*, when the set countably infinite, are used instead.
- For an orthonormal set $\{e_i\}_{i \in I}$,

$$(e_i, e_j) = \delta_{ij}, \ i, j \in I,$$

where δ_{ij} is the *Kronecker delta*.

- Any orthogonal set $\{x_i\}_{i \in I}$ with nonzero elements can be transformed into an orthonormal set $\{e_i\}_{i \in I}$ with the *same span* via the following simple *normalization procedure*:

$$e_i := \frac{x_i}{\|x_i\|}, \ i \in I.$$

Examples 6.11.

1. In the (real or complex) space $l_2^{(n)}$ ($n \in \mathbb{N}$), the *standard unit basis*

$$\{e_1 := (1, 0, 0, \dots, 0), e_2 := (0, 1, 0, \dots, 0), \dots, e_n := (0, 0, 0, \dots, 1)\}$$

   is an *orthonormal set* (see Examples 6.4).

2. In the (real or complex) space l_2, the *standard Schauder basis*

$$\{e_n := (\delta_{nk})_{k \in \mathbb{N}}\}_{n \in \mathbb{N}},$$

   where δ_{nk} is the Kronecker delta, is an *orthonormal set* (see Examples 6.4).

3. In the complex space $(C[0, 2\pi], (\cdot, \cdot)_2)$, the set

$$\{x_k(t) := e^{ikt} \mid k \in \mathbb{Z}, t \in [0, 2\pi]\}$$

   is *orthogonal* (see Examples 6.4) and, in view of

$$\|x_k\|_2 = \sqrt{2\pi}, \ k \in \mathbb{Z},$$

   it can be normalized into the following *orthonormal* one:

$$\left\{ e_k(t) := \frac{e^{ikt}}{\sqrt{2\pi}} \,\middle|\, k \in \mathbb{Z}, t \in [0, 2\pi] \right\}.$$

   In the real space $(C[0, 2\pi], (\cdot, \cdot)_2)$, the set

$$\{1, \cos nt, \ \sin nt \mid n \in \mathbb{N}, t \in [0, 2\pi]\}$$

   is *orthogonal* (see Examples 6.4) and, in view of

$$\|1\|_2 = \sqrt{2\pi}, \ \| \cos n \cdot \|_2 = \| \sin n \cdot \|_2 = \sqrt{\pi}, \ n \in \mathbb{N},$$

   it can be normalized into the following *orthonormal* one:

$$\left\{ \frac{1}{\sqrt{2\pi}}, \frac{\cos nt}{\sqrt{\pi}}, \frac{\sin nt}{\sqrt{\pi}} \,\middle|\, n \in \mathbb{N}, t \in [0, 2\pi] \right\}.$$

4. The sets of the prior example are also orthogonal and orthonormal in the space $L_2(0, 2\pi)$ (complex or real, respectively), their elements understood as the equivalence classes represented by the corresponding functions.

Exercise 6.30. Verify.

Proposition 6.8 (Linear Independence of Orthogonal Sets). *In an inner product space* $(X, (\cdot, \cdot), \| \cdot \|)$, *an orthogonal set* $\{x_i\}_{i \in I}$ *with nonzero elements, in particular an orthonormal set, is linearly independent.*

Exercise 6.31. Prove.

Definition 6.9 (Complete Orthonormal Set). In an inner product space $(X, (\cdot, \cdot), \| \cdot \|)$, an orthonormal set $\{e_i\}_{i \in I}$, which is not a proper subset of any orthonormal set, i. e., a maximal orthonormal set relative to the set-theoretic inclusion $\subseteq$, is called *complete*.

The following statement resembles the *Basis Theorem* (Theorem 3.2). The same is true for the proofs.

Theorem 6.9 (Existence of a Complete Orthonormal Set). *In a nonzero inner product space* $(X, (\cdot, \cdot), \| \cdot \|)$, *each orthonormal set* S *can be extended to a complete orthonormal set* S'.

Proof. Let S be an arbitrary orthonormal set in $(X, (\cdot, \cdot), \| \cdot \|)$.

In the collection $(\mathcal{O}, \subseteq)$ of all orthonormal subsets in $(X, (\cdot, \cdot), \| \cdot \|)$ partially ordered by the set-theoretic inclusion $\subseteq$, consider an arbitrary *chain* $\mathscr{C}$.

The set

$$U := \bigcup_{C \in \mathscr{C}} C,$$

is also *orthonormal* and is an *upper bound* of $\mathscr{C}$ in $(\mathcal{O}, \subseteq)$.

Exercise 6.32. Verify.

By *Zorn's Lemma (Precise Version)* (Theorem A.6), there exists a *maximal element* S' in $(\mathcal{O}, \subseteq)$, i. e., a *complete orthonormal set* in X, such that $S \subseteq S'$. $\quad\square$

Proposition 6.9 (Characterization of Complete Orthonormal Sets). *In an inner product space* $(X, (\cdot, \cdot), \| \cdot \|)$, *an orthonormal set* $\{e_i\}_{i \in I}$ *is complete iff*

$$\{e_i\}_{i \in I}^{\perp} = \{0\}.$$

Exercise 6.33. Prove.

Remarks 6.15.
- The *Existence of a Complete Orthonormal Set Theorem* (Theorem 6.9), in particular, establishes the *existence* of a complete orthonormal set in a nonzero inner product space $(X, (\cdot, \cdot), \| \cdot \|)$.

Indeed, one can take a singleton $S := \{\frac{x}{\|x\|}\}$, where $x \in X\backslash\{0\}$ is an arbitrary nonzero element, and extend it to a complete orthonormal set S' in $(X, (\cdot, \cdot), \|\cdot\|)$.

– A complete orthonormal set in an inner product space, although *existent* by the prior theorem, *need not be unique*. If $S := \{e_j\}_{j \in I}$ is a complete orthonormal set in a complex inner product space $(X, (\cdot, \cdot), \|\cdot\|)$, for any $\theta_j \in \mathbb{R}, j \in I$, the set of rotations $S' := \{e^{i\theta_j}e_j\}_{j \in I}$ (i in the exponent is the *imaginary unit*) is a complete orthonormal set in $(X, (\cdot, \cdot), \|\cdot\|)$ as well.

In particular, in the (real or complex) space $l_2^{(2)}$, the *standard unit basis*

$$\{e_1 := (1, 0), e_2 := (0, 1)\}$$

is a *complete orthonormal set* as well as its clockwise rotation by $\pi/4$ (radians):

$$\left\{e_1' := \frac{1}{\sqrt{2}}(1, -1), e_2' := \frac{1}{\sqrt{2}}(1, 1)\right\}$$

Exercise 6.34. Verify.

Hint. Use the prior characterization.

Examples 6.12.

1. In the (real or complex) space $l_2^{(n)}$ ($n \in \mathbb{N}$), the *standard unit basis*

$$\{e_1 := (1, 0, 0, \ldots, 0), e_2 := (0, 1, 0, \ldots, 0), \ldots, e_n := (0, 0, 0, \ldots, 1)\}$$

is a *complete orthonormal set*.

2. In the (real or complex) space l_2, the *standard Schauder basis*

$$\{e_n := (\delta_{nk})_{k \in \mathbb{N}}\}_{n \in \mathbb{N}},$$

where δ_{nk} is the Kronecker delta, is a *complete orthonormal sequence*.

3. In the *complex* space $(C[0, 2\pi], (\cdot, \cdot)_2)$, the set

$$\left\{\frac{e^{ikt}}{\sqrt{2\pi}} \,\middle|\, k \in \mathbb{Z}, t \in [0, 2\pi]\right\}$$

is a *complete orthonormal sequence*.
In the *real* space $(C[0, 2\pi], (\cdot, \cdot)_2)$,

$$\left\{\frac{1}{\sqrt{2\pi}}, \frac{\cos kt}{\sqrt{\pi}}, \frac{\sin kt}{\sqrt{\pi}} \,\middle|\, k \in \mathbb{Z}, t \in [0, 2\pi]\right\}$$

is a *complete orthonormal sequence*.

4. The sets of the prior example are also *complete orthonormal sequences* in the Hilbert space $L_2(0, 2\pi)$ (complex or real, respectively), their elements understood as the equivalence classes represented by the corresponding functions.

5. The set

$$S := \{e^{i\lambda t} \mid t \in \mathbb{R}\}_{\lambda \in \mathbb{R}}$$

is an *uncountable complete orthonormal set* in a complex inner product space $(X, (\cdot, \cdot)_X)$ constructed as follows:

- the vector space $X := \text{span}(S)$, i. e., the set of all *"exponential polynomials"* of the form

$$x(t) := \sum_{k=1}^{n} a_k e^{i\lambda_k t},$$

where $n \in \mathbb{N}$, $a_1, \ldots, a_n \in \mathbb{C}$, and $\lambda_1, \ldots, \lambda_n \in \mathbb{R}$;

- for arbitrary

$$x(t) := \sum_{k=1}^{n} a_k e^{i\lambda_k t}, \ y(t) := \sum_{l=1}^{m} b_l e^{i\mu_l t} \in X,$$

the inner product (x, y) is defined by

$$(x, y) = \lim_{T \to \infty} \frac{1}{2T} \int_{-T}^{T} x(t)\overline{y(t)} \, dt = \lim_{T \to \infty} \frac{1}{2T} \sum_{k=1}^{n} \sum_{l=1}^{m} a_k \overline{b_l} \int_{-T}^{T} e^{i\lambda_k t} e^{-i\mu_l t} \, dt$$

$$= \sum_{k=1}^{n} \sum_{l=1}^{m} a_k \overline{b_l} \lim_{T \to \infty} \frac{1}{2T} \int_{-T}^{T} e^{i(\lambda_k - \mu_l)t} \, dt = \sum_{k=1}^{n} \sum_{l=1}^{m} a_k \overline{b_l} \delta(\lambda_k, \mu_l),$$

where $\delta(\lambda_k, \mu_l)$ is the *Kronecker delta*.

Exercise 6.35. Verify 1, 2, and 5.

One can also characterize complete orthonormal sets in a Hilbert space setting as follows:

Proposition 6.10 (Characterization of Complete Orthonormal Sets). *In a Hilbert space* $(X, (\cdot, \cdot), \| \cdot \|)$*, an orthonormal set* $\{e_i\}_{i \in I}$ *is complete iff it is fundamental, i. e.,*

$$\overline{\text{span}(\{e_i\}_{i \in I})} = X.$$

Proof. Since, in an inner product space, by the *Coincidence of Orthogonal Complements Proposition* (Proposition 6.22, Section 6.12, Problem 10),

$$\overline{\text{span}(\{e_i\}_{i \in I})}^{\perp} = \{e_i\}_{i \in I}^{\perp},$$

by the *Characterization of Complete Orthonormal Sets* (Proposition 6.9), the *completeness* of $\{e_i\}_{i \in I}$ is *equivalent* to

$$\overline{\text{span}(\{e_i\}_{i \in I})}^{\perp} = \{e_i\}_{i \in I}^{\perp} = \{0\}.$$

The latter, in a Hilbert space, by the *Projection Theorem* (Theorem 6.7), is *equivalent* to

$$\overline{\text{span}(\{e_i\}_{i \in I})} = X,$$

i. e., to the *fundamentality* of $\{e_i\}_{i \in I}$. □

6.6 Gram–Schmidt Process

The *Gram²–Schmidt³ process* transforms a *linearly independent countable set* in an inner product space into an *orthogonal* or *orthonormal set* with the *same span*.

Given a linearly independent countable set $\{x_n\}_{n \in I}$, where $I = \{1, \ldots, N\}$ with some $N \in \mathbb{N}$ or $I = \mathbb{N}$, in an inner product space $(X, (\cdot, \cdot), \| \cdot \|)$, one can inductively build new sets $\{y_n\}_{n \in I}$ and $\{e_n\}_{n \in I}$ as follows:

$$y_1 := x_1, \ e_1 := \frac{y_1}{\|y_1\|},$$

and

$$y_n := x_n - \sum_{k=1}^{n-1} (x_n, e_k) e_k, \ e_n := \frac{y_n}{\|y_n\|}, \ n \in I, n \geq 2.$$

The output sets $\{y_n\}_{n \in I}$ and $\{e_n\}_{n \in I}$ are *orthogonal* and *orthonormal*, respectively, and

$$\forall n \in I: \ \text{span}(\{e_1, \ldots, e_n\}) = \text{span}(\{y_1, \ldots, y_n\}) = \text{span}(\{x_1, \ldots, x_n\}).$$

Exercise 6.36. Verify.

The latter implies that

$$\text{span}(\{e_n\}_{n \in I}) = \text{span}(\{y_n\}_{n \in I}) = \text{span}(\{x_n\}_{n \in I}).$$

The process of building the orthogonal set $\{y_n\}_{n \in I}$ and the orthonormal set $\{e_n\}_{n \in I}$ is called the *Gram–Schmidt orthogonalization* and the *Gram–Schmidt orthonormalization*, respectively.

Example 6.13. Applying the *Gram–Schmidt orthonormalization* in the (real or complex) Hilbert space l_2 to the linearly independent set

$$\{x_1 := (1, 0, 0, \ldots), x_2 := (-1, -1, 0, \ldots)\},$$

---

2 Jørgen Pedersen Gram (1850–1916).
3 Erhard Schmidt (1876–1959).

we obtain $e_1 := \frac{x_1}{\|x_1\|_2} = x_1 = (1, 0, 0, \ldots)$ and

$$e_2 := \frac{x_2 - (x_2, e_1)e_1}{\|x_2 - (x_2, e_1)e_1\|_2} = \frac{(-1, -1, 0, \ldots) + (1, 0, 0, \ldots)}{\|(-1, -1, 0, \ldots) + (1, 0, 0, \ldots)\|_2} = (0, -1, 0, \ldots),$$

i. e., the orthonormal set $\{e_1 = (1, 0, 0, \ldots), e_2 = (0, -1, 0, \ldots)\}$.

Examples 6.14.

1. In the Hilbert space $L_2(-1, 1)$, the *Gram–Schmidt orthonormalization* applied to the set

$$\{t^n\}_{n \in \mathbb{Z}_+}$$

yields the *complete orthonormal sequence* of the well-known *Legendre*[4] *polynomials*

$$\left\{ P_n(t) = \frac{\sqrt{n + 1/2}}{2^n n!} \frac{d^n}{dt^n} [(t^2 - 1)^n] \right\}_{n \in \mathbb{Z}_+}.$$

2. In the Hilbert space $L_2((-\infty, \infty), e^{-t^2} dt)$, the *Gram–Schmidt orthonormalization* applied to the set

$$\{t^n\}_{n \in \mathbb{Z}_+}$$

yields the *complete orthonormal sequence* of *Hermite*[5] *polynomials*

$$\left\{ H_n(t) = \frac{(-1)^n}{\sqrt{\sqrt{\pi} 2^n n!}} e^{t^2} \frac{d^n}{dt^n} e^{-t^2} \right\}_{n \in \mathbb{Z}_+}.$$

3. In the Hilbert space $L_2((0, \infty), e^{-t} dt)$, the *Gram–Schmidt orthonormalization* applied to the set

$$\{t^n\}_{n \in \mathbb{Z}_+}$$

yields the *complete orthonormal sequence* of *Laguerre*[6] *polynomials*

$$\left\{ L_n(t) = (-1)^n e^t \frac{d^n}{dt^n} [t^n e^{-t}] \right\}_{n \in \mathbb{Z}_+}.$$

See, e. g., [13].

Remark 6.16. To be precise, in L_2 spaces, one should talk about the equivalence classes represented by the corresponding polynomials (see Examples 6.1).

---

4 Adrien-Marie Legendre (1752–1833).
5 Charles Hermite (1822–1901).
6 Edmond Laguerre (1834–1886).

6.7 Generalized Fourier Series

We begin our treatment of *generalized Fourier series* in a Hilbert space with studying the case of a finite orthonormal set.

6.7.1 Finite Orthonormal Set

Theorem 6.10 (Finite Orthonormal Set). *Let $\{e_1, \ldots, e_n\}$ ($n \in \mathbb{N}$) be a finite orthonormal set in an inner product space $(X, (\cdot, \cdot), \|\cdot\|)$. Then, for each $x \in X$, the nearest point to x in the n-dimensional subspace $Y := \mathrm{span}(\{e_1, \ldots, e_n\})$ is*

$$y := \sum_{i=1}^{n} (x, e_i) e_i$$

with

$$x - y \in Y^{\perp} \quad \text{and} \quad \|y\|^2 = \sum_{i=1}^{n} |(x, e_i)|^2$$

and, the point y being the nearest to x in Y,

$$\rho^2(x, Y) = \|x - y\|^2 = \|x\|^2 - \|y\|^2 = \|x\|^2 - \sum_{i=1}^{n} |(x, e_i)|^2,$$

which implies that

$$\sum_{i=1}^{n} |(x, e_i)|^2 \le \|x\|^2.$$

Furthermore,

$$x \in Y \iff x = y \iff \|x - y\| = 0 \iff \sum_{i=1}^{n} |(x, e_i)|^2 = \|x\|^2.$$

Proof. Let

$$y := \sum_{i=1}^{n} (x, e_i) e_i.$$

As is easily seen $y \in Y := \mathrm{span}(\{e_1, \ldots, e_n\})$ and $z := x - y \in Y^{\perp}$.

Exercise 6.37. Verify the latter.

Hint. Show that

$$z \perp e_i, \ i = 1, \ldots, n,$$

and apply the *Coincidence of Orthogonal Complements Proposition* (Proposition 6.22, Section 6.12, Problem 10).

Now, let us show that y is the *nearest point* to x in Y, i. e., that, for any $y' \in Y \setminus \{y\}$,

$$\|x - y'\| > \|x - y\|.$$

Indeed,

$$\|x - y'\|^2 = \|y + z - y'\|^2 = \|z + (y - y')\|^2$$

$$\qquad \text{since } y - y' \in Y \text{ and } z \perp y - y', \text{ by the } \textit{Pythagorean Theorem} \text{ (Theorem 6.4)},$$

$$= \|z\|^2 + \|y - y'\|^2 \qquad\qquad\qquad\qquad \text{since } y \neq y' \in Y, \|y - y'\|^2 > 0;$$

$$> \|z\|^2 = \|x - y\|^2.$$

By the *Generalized Pythagorean Theorem* (Theorem 6.27, Section 6.12, Problem 16), in view of *absolute scalability* of norm and considering that $\|e_i\| = 1$, $i = 1, \ldots, n$,

$$\|y\|^2 = \sum_{i=1}^{n} \|(x, e_i)e_i\|^2 = \sum_{i=1}^{n} |(x, e_i)|^2.$$

Furthermore, since $x - y \in Y^{\perp}$, by the *Pythagorean Theorem* (Theorem 6.4),

$$\rho^2(x, Y) = \|x - y\|^2 = \|x\|^2 - \|y\|^2 = \|x\|^2 - \sum_{i=1}^{n} |(x, e_i)|^2,$$

which implies that

$$\sum_{i=1}^{n} |(x, e_i)|^2 \leq \|x\|^2$$

and

$$x \in Y \iff x = y \iff \|x - y\| = 0 \iff \sum_{i=1}^{n} |(x, e_i)|^2 = \|x\|^2. \qquad \square$$

From the prior theorem, the *Closedness of Finite-Dimensional Subspaces Theorem* (Theorem 3.12), and the *Projection Theorem* (Theorem 6.7), we immediately obtain the following:

Corollary 6.4 (Finite Orthonormal Set). *Let $\{e_1, \ldots, e_n\}$ ($n \in \mathbb{N}$) be a finite orthonormal set in a Hilbert space $(X, (\cdot, \cdot), \| \cdot \|)$. Then, for each $x \in X$, the orthogonal projection of x onto the n-dimensional subspace $Y := \mathrm{span}(\{e_1, \ldots, e_n\})$ is*

$$y := \sum_{i=1}^{n} (x, e_i)e_i$$

with

$$x - y \in Y^{\perp} \quad \text{and} \quad \|y\|^2 = \sum_{i=1}^{n} |(x, e_i)|^2$$

and, the point y being the nearest to x in Y,

$$\rho^2(x, Y) = \|x - y\|^2 = \|x\|^2 - \|y\|^2 = \|x\|^2 - \sum_{i=1}^{n} |(x, e_i)|^2,$$

which implies that

$$\sum_{i=1}^{n} |(x, e_i)|^2 \le \|x\|^2.$$

Furthermore,

$$x \in Y \iff x = y \iff \|x - y\| = 0 \iff \sum_{i=1}^{n} |(x, e_i)|^2 = \|x\|^2.$$

Examples 6.15.

1. Thus, when applying the *Gram–Schmidt process* to linearly independent count-
    able set $\{x_n\}_{n \in I}$, where $I = \{1, \dots, N\}$ with some $N \in \mathbb{N}$ or $I = \mathbb{N}$, in a Hilbert space
    $(X, (\cdot, \cdot), \| \cdot \|)$, we set

    $$y_1 := x_1, \ e_1 := \frac{y_1}{\|y_1\|}$$

    and, to obtain y_n for $n \in I$, $n \ge 2$, we subtract from x_n its orthogonal projection
    onto the $(n-1)$-dimensional subspace $Y_{n-1} := \text{span}(\{e_1, \dots, e_{n-1}\})$:

    $$y_n := x_n - \sum_{k=1}^{n-1} (x_n, e_k) e_k \in Y_{n-1}^{\perp}, \ e_n := \frac{y_n}{\|y_n\|}, \ n \in I, n \ge 2$$

    (see Section 6.6).

2. In the (real or complex) space $l_2^{(n)}$ ($n \in \mathbb{N}$), the *orthogonal projection* of an arbitrary
    n-tuple $x := (x_1, \dots, x_n)$ on the k-dimensional subspace Y_k spanned by the first k
    vectors of the *standard unit basis* $\{e_1, \dots, e_n\}$ $(1 \le k \le n)$ is

    $$y = \sum_{i=1}^{k} (x, e_i) e_i = \sum_{i=1}^{k} x_i e_i = (x_1, \dots, x_k, 0, \dots, 0).$$

    In particular, for $k = n$, $Y_n = l_2^{(n)}$ and

    $$y = \sum_{i=1}^{n} (x, e_i) e_i = \sum_{i=1}^{n} x_i e_i = (x_1, \dots, x_n) = x.$$

3. In the (real or complex) space l_2, for any $n \in \mathbb{N}$, the *orthogonal projection* of an
    arbitrary sequence $x := (x_k)_{k \in \mathbb{N}}$ on the n-dimensional subspace Y_n spanned by
    the first n vectors of the standard Schauder basis $\{e_i\}_{i \in \mathbb{N}}$ $(n \in \mathbb{N})$ is

    $$y = \sum_{i=1}^{n} (x, e_i) e_i = \sum_{i=1}^{n} x_i e_i = (x_1, \dots, x_n, 0, \dots).$$

6.7.2 Arbitrary Orthonormal Set

Let us now proceed to the case of an arbitrary orthonormal set.

Theorem 6.11 (Arbitrary Orthonormal Set). *Let* $S := \{e_i\}_{i \in I}$ *be an orthonormal set in a Hilbert space* $(X, (\cdot, \cdot), \|\cdot\|)$. *Then, for each* $x \in X$, *the set of all indices* $i \in I$, *for which the generalized Fourier coefficients* (x, e_i), $i \in I$, *of x relative to S do not vanish, i. e.,*

$$N(x) := \{i \in I \mid (x, e_i) \neq 0\},$$

is countable, and hence, the summation

$$\sum_{i \in I} (x, e_i) e_i$$

is either a finite sum or an infinite series, called the generalized Fourier series of x relative to S.

   If $\{i(n)\}_{n=1}^{N}$ ($N \in \mathbb{N}$ *or* $N := \infty$) *is an arbitrary countable arrangement of the set* $N(x)$, *then*

$$\sum_{k=1}^{N} |(x, e_{i(k)})|^2 \leq \|x\|^2 \quad (\text{Bessel's Inequality}),$$

and

$$y := \sum_{k=1}^{N} (x, e_{i(k)}) e_{i(k)}$$

is the orthogonal projection of x onto the closed subspace $Y := \overline{\text{span}(S)}$ *with*

$$x - y \in Y^{\perp} \quad \text{and} \quad \|y\|^2 = \sum_{k=1}^{N} |(x, e_{i(k)})|^2.$$

The point y is the nearest to x in Y with

$$\rho^2(x, Y) = \|x - y\|^2 = \|x\|^2 - \|y\|^2 = \|x\|^2 - \sum_{k=1}^{N} |(x, e_{i(k)})|^2.$$

Furthermore,

$$x \in Y \Leftrightarrow x = y \Leftrightarrow \|x - y\| = 0 \Leftrightarrow \sum_{k=1}^{N} |(x, e_{i(k)})|^2 = \|x\|^2 \quad (\text{Parseval's Identity}).$$

Proof. Since, by the *Closure of a Subspace Proposition* (Proposition 3.13, Section 3.5, Problem 6), $Y := \overline{\text{span}(S)}$ is a *closed subspace* in the Hilbert space $(X, (\cdot, \cdot), \|\cdot\|)$, by the *Projection Theorem* (Theorem 6.7),

$$X = Y \oplus Y^{\perp},$$

i. e., each $x \in X$ can be uniquely represented as

$$x = y + z$$

with some $y \in Y$ and $z := x - y \in Y^{\perp}$, where y is the *orthogonal projection* of x onto Y.

For each $x \in X$,

$$N(x) := \{i \in I \mid (x, e_i) \neq 0\} = \bigcup_{n=1}^{\infty} N_n(x),$$

where

$$N_n(x) := \{i \in N(x) \mid |(x, e_i)| > 1/n\}, \ n \in \mathbb{N}.$$

By the *Finite Orthonormal Set Corollary* (Corollary 6.4), for each $n \in \mathbb{N}$, the set $N_n(x)$ is *finite*.

Exercise 6.38. Prove *by contradiction*.

Hence, by the *Properties of Countable Sets* (Proposition 1.3), the set $N(x)$ is *countable*.

Let $\{i(n)\}_{n=1}^{N}$ ($N \in \mathbb{N}$ or $N := \infty$) be an arbitrary countable arrangement of the set $N(x)$.

The case of a *finite* $N(x)$, i. e., $N \in \mathbb{N}$, immediately follows from the *Finite Orthonormal Set Corollary* (Corollary 6.4).

Suppose that the set $N(x)$ is *countably infinite*, i. e., $N := \infty$. Then, by the *Finite Orthonormal Set Corollary* (Corollary 6.4),

$$\forall n \in \mathbb{N} : \ \sum_{k=1}^{n} |(x, e_{i(k)})|^2 \leq \|x\|^2,$$

whence, letting $n \to \infty$, we obtain *Bessel's*[7] *inequality*:

$$\sum_{k=1}^{\infty} |(x, e_{i(k)})|^2 \leq \|x\|^2,$$

which, since, for any $n, p \in \mathbb{N}$, by the *inner product axioms*, implies that

$$\left\| \sum_{k=n+1}^{n+p} (x, e_{i(k)}) e_{i(k)} \right\|^2 = \left(\sum_{k=n+1}^{n+p} (x, e_{i(k)}) e_{i(k)}, \sum_{l=n+1}^{n+p} (x, e_{i(l)}) e_{i(l)} \right)$$

$$= \sum_{k=n+1}^{n+p} \sum_{l=n+1}^{n+p} (x, e_{i(k)}) \overline{(x, e_{i(l)})} (e_{i(k)}, e_{i(l)}) \qquad \text{since } (e_{i(k)}, e_{i(l)}) = \delta_{i(k)i(l)};$$

$$= \sum_{k=n+1}^{n+p} |(x, e_{i(k)})|^2 \to 0, \ n \to \infty.$$

---

7 Friedrich Wilhelm Bessel (1784–1846).

Whence, in view of the *completeness* of $(X, (\cdot, \cdot), \|\cdot\|)$, by *Cauchy's Convergence Test for Series* (Theorem 3.7), we infer that the series

$$\sum_{k=1}^{\infty} (x, e_{i(k)}) e_{i(k)},$$

converges in $(X, (\cdot, \cdot), \|\cdot\|)$.

The fact that

$$y := \sum_{k=1}^{\infty} (x, e_{i(k)}) e_{i(k)},$$

is the orthogonal projection of x onto the subspace $Y := \overline{\text{span}(S)}$, by the *Coincidence of Orthogonal Complements Proposition* (Proposition 6.22, Section 6.12, Problem 10), follows from the fact that

$$\forall n \in \mathbb{N} : x - y \perp e_{i(n)}.$$

Exercise 6.39. Verify.

By *continuity* of norm, the *Generalized Pythagorean Theorem* (Theorem 6.27, Section 6.12, Problem 16), and in view of $\|e_i\| = 1$, $i \in I$,

$$\|y\|^2 = \lim_{n \to \infty} \left\| \sum_{k=1}^{n} (x, e_{i(k)}) e_{i(k)} \right\|^2 = \lim_{n \to \infty} \sum_{k=1}^{n} \|(x, e_{i(k)}) e_{i(k)}\|^2$$

$$= \lim_{n \to \infty} \sum_{k=1}^{n} |(x, e_{i(k)})|^2 = \sum_{k=1}^{\infty} |(x, e_{i(k)})|^2.$$

By the *Projection Theorem* (Theorem 6.7), y is the *nearest point* to x in Y, which, by the *Pythagorean Theorem* (Theorem 6.4), implies that

$$\rho^2(x, Y) = \|x - y\|^2 = \|x\|^2 - \|y\|^2 = \|x\|^2 - \sum_{k=1}^{\infty} |(x, e_{i(k)})|^2,$$

and hence,

$$x \in Y \iff x = y \iff \|x - y\| = 0 \iff \sum_{k=1}^{\infty} |(x, e_{i(k)})|^2 = \|x\|^2,$$

the latter called *Parseval's*[8] *identity*. □

---

[8] Marc-Antoine Parseval (1755–1836).

Remarks 6.17.

– Thus, under the conditions of the prior theorem, for an orthonormal set $S := \{e_i\}_{i \in I}$ in a Hilbert space $(X, (\cdot, \cdot), \|\cdot\|)$ and an arbitrary $x \in X$, without specifying a countable arrangement of the set

$$N(x) := \{i \in I \mid (x, e_i) \neq 0\},$$

we can write

$$\sum_{i \in N(x)} |(x, e_i)|^2 \leq \|x\|^2 \quad (\textit{Bessel's Inequality}),$$

$$y := \sum_{i \in N(x)} (x, e_i)e_i \in Y \text{ with } x - y \in Y^{\perp} \text{ and } \|y\|^2 = \sum_{i \in N(x)} |(x, e_i)|^2,$$

where $Y := \overline{\text{span}(S)}$,

$$\rho^2(x, Y) = \|x - y\|^2 = \|x\|^2 - \|y\|^2 = \|x\|^2 - \sum_{i \in N(x)} |(x, e_i)|^2,$$

and

$$x \in Y \Leftrightarrow x = y \Leftrightarrow \|x - y\| = 0 \Leftrightarrow \sum_{i \in N(x)} |(x, e_i)|^2 = \|x\|^2 \quad (\textit{Parseval's Identity}).$$

– *Parseval's identity*, which can be regarded as a *Pythagorean Theorem*, has the following equivalent inner-product form:

$$\forall x, y \in Y := \overline{\text{span}(S)} : \quad (x, y) = \sum_{i \in N(x) \cap N(y)} (x, e_i)\overline{(y, e_i)}. \tag{6.8}$$

Exercise 6.40. Verify.

– As follows from the proof of the prior theorem, for an orthonormal set $S := \{e_i\}_{i \in I}$ in a Hilbert space $(X, (\cdot, \cdot), \|\cdot\|)$ over $\mathbb{F}$ and an arbitrary numeric I-tuple $(c_i)_{i \in I} \in \mathbb{F}^I$ such that

$$c_i \neq 0 \text{ for } \textit{countably many } i\text{'s} \quad \text{and} \quad \sum_{i \in I} |c_i|^2 < \infty,$$

the series

$$\sum_{i \in I} c_i e_i,$$

converges in $(X, (\cdot, \cdot), \|\cdot\|)$ and

$$\left\| \sum_{i \in I} c_i e_i \right\|^2 = \sum_{i \in I} |c_i|^2.$$

6.7.3 Orthonormal Sequence

As an important particular case, we obtain that of an orthonormal sequence ($I = \mathbb{N}$).

Corollary 6.5 (Orthonormal Sequence). *Let $\{e_n\}_{n\in\mathbb{N}}$ be a countably infinite orthonormal set (an orthonormal sequence) in a Hilbert space $(X, (\cdot, \cdot), \|\cdot\|)$. Then, for each $x \in X$,*

$$\sum_{n=1}^{\infty} |(x, e_n)|^2 \le \|x\|^2 \quad (\textit{Bessel's Inequality}),$$

and the generalized Fourier series $\sum_{n=1}^{\infty}(x, e_n)e_n$ of x relative to $\{e_n\}_{n\in\mathbb{N}}$ converges in $(X, (\cdot, \cdot), \|\cdot\|)$ to the orthogonal projection

$$y := \sum_{n=1}^{\infty}(x, e_n)e_n$$

of x onto the closed subspace $Y := \overline{\mathrm{span}(\{e_n\}_{n\in\mathbb{N}})}$ with

$$x - y \in Y^{\perp} \quad \text{and} \quad \|y\|^2 = \sum_{n=1}^{\infty} |(x, e_n)|^2,$$

which is the nearest point to x in Y with

$$\rho^2(x, Y) = \|x - y\|^2 = \|x\|^2 - \|y\|^2 = \|x\|^2 - \sum_{n=1}^{\infty} |(x, e_n)|^2.$$

Furthermore,

$$x \in Y \Leftrightarrow x = y \Leftrightarrow \|x - y\| = 0 \Leftrightarrow \sum_{n=1}^{\infty} |(x, e_n)|^2 = \|x\|^2 \quad (\textit{Parseval's Identity}).$$

From *Bessel's inequality*, we immediately obtain the following statement:

Corollary 6.6 (Generalized Fourier Coefficients Sequence). *Let $\{e_n\}_{n\in\mathbb{N}}$ be an orthonormal sequence in a Hilbert space $(X, (\cdot, \cdot), \|\cdot\|)$. Then*

$$(x, e_n) \to 0, \; n \to \infty.$$

Exercise 6.41. Prove.

6.8 Orthonormal Bases and Orthogonal Dimension

In addition to the *three* meanings the notion of *basis* has in a Banach space, in a Hilbert space setting (see Sections 3.1.5.2 and 3.2.5), due to the presence of *orthogonality*, it acquires the extra one discussed below.

Definition 6.10 (Orthonormal Basis). An orthonormal set $S := \{e_i\}_{i \in I}$ in a nonzero Hilbert space $(X, (\cdot, \cdot), \| \cdot \|)$ is called an *orthonormal basis* for X if the generalized Fourier series representation

$$x = \sum_{i \in N(x)} (x, e_i) e_i,$$

where

$$N(x) := \{i \in I \mid (x, e_i) \neq 0\},$$

holds for each $x \in X$.

Theorem 6.12 (Orthonormal Basis Characterizations). *An orthonormal set $S := \{e_i\}_{i \in I}$ in a nonzero Hilbert space $(X, (\cdot, \cdot), \| \cdot \|)$ is an orthonormal basis for X iff any of the following equivalent conditions is satisfied.*
1. *S is complete, i. e., $S^\perp = \{0\}$.*
2. *S is fundamental, i. e., $\overline{\operatorname{span}(S)} = X$.*
3. *Parseval's identity*

$$\sum_{i \in N(x)} |(x, e_i)|^2 = \|x\|^2,$$

where

$$N(x) := \{i \in I \mid (x, e_i) \neq 0\},$$

holds for each $x \in X$.

Proof. A rather effortless proof immediately follows from the definition, the *Characterization of Complete Orthonormal Sets* (Proposition 6.10), and the *Arbitrary Orthonormal Set Theorem* (Theorem 6.11).

Exercise 6.42. Fill in the details. □

By the prior statement and the *Existence of a Complete Orthonormal Set Theorem* (Theorem 6.9), we obtain the following analogue of the *Basis Theorem* (Theorem 3.2):

Theorem 6.13 (Orthonormal Basis Theorem). *In a nonzero Hilbert space $(X, (\cdot, \cdot), \| \cdot \|)$, each orthonormal set S can be extended to an orthonormal basis S' of X.*

Remark 6.18. An orthonormal basis in a Hilbert space, although *existent* by the prior theorem, *need not be unique* (see Remarks 6.15) (see Section 6.12, Problem 19).

Examples 6.16. The complete orthonormal sets from Examples 6.12 and the complete orthonormal sequences of *Legendre, Hermite,* and *Laguerre polynomials* from Examples 6.14 are *orthonormal bases* in the corresponding Hilbert spaces.

Furthermore, the following statement is an analogue of the *Dimension Theorem* (Theorem 3.4).

Theorem 6.14 (Dimension Theorem for Hilbert Spaces). *All orthonormal bases of a nonzero Hilbert space have equally many elements.*

Proof. Let $S := \{e_i\}_{i \in I}$ and $S' := \{e'_j\}_{j \in J}$ be two orthonormal bases for a Hilbert space $(X, (\cdot, \cdot), \|\cdot\|)$, with $|I|$ and $|J|$ being their cardinalities, respectively.

If S is *finite*, i. e., $|I| \in \mathbb{N}$, then S is a *Hamel basis* for X, and the *algebraic dimension* of X is $|I|$. Since, by the *Linear Independence of Orthogonal Sets Proposition* (Proposition 6.8), the orthonormal set S' is *linearly independent*, it is also finite and

$$|J| \le |I|.$$

Hence, $|J| \in \mathbb{N}$ and symmetrically, we have

$$|I| \le |J|,$$

which implies that

$$|I| = |J|.$$

Suppose that S is *infinite*, i. e., $|I| \ge \aleph_0$ (see Examples 1.1). This immediately implies that $|J| \ge \aleph_0$.

Exercise 6.43. Explain.

For each $i \in I$, the set

$$N'_i = \{j \in J \mid (e_i, e'_j) \ne 0\}$$

is *nonempty*, since otherwise the orthonormal set S' would be *incomplete*.

Exercise 6.44. Explain.

By the *Arbitrary Orthonormal Set Theorem* (Theorem 6.11), the nonempty set N'_i, $i \in I$, is *countable*.
Furthermore,

$$\forall j \in J \, \exists i \in I : j \in N'_i,$$

since otherwise the orthonormal set S would be *incomplete*.

Exercise 6.45. Explain.

Hence,

$$J := \bigcup_{i \in I} N'_i,$$

which, by the arithmetic of cardinals (see, e. g., [29, 33, 41, 52]), implies that

$$|J| \leq \aleph_0 |I| = |I|.$$

Symmetrically,

$$|I| \leq \aleph_0 |J| = |J|,$$

and hence,

$$|I| = |J|,$$

which completes the proof. $\square$

By the *Dimension Theorem for Hilbert Spaces*, the following notion is well defined:

Definition 6.11 (Orthogonal Dimension of a Hilbert Space). The *orthogonal dimension* of a nonzero Hilbert space $(X, (\cdot, \cdot), \| \cdot \|)$ is the *common cardinality* of all orthonormal bases of X.

The dimension of a zero space is naturally defined to be 0.

Remark 6.19. The symbol $\dim X$ can be used contextually.

The case of a separable Hilbert space deserves special attention.

Theorem 6.15 (Orthogonal Dimension of a Separable Hilbert Space). *A Hilbert space* $(X, (\cdot, \cdot), \| \cdot \|)$ *is separable iff the orthogonal dimension of X does not exceed $\aleph_0$, i. e., every orthonormal basis of X is countable.*

Proof. The case of $X = \{0\}$ being trivial, let us assume that $X \neq \{0\}$.

"*If*" part. Suppose that X has a *countable* orthonormal basis. Then all finite linear combinations of the basis vectors with rational/complex rational coefficients form a *countable dense set* in X, which implies that the space X is *separable*.

Exercise 6.46. Explain.

"*Only if*" part. Suppose that X is *separable*. Then it has a *countably infinite dense subset* $M := \{x_n\}_{n \in \mathbb{N}}$, which is *fundamental* since

$$\overline{\text{span}(M)} \supseteq \overline{M} = X.$$

Let us inductively construct a linearly independent subset M' of M with

$$\text{span}(M') = \text{span}(M).$$

Letting

$$n(1) := \min\{n \in \mathbb{N} \mid x_n \neq 0\},$$

we choose our first nonzero element $x_{n(1)} \in M$.

Letting

$$n(2) := \min\{n \in \mathbb{N} \mid x_{n(1)}, x_n \text{ are } linearly\ independent\},$$

we choose our second element $x_{n(2)} \in M$ linearly independent of $x_{n(1)}$, if any.

Continuing inductively in this manner (using the *Axiom of Choice* (see Appendix A)), we obtain a *countable linearly independent subset* $M' := \{x_{n(i)}\}_{i \in I}$ ($I = \{1, \ldots, N\}$ with some $N \in \mathbb{N}$ or $I = \mathbb{N}$) of M such that

$$\text{span}(M') = \text{span}(M).$$

Exercise 6.47. Explain.

Hence,

$$\overline{\text{span}(M')} = \overline{\text{span}(M)} = X,$$

i. e., the set M' is *fundamental*.

Applying to $M' := \{x_{n(i)}\}_{i \in I}$ the *Gram–Schmidt orthonormalization* (see Section 6.6), we obtain a *countable orthonormal set* $S := \{e_i\}_{i \in I}$, which is *fundamental* as well as M' since

$$\text{span}(S) = \text{span}(M'),$$

and hence, by the *Orthonormal Basis Characterizations* (Theorem 6.12), is an *orthonormal basis* of X. □

Remarks 6.20.

– The *orthogonal dimension* of a separable Hilbert space is *equal* to its *algebraic dimension*, provided the space is finite-dimensional, and is *less* than its *algebraic dimension*, provided the space is infinite-dimensional.

Exercise 6.48. Explain.

– An orthonormal basis $\{e_n\}_{n \in \mathbb{N}}$ of an infinite-dimensional separable Hilbert space is also its *Schauder basis* (see Section 3.2.5), as is, in l_2, the *standard orthonormal basis*

$$\{e_n := (\delta_{nk})_{k \in \mathbb{N}}\}_{n \in \mathbb{N}},$$

where δ_{nk} is the Kronecker delta (see Examples 6.12).

Finally, the following statement is an analogue of the *Isomorphism Theorem* (Theorem 3.5):

Theorem 6.16 (Isomorphism Theorem for Hilbert Spaces). *Two nonzero Hilbert spaces* $(X, (\cdot, \cdot)_X, \| \cdot \|_X)$ *and* $(Y, (\cdot, \cdot)_Y, \| \cdot \|_Y)$ *are isometrically isomorphic iff they have the same orthogonal dimension.*

Proof. "*Only if*" part. Let $T : X \to Y$ be an *isometric isomorphism* between X and Y.

By the *Polarization Identities* (Proposition 6.2), along with inner-product norm, T preserves inner product, i. e.,

$$(x, y)_X = (Tx, Ty)_Y, \quad x, y \in X,$$

and hence, T preserves *orthogonality*, i. e.,

$$x \perp y \text{ in } X \iff Tx \perp Ty \text{ in } Y.$$

Exercise 6.49. Explain.

Therefore, a set S is an *orthonormal basis* in X iff $T(S)$ is an *orthonormal basis* in Y, which, since T is a *bijection*, implies that X and Y have the same orthogonal dimension.

"*If*" part. Suppose that X and Y have the same orthogonal dimension. Choosing orthonormal bases $S := \{e_i\}_{i \in I}$ for X and $S' := \{e'_i\}_{i \in I}$ for Y sharing the *indexing set* I, we can establish an *isometric isomorphism* T between X and Y by matching the vectors with the *identical Fourier series representations* relative to the bases S and S', respectively, as follows:

$$X \ni x = \sum_{i \in I}(x, e_i)e_i \mapsto Tx := \sum_{i \in I}(x, e_i)e'_i \in Y.$$

In particular, $Te_i = e'_i$, $i \in I$.

The mapping $T : X \to Y$ is *well defined* since, for each $x \in X$, by *Parseval's identity* (see the *Orthonormal Sequence Corollary* (Corollary 6.5)),

$$\sum_{i \in I}|(x, e_i)|^2 = \|x\|^2 < \infty,$$

which implies the convergence for the series

$$\sum_{i \in I}(x, e_i)e'_i$$

in $(Y, (\cdot, \cdot)_Y, \| \cdot \|_Y)$ and the fact that

$$\|Tx\|^2 = \sum_{i \in I}|(x, e_i)|^2 = \|x\|^2$$

(see Remarks 6.17).

Thus, $T : X \to Y$ is *norm preserving*, and hence, isometric.

The mapping $T : X \to Y$ is, obviously, *linear* and also *onto* (i. e., *surjective*). Since, for each

$$y = \sum_{i \in I}(y, e'_i)e'_i \in Y,$$

we can choose

$$x := \sum_{i \in I} (y, e_i')e_i \in X$$

so that $y = Tx$.

Thus, $T : X \to Y$ is an *isometric isomorphism* between X and Y. $\quad\square$

Remark 6.21. Therefore, two Hilbert spaces differ from each other only in their *orthogonal dimension*.

From the *Orthogonal Dimension of a Separable Hilbert Space Theorem* (Theorem 6.15) and the *Isomorphism Theorem for Hilbert Spaces* (Theorem 6.16), we obtain the following direct corollary:

Corollary 6.7 (Isomorphism of Separable Hilbert Spaces). *A separable Hilbert space* $(X, (\cdot, \cdot), \| \cdot \|)$ *over* $\mathbb{F}$ *is isometrically isomorphic to either* $l_2^{(n)}(\mathbb{F})$ *with some* $n \in \mathbb{Z}_+$ *or to* $l_2(\mathbb{F})$.

Remarks 6.22.
– More generally, a Hilbert space $(X, (\cdot, \cdot), \| \cdot \|)$ over $\mathbb{F}$ with an orthonormal basis $S := \{e_i\}_{i \in I}$ is isometrically isomorphic to the Hilbert space defined as follows:

$$l_2(I, \mathbb{F}) := \left\{ x := (x_i)_{i \in I} \in \mathbb{F}^I \,\middle|\, x_i \neq 0 \text{ for } countably\ many\ i\text{'s and } \sum_{i \in I} |x_i|^2 < \infty \right\}$$

with the inner product

$$(x, y) = \sum_{i \in I} x_i \overline{y_i}.$$

Exercise 6.50. Describe an *orthonormal basis* in $l_2(I, \mathbb{F})$.

– For a set I of an arbitrary cardinality $|I|$, there is a Hilbert space $l_2(I, \mathbb{F})$ of orthogonal dimension $|I|$.

6.9 Adjoint Operators

The notion of *adjoint operator* is rooted in the *self-duality* of Hilbert spaces, as described in the *Riesz Representation Theorem* (Theorem 6.8).

6.9.1 Definition, Linearity, Examples

Definition 6.12 (Adjoint Operator). Let $(A, D(A))$ be a densely defined (i. e., $\overline{D(A)} = X$) linear operator in a Hilbert space $(X, (\cdot, \cdot), \| \cdot \|)$. The operator *adjoint to A* is a mapping in X defined as follows:

$$D(A^*) := \{y \in X \mid \exists! z \in X \,\forall x \in D(A) : (Ax, y) = (x, z)\} \ni y \mapsto A^*y := z.$$

Remarks 6.23.

– The *denseness* of the domain $D(A)$ of the operator A $(\overline{D(A)} = X)$ fundamentally underlies the fact that the *adjoint operator* $(A^*, D(A^*))$ is *well defined*. Indeed, $D(A^*) \neq \emptyset$ since $0 \in D(A^*)$ with $A^*0 = 0$.

Exercise 6.51. Verify.

Furthermore, if, for some $y \in X$,

$$\exists z_1, z_2 \in X \; \forall x \in D(A) : \; (x, z_1) = (Ax, y) = (x, z_2),$$

then

$$(x, z_1 - z_2) = 0, \; x \in D(A),$$

and hence, by the *Characterization of Denseness of Subspace* (Proposition 6.7),

$$z_1 - z_2 \in D(A)^{\perp} = \{0\},$$

i. e., $z_1 = z_2$. This implies that, for any $y \in D(A^*)$, the vector $z \in X$ such that

$$\forall x \in D(A) : \; (Ax, y) = (x, z)$$

is *unique*, and hence, the value $A^*y := z$ is *uniquely defined*.
More precisely, the denseness of the domain $D(A)$ of the operator A is *equivalent* to the fact that the adjoint to A operator A^* can be well defined.
Indeed, if $\overline{D(A)} \neq X$, then, by the *Characterization of Denseness of Subspace* (Proposition 6.7),

$$D(A)^{\perp} \neq \{0\},$$

which implies that there exists a $z \in D(A)^{\perp} \setminus \{0\}$, and hence,

$$\forall x \in D(A) : \; (x, 0) = (Ax, 0) = (x, z),$$

i. e., the value for the adjoint operator at 0 is not uniquely defined: it can be equal to 0 or to $z \neq 0$.

– By the *Riesz Representation Theorem* (Theorem 6.8), we also infer that, for a densely defined linear operator in a Hilbert space $(X, (\cdot, \cdot), \| \cdot \|)$,

$$y \in D(A^*) \iff f(x) := (Ax, y), \; x \in D(A), \text{ is a } \textit{bounded linear functional} \text{ on } D(A).$$

Exercise 6.52. Explain.

The following statement answers the natural question whether the adjoint operator is *linear*.

Proposition 6.11 (Linearity of Adjoint Operator). *Let $(A, D(A))$ be a densely defined linear operator in a Hilbert space $(X, (\cdot, \cdot), \| \cdot \|)$ over $\mathbb{F}$. Then the adjoint $(A^*, D(A^*))$ is a linear operator in X.*

Proof. For arbitrary $y_1, y_2 \in D(A^*)$ and any $\lambda_1, \lambda_2 \in \mathbb{F}$, since

$$\forall x \in D(A) : (Ax, \lambda_1 y_1 + \lambda_2 y_2) = \bar{\lambda}_1 (Ax, y_1) + \bar{\lambda}_2 (Ax, y_2)$$
$$= \bar{\lambda}_1 (x, A^* y_1) + \bar{\lambda}_2 (x, A^* y_2)$$
$$= (x, \lambda_1 A^* y_1 + \lambda_2 A^* y_2),$$

we infer that

$$\lambda_1 y_1 + \lambda_2 y_2 \in D(A^*), \quad \text{and} \quad A^* (\lambda_1 y_1 + \lambda_2 y_2) = \lambda_1 A^* y_1 + \lambda_2 A^* y_2,$$

which shows that $D(A^*)$ is a subspace of X, and $(A^*, D(A^*))$ is a linear operator in $(X, (\cdot, \cdot), \| \cdot \|)$. □

Examples 6.17.
1. For a Hilbert space $(X, (\cdot, \cdot), \| \cdot \|)$ over $\mathbb{F}$, the adjoint to the bounded linear operator A of multiplication by an arbitrary number $\lambda \in \mathbb{F}$ (see Example 4.2) is the operator A^* of multiplication by the conjugate number $\bar{\lambda}$ and $D(A^*) = D(A) = X$, the conjugation being superfluous when the space is real.
   In particular, $0^* = 0$ ($\lambda = 0$), and $I^* = I$ ($\lambda = 1$).
2. For the Hilbert space $l_2^{(n)}$ ($n \in \mathbb{N}$), the adjoint to the bounded linear operator A of multiplication by an $n \times n$ matrix $[a_{ij}]$ with entries from $\mathbb{F}$ is the operator A^* of multiplication by its *conjugate transpose* $[\overline{a_{ij}}]^T$ and $D(A^*) = D(A) = X$, the conjugation being superfluous when the space is real.
3. For the Hilbert space l_2,
   (a) the adjoint to the closed densely defined linear operator of multiplication by a numeric sequence $(a_n)_{n \in \mathbb{N}}$,

$$x := (x_n)_{n \in \mathbb{N}} \mapsto Ax := (a_n x_n)_{n \in \mathbb{N}}$$

   with the maximal domain

$$D(A) := \{(x_n)_{n \in \mathbb{N}} \in l_2 \mid (a_n x_n)_{n \in \mathbb{N}} \in l_2\}$$

   (see Examples 4.7), is the operator of multiplication by the conjugate sequence $(\overline{a_n})_{n \in \mathbb{N}}$,

$$y := (y_n)_{n \in \mathbb{N}} \mapsto A^* y = (\overline{a_n} y_n)_{n \in \mathbb{N}}$$

   with

$$D(A^*) = \{(y_n)_{n \in \mathbb{N}} \in l_2 \mid (\overline{a_n} y_n)_{n \in \mathbb{N}} \in l_2\} = D(A);$$

(b) the adjoint to the bounded linear *right shift operator*

$$l_2 \ni x := (x_1, x_2, \ldots) \mapsto Ax := (0, x_1, x_2, \ldots) \in l_2$$

is the bounded linear *left shift operator*

$$l_2 \ni y := (y_1, y_2, \ldots) \mapsto By := (y_2, y_3, y_4, \ldots) \in l_2,$$

i. e., $A^* = B$.

Also, $B^* = A$.

4. For the Hilbert space $L_2(a, b)$ $(-\infty \le a < b \le \infty)$ over $\mathbb{F}$, the adjoint to the closed densely defined linear operator A of multiplication by a *continuous function* $m :$ $(a, b) \to \mathbb{F}$ with the maximal domain

$$D(A) = \left\{ x \in L_2(a, b) \;\middle|\; \int_a^b |m(t)x(t)|^2 \, dt < \infty \right\}$$

is the operator A^* of multiplication by the conjugate function $\overline{m(\cdot)}$ with

$$D(A^*) = \left\{ y \in L_2(a, b) \;\middle|\; \int_a^b |\overline{m(t)}y(t)|^2 \, dt < \infty \right\} = D(A).$$

Exercise 6.53. Verify 1–3.

6.9.2 Existence, Graph, and Closedness

The following two linear operators, henceforth intuitively referred to as *reflections*, allow us to implement a convenient approach based on operators' graphs, which turns out to be very useful for proving several subsequent statements:

Proposition 6.12 (Reflections). *Let* $(X, (\cdot, \cdot), \| \cdot \|)$ *be a Hilbert space. On the product space* $(X \times X, (\cdot, \cdot)_{X \times X}, \| \cdot \|_{X \times X})$, *which is also a Hilbert space (see Section 6.12, Problem 5), the following mappings, subsequently referred to as reflections,*

$$X \times X \ni \langle x_1, x_2 \rangle \mapsto \mathbb{U} \langle x_1, x_2 \rangle := \langle x_2, x_1 \rangle$$

and

$$X \times X \ni \langle x_1, x_2 \rangle \mapsto \mathbb{O} \langle x_1, x_2 \rangle := \langle -x_2, x_1 \rangle$$

are isometric automorphisms of $(X \times X, (\cdot, \cdot)_{X \times X}, \| \cdot \|_{X \times X})$, *and*

$$\mathbb{U}^2 = \mathbb{I}, \quad \mathbb{O}^2 = -\mathbb{I}, \quad \mathbb{OU} = -\mathbb{UO}, \tag{6.9}$$

where $\mathbb{I}$ *is the identity operator on* $X \times X$.

Exercise 6.54. Prove.

Remark 6.24. By the *Graph Characterization of Invertibility* (Proposition 4.12, Section 4.7, Problem 5), the inverse operator A^{-1} exists *iff* the subspace

$$\mathbb{U}G_A = \{\langle Ax, x \rangle \in X \times X \mid x \in D(A)\}$$

of the product space $X \times X$ is the graph of a linear operator in X, in which case

$$G_{A^{-1}} = \mathbb{U}G_A.$$

Theorem 6.17 (Existence and Graph of Adjoint Operator). *Let $(A, D(A))$ be a linear operator in a Hilbert space $(X, (\cdot, \cdot), \| \cdot \|)$. The adjoint to A operator A^* exists iff the closed subspace $(\mathbb{O}G_A)^\perp$ of the product space $(X \times X, (\cdot, \cdot)_{X \times X}, \| \cdot \|_{X \times X})$ is the graph of a linear operator in X, in which case,*

$$G_{A^*} = (\mathbb{O}G_A)^\perp.$$

Proof. By the *Characterization of the Graph of a Linear Operator* (Proposition 4.10, Section 4.7, Problem 2), the *closed subspace* $(\mathbb{O}G_A)^\perp$ of the product space $(X \times X, (\cdot, \cdot)_{X \times X}, \| \cdot \|_{X \times X})$ (see the *Orthogonal Complement Proposition* (Proposition 6.4)) is the graph of a linear operator in X *iff*

$$\langle 0, z \rangle \in (\mathbb{O}G_A)^\perp$$

implies that $z = 0$, which is *equivalent* to the fact that the identity

$$\forall x \in D(A) : \quad (x, z) = (-Ax, 0) + (x, z) = (\langle -Ax, x \rangle, \langle 0, z \rangle)_{X \times X}$$
$$= (\mathbb{O}\langle x, Ax \rangle, \langle 0, z \rangle)_{X \times X} = 0,$$

implies that $z = 0$, i. e., that

$$D(A)^\perp = \{0\},$$

which, by the *Characterization of Denseness of Subspace* (Proposition 6.7), is *equivalent* to the fact that

$$\overline{D(A)} = X,$$

which, in its turn, is *equivalent* to the existence of the adjoint operator A^* (see Remarks 6.23).

Furthermore, for $y, z \in X$,

$$\langle y, z \rangle \in G_{A^*} \quad \Leftrightarrow \quad \forall x \in D(A) : (Ax, y) = (x, z),$$

which is *equivalent* to the identity

$$\forall x \in D(A) : \ (-Ax, y) + (x, z) = 0,$$

which, in its turn, is *equivalent* to the identity

$$\forall x \in D(A) : \ (\langle -Ax, x \rangle, \langle y, z \rangle)_{X \times X} = 0,$$

the latter being true *iff* $\langle y, z \rangle \in (\mathbb{O}G_A)^{\perp}$.

This shows that

$$G_{A^*} = (\mathbb{O}G_A),^{\perp}$$

and completes the proof. $\qquad\qquad\qquad\qquad\qquad\qquad\qquad\qquad\qquad\qquad$ □

Considering that, whenever the adjoint operator A^* exists, i. e., $\overline{D(A)} = X$ (see Remarks 6.23), its graph $G_{A^*} = (\mathbb{O}G_A)^{\perp}$ is a *closed subspace* of the *product space* $(X \times X, (\cdot, \cdot)_{X \times X}, \|\cdot\|_{X \times X})$, we immediately obtain the following

Corollary 6.8 (Closedness of Adjoint Operator). *For any densely defined linear operator $(A, D(A))$ in a Hilbert space $(X, (\cdot, \cdot), \|\cdot\|)$, the adjoint operator $(A^*, D(A^*))$ is closed.*

6.9.3 Second Adjoint

Theorem 6.18 (Second Adjoint Operator). *A densely defined linear operator $(A, D(A))$ in a Hilbert space $(X, (\cdot, \cdot), \|\cdot\|)$ is closed iff the adjoint operator $(A^*, D(A^*))$ is densely defined and the second adjoint $A^{**} := (A^*)^*$ coincides with A, i. e.,*

$$A^{**} = A.$$

Proof. "*Only if*" part. Suppose that a densely defined linear operator $(A, D(A))$ in a Hilbert space $(X, (\cdot, \cdot), \|\cdot\|)$ is *closed*.

By the *closedness* of the operator A, its graph G_A is a *closed subspace* of the *product space* $(X \times X, (\cdot, \cdot)_{X \times X}, \|\cdot\|_{X \times X})$, and hence, so is its image $\mathbb{O}G_A$ under the *reflection* $\mathbb{O}$, which is an *isometric automorphism* of the product space $X \times X$ (see the *Reflections Proposition* (Proposition 6.12)).

By the *Projection Theorem* (Theorem 6.7) and the *Existence and Graph of Adjoint Operator Theorem* (Theorem 6.17),

$$X \times X = \mathbb{O}G_A \oplus (\mathbb{O}G_A)^{\perp} = \mathbb{O}G_A \oplus G_{A^*}.$$

In view of the fact that the *reflection* $\mathbb{O}$, being an *isometric automorphism* of the product space $X \times X$, preserves *orthogonality* in it (see the proof of the *Isomorphism Theorem for Hilbert Spaces* (Theorem 6.16)), we have

$$X \times X = \mathbb{O}(X \times X) = \mathbb{O}^2 G_A \oplus \mathbb{O}G_{A^*}.$$

By (6.9), considering that G_A is a subspace of $X \times X$, we also have

$$\mathbb{O}^2 G_A = -\mathbb{I} G_A = -G_A = G_A.$$

Thus,

$$X \times X = G_A \oplus \mathbb{O} G_{A^*},$$

which, by the *Projection Theorem* (Theorem 6.7), implies that

$$G_A = (\mathbb{O} G_{A^*})^\perp.$$

Whence, by the *Existence and Graph of Adjoint Operator Theorem* (Theorem 6.17) applied to the adjoint operator A^*, we infer that the *second adjoint* $A^{**} := (A^*)^*$ exists, i. e., $\overline{D(A^*)} = X$ (see Remarks 6.23), and

$$G_{A^{**}} = G_A,$$

which implies that $A^{**} = A$.

"*If*" part. This part follows directly from the *Closedness of Adjoint Operator Corollary* (Corollary 6.8). □

6.9.4 Properties

Theorem 6.19 (Properties of Adjoint Operators). *Let $(A, D(A))$ be a densely defined linear operator in a Hilbert space $(X, (\cdot, \cdot), \| \cdot \|)$ over $\mathbb{F}$. Then*
1. *if $A \in L(X)$, $A^* \in L(X)$ and*

$$\|A^*\| = \|A\|,$$

where the notation $\| \cdot \|$ is used to designate the operator norm;
2. *if $A \in K(X)$, $A^* \in K(X)$;*
3. *if $B \in L(X)$,*
   (a) *for any $\lambda, \mu \in \mathbb{F}$, $\lambda \neq 0$, $(\lambda A + \mu B)^* = \bar{\lambda} A^* + \bar{\mu} B^*$,*
   (b) *provided $A \in L(X)$, $(AB)^* = B^* A^*$.*

Proof.
1. If $A \in L(X)$, as is easily seen, $D(A^*) = X$.

    Exercise 6.55. Verify (see Remarks 6.23).

    This, in view of the *closedness* of A^*, by the *Closed Graph Theorem* (Theorem 4.10), implies that $A^* \in L(X)$.

By the *Norm of a Bounded Linear Operator Proposition* (Proposition 6.26, Section 6.12, Problem 20) and the *Second Adjoint Operator Theorem* (Theorem 6.18),

$$\|A^*\| = \sup_{\|x\|=1, \|y\|=1} |(A^*x, y)| = \sup_{\|x\|=1, \|y\|=1} |(x, A^{**}y)| = \sup_{\|x\|=1, \|y\|=1} |(x, Ay)|$$

$$= \sup_{\|x\|=1, \|y\|=1} |(x, Ay)| = \sup_{\|x\|=1, \|y\|=1} |(Ay, x)| = \|A\|.$$

2. Suppose that $A \in K(X)$ and let $(y_n)_{n \in \mathbb{N}}$ be an arbitrary *bounded* sequence in $(X, (\cdot, \cdot), \| \cdot \|)$, i. e.,

$$\exists C > 0 \; \forall n \in \mathbb{N} : \|y_n\| \le C.$$

By the *Riesz Representation Theorem* (Theorem 6.8), for each $n \in \mathbb{N}$,

$$f_n(x) := (x, y_n), \; n \in \mathbb{N}, x \in X,$$

is a *bounded linear functional* on $(X, (\cdot, \cdot), \| \cdot \|)$ (i. e., $f \in X^*$) with $\|f_n\| = \|y_n\|$, where the same notation $\| \cdot \|$ is used to designate the norm in the *dual space* X^*.
Since the operator A is *compact*, the set $T := \overline{AB}(0, 1)$, where $A\overline{B}(0, 1)$ is the image under A of the closed unit ball

$$\overline{B}(0, 1) := \{x \in X \mid \|x\| \le 1\},$$

is *compact* in $(X, (\cdot, \cdot), \| \cdot \|)$.
The set $\{f_n\}_{n \in \mathbb{N}}$ of functions continuous on T is *uniformly bounded* and *equicontinuous* on T (see Section 2.17.2).

Exercise 6.56. Verify.

By the *Arzelà–Ascoli Theorem* (Theorem 2.59), the set $\{f_n\}_{n \in \mathbb{N}}$ is *precompact* in the space $(C(T), \rho_\infty)$, and hence, by the *Sequential Characterization of Precompactness* (Theorem 2.51), the sequence $(f_n)_{n \in \mathbb{N}}$ contains a subsequence $(f_{n(k)})_{k \in \mathbb{N}}$ *uniformly convergent* on T to a function $f \in C(T)$.
Therefore, (see Section 6.12, Problem 2),

$$\|A^*y_{n(i)} - A^*y_{n(j)}\| = \sup_{\|x\| \le 1} |(x, A^*y_{n(i)} - A^*y_{n(j)})|$$

$$= \sup_{\|x\| \le 1} |(x, A^*(y_{n(i)} - y_{n(j)}))| = \sup_{\|x\| \le 1} |(Ax, y_{n(i)} - y_{n(j)})|$$

$$= \sup_{\|x\| \le 1} |(Ax, y_{n(i)}) - (Ax, y_{n(j)})| = \sup_{\|x\| \le 1} |f_{n(i)}(Ax) - f_{n(j)}(Ax)|$$

$$\le \sup_{\|x\| \le 1} |f_{n(i)}(Ax) - f(Ax)| + \sup_{\|x\| \le 1} |f(Ax) - f_{n(j)}(Ax)| \to 0, \; i, j \to \infty,$$

which implies that the subsequence $(Ay_{n(k)})_{k \in \mathbb{N}}$ is *fundamental* in $(X, (\cdot, \cdot), \| \cdot \|)$, and hence, by the *completeness* of the space $(X, (\cdot, \cdot), \| \cdot \|)$, is *convergent*.
By the *Sequential Characterization of Compact Operators* (Proposition 5.2), we conclude that the adjoint operator is *compact*, i. e., $A^* \in K(X)$.

3. Let $B \in L(X)$. Then, by part 1, $B^* \in L(X)$.

    (a) For any $\lambda, \mu \in \mathbb{F}$ with $\lambda \neq 0$, since $D(\lambda A + \mu B) = D(A)$, $\overline{D(\lambda A + \mu B)} = \overline{D(A)} = X$, and hence, $(\lambda A + \mu B)^*$ exists (see Remarks 6.23).

    Let $y \in D(\bar{\lambda} A^* + \bar{\mu} B^*) = D(A^*)$ be arbitrary. Then, for any $x \in D(\lambda A + \mu B) = D(A)$,

    $$((\lambda A + \mu B)x, y) = \lambda(Ax, y) + \mu(Bx, y) = \lambda(x, A^* y) + \mu(x, B^* y)$$
    $$= (x, \bar{\lambda} A^* y + \bar{\mu} B^* y) = (x, (\bar{\lambda} A^* + \bar{\mu} B^*)y),$$

    which implies that

    $$y \in D((\lambda A + \mu B)^*) \quad \text{and} \quad (\lambda A + \mu B)^* y = (\bar{\lambda} A^* + \bar{\mu} B^*)y.$$

    On the other hand, let $y \in D((\lambda A + \mu B)^*)$ be arbitrary, then, for any $x \in D(\lambda A + \mu B) = D(A)$,

    $$((\lambda A + \mu B)x, y) = (x, (\lambda A + \mu B)^* y),$$

    and hence,

    $$\forall x \in D(A) : \lambda(Ax, y) + (x, \bar{\mu} B^* y) = (x, (\lambda A + \mu B)^* y).$$

    Therefore,

    $$\forall x \in D(A) : (Ax, y) = \left(x, \frac{1}{\lambda}[(\lambda A + \mu B)^* y - \bar{\mu} B^* y] \right),$$

    which implies that $y \in D(A^*) = D(\bar{\lambda} A^* + \bar{\mu} B^*)$, and

    $$A^* y = \frac{1}{\lambda}[(\lambda A + \mu B)^* y - \bar{\mu} B^* y].$$

    Whence, we conclude that

    $$y \in D(\bar{\lambda} A^* + \bar{\mu} B^*) \quad \text{and} \quad (\lambda A + \mu B)^* y = (\bar{\lambda} A^* + \bar{\mu} B^*)y,$$

    which completes the proof of part 3 (a). □

Exercise 6.57. Prove part 3 (b).

Remarks 6.25.
–   Provided $A \in L(X)$, part 3(a) also holds when $\lambda = 0$ (see Examples 5.1).
–   As follows from part 3(a), for an arbitrary densely defined linear operator $(A, D(A))$ in a Hilbert space $(X, (\cdot, \cdot), \| \cdot \|)$ over $\mathbb{F}$ and any $\mu \in \mathbb{F}$,

    $$(A + \mu I)^* = A^* + \bar{\mu} I$$

    (see Examples 6.17).
–   As follows from part 1, for a Hilbert space $(X, (\cdot, \cdot), \| \cdot \|)$, the operation

    $$L(X) \ni A \mapsto A^* \in L(X)$$

    is an *isometric involution* on the Banach algebra $L(X)$ (see Remark 4.9).

6.9.5 Inverse of Adjoint, Orthogonal Sum Decompositions

The following two statements are instrumental for establishing connections between the spectrum of a closed densely defined linear operator in a complex Hilbert space and its adjoint as stated in the *Spectrum of Adjoint Operator Theorem* (Theorem 6.22):

Theorem 6.20 (Inverse of Adjoint Operator). *If, for a densely defined linear operator $(A, D(A))$ in a Hilbert space $(X, (\cdot, \cdot), \|\cdot\|)$ with a dense range (i. e., $\overline{R(A)} = X$), there exists an inverse operator $A^{-1} : R(A) \to D(A)$, then there exists an inverse $(A^*)^{-1}$ to the adjoint operator A^* and*

$$\left(A^*\right)^{-1} = \left(A^{-1}\right)^*,$$

i. e., the inverse of the adjoint is the adjoint of the inverse.

Proof. By hypothesis, an inverse operator $A^{-1} : R(A) \to D(A)$ exists and its domain $D(A^{-1}) = R(A)$ is *dense* in $(X, (\cdot, \cdot), \|\cdot\|)$. Hence, there exists the adjoint $(A^{-1})^*$ (see Remarks 6.23).

Now, we are going to make use of the *reflections* $\mathbb{U}$ and $\mathbb{O}$, introduced in the *Reflections Proposition* (Proposition 6.12).

Considering that, by the *Existence and Graph of Adjoint Operator Theorem* (Theorem 6.17), $(\mathbb{O}G_A)^\perp = G_{A^*}$, we have the following:

$$G_{(A^{-1})^*} = (\mathbb{O}G_{A^{-1}})^\perp \qquad \text{since } G_{A^{-1}} = \mathbb{U}G_A \text{ (see Remark 6.24);}$$
$$= (\mathbb{O}\mathbb{U}G_A)^\perp \qquad\qquad \text{since } \mathbb{O}\mathbb{U} = -\mathbb{U}\mathbb{O} \text{ (see (6.9));}$$
$$= (-\mathbb{U}\mathbb{O}G_A)^\perp$$

$\qquad$ since $\mathbb{U}\mathbb{O}G_A$ is a *subspace* in the product space $(X \times X, (\cdot, \cdot)_{X \times X}, \|\cdot\|_{X \times X})$;

$$= (\mathbb{U}\mathbb{O}G_A)^\perp$$

$\qquad\qquad$ since $\mathbb{U}$ is an *isometric automorphism* of $X \times X$ (see Proposition 6.12);

$$= \mathbb{U}(\mathbb{O}G_A)^\perp = \mathbb{U}G_{A^*},$$

which implies that $(A^*)^{-1}$ exists, and

$$G_{(A^*)^{-1}} = \mathbb{U}G_{A^*} = G_{(A^{-1})^*}$$

(see Remark 6.24), and hence,

$$\left(A^*\right)^{-1} = \left(A^{-1}\right)^*. \qquad\qquad \square$$

Theorem 6.21 (Orthogonal Sum Decompositions). *For a densely defined closed linear operator $(A, D(A))$ in a Hilbert space $(X, (\cdot, \cdot), \|\cdot\|)$, the orthogonal sum decompositions*

$$X = \ker A^* \oplus \overline{R(A)} \quad \text{and} \quad X = \ker A \oplus \overline{R(A^*)}$$

hold.

Exercise 6.58. Prove.

Hint. To prove the first decomposition, show that $R(A)^\perp = \ker A^*$ and apply the *Projection Theorem* (Theorem 6.7). To prove the second decomposition, use the first decomposition for A^* in place of A and apply the *Second Adjoint Operator Theorem* (Theorem 6.18).

6.9.6 Spectrum of Adjoint Operator

Now, we are ready to prove the following important statement:

Theorem 6.22 (Spectrum of Adjoint Operator). *For a densely defined closed linear operator $(A, D(A))$ in a complex Hilbert space $(X, (\cdot, \cdot), \|\cdot\|)$,*

$$\overline{\sigma(A)} := \{\overline{\lambda} \mid \lambda \in \sigma(A)\} = \sigma(A^*)$$

with

$$\overline{\sigma_r(A)} \subseteq \sigma_p(A^*), \ \overline{\sigma_p(A)} \subseteq \sigma_p(A^*) \cup \sigma_r(A^*), \ and \ \overline{\sigma_c(A)} = \sigma_c(A^*)$$

(the bar is understood as conjugation in the complex plane).

Proof. Let $\lambda \in \rho(A)$ be arbitrary. Then

$$\exists (A - \lambda I)^{-1} \in L(X),$$

and hence, by the *Inverse of Adjoint Theorem* (Theorem 6.20) and the *Properties of Adjoint Operators* (Theorem 6.19),

$$\exists (A^* - \overline{\lambda} I)^{-1} = ((A - \lambda I)^*)^{-1} = ((A - \lambda I)^{-1})^* \in L(X),$$

which implies that $\overline{\lambda} \in \rho(A^*)$.

Therefore, we have the inclusion

$$\overline{\rho(A)} \subseteq \rho(A^*). \tag{6.10}$$

Since, by the *Second Adjoint Operator Theorem* (Theorem 6.18), $A^{**} := (A^*)^* = A$, we also have the inclusion

$$\overline{\rho(A^*)} \subseteq \rho(A),$$

or equivalently,

$$\rho(A^*) \subseteq \overline{\rho(A)}. \tag{6.11}$$

Inclusions (6.10) and (6.11) jointly imply that

$$\rho(A^*) = \overline{\rho(A)},$$

which is equivalent to

$$\sigma(A^*) = \rho(A^*)^c = \overline{\rho(A)}^c = \overline{\sigma(A)}.$$

For each $\lambda \in \sigma_r(A)$, $\overline{R(A - \lambda I)} \neq X$ (see Section 5.2.2), which, by the *Orthogonal Sum Decompositions Theorem* (Theorem 6.21), the *Properties of Adjoint Operators* (Theorem 6.19), and the *Characterization of Denseness of Subspace* (Proposition 6.7), implies that

$$\ker(A^* - \overline{\lambda} I) = \ker((A - \lambda I)^*) = \overline{R(A - \lambda I)}^\perp \neq \{0\},$$

i. e., $\overline{\lambda} \in \sigma_p(A^*)$, and hence, we have the inclusion

$$\overline{\sigma_r(A)} \subseteq \sigma_p(A^*). \tag{6.12}$$

For each $\lambda \in \sigma_p(A)$, $\ker(A - \lambda I) \neq \{0\}$, which, by the *Orthogonal Sum Decompositions Theorem* (Theorem 6.21) and the *Properties of Adjoint Operators* (Theorem 6.19), implies that

$$\overline{R(A^* - \overline{\lambda} I)} = \overline{R((A - \lambda I)^*)} = \ker(A - \lambda I)^\perp \neq X,$$

i. e., $\overline{\lambda} \in \sigma_p(A^*) \cup \sigma_r(A^*)$, and hence, we have the inclusion

$$\overline{\sigma_p(A)} \subseteq \sigma_p(A^*) \cup \sigma_r(A^*). \tag{6.13}$$

By the *Second Adjoint Operator Theorem* (Theorem 6.18), inclusions (6.12) and (6.13) imply that

$$\overline{\sigma_c(A)} = \sigma_c(A^*).$$

Exercise 6.59. Explain. □

Examples 6.18.
1. For a complex Hilbert space $(X, (\cdot, \cdot), \| \cdot \|)$, the adjoint to the bounded linear operator A of multiplication by an arbitrary number $\lambda \in \mathbb{C}$ is the operator A^* of multiplication by the conjugate number $\overline{\lambda}$ with $D(A^*) = D(A) = X$ (see Examples 6.17) and, consistently with the prior theorem and Examples 5.2,

$$\sigma(A^*) = \sigma_p(A^*) = \{\overline{\lambda}\} = \overline{\sigma_p(A)} = \overline{\sigma(A)}$$

(the bar is understood as conjugation in the complex plane).

2. For the complex Hilbert space $l_2^{(n)}$ ($n \in \mathbb{N}$), the conjugate to the bounded linear operator A of multiplication by an $n \times n$ matrix $[a_{ij}]$ with complex entries is the operator A^* of multiplication by its conjugate transpose $[\overline{a_{ij}}]^T$ with $D(A^*) = D(A) = X$ (see Examples 6.17) and, consistently with the prior theorem and Examples 5.2,

$$\sigma(A^*) = \sigma_p(A^*) = \overline{\sigma_p(A)} = \overline{\sigma(A)}$$

(the bar is understood as conjugation in the complex plane).

3. For the complex Hilbert space l_2,

   (a) the adjoint to the closed densely defined linear operator of multiplication by a numeric sequence $(a_n)_{n \in \mathbb{N}}$

   $$x := (x_n)_{n \in \mathbb{N}} \mapsto Ax := (a_n x_n)_{n \in \mathbb{N}}$$

   with the maximal domain

   $$D(A) := \{(x_n)_{n \in \mathbb{N}} \in l_2 \mid (a_n x_n)_{n \in \mathbb{N}} \in l_2\}$$

   is the operator of multiplication by the conjugate sequence $(\overline{a_n})_{n \in \mathbb{N}}$

   $$y := (y_n)_{n \in \mathbb{N}} \mapsto A^* y = (\overline{a_n} y_n)_{n \in \mathbb{N}}$$

   with

   $$D(A^*) = \{(y_n)_{n \in \mathbb{N}} \in l_2 \mid (\overline{a_n} y_n)_{n \in \mathbb{N}} \in l_2\} = D(A)$$

   (see Examples 6.17) and, consistently with the prior theorem and Examples 5.2,

   $$\sigma_r(A^*) = \sigma_r(A) = \emptyset, \ \sigma_p(A^*) = \overline{\{a_n\}_{n \in \mathbb{N}}} = \overline{\sigma_p(A)}, \text{ and } \sigma_c(A^*) = \overline{\sigma_c(A)}$$

   (the bar is understood as conjugation in the complex plane);

   (b) the adjoint to the bounded linear *right shift operator*

   $$l_2 \ni x := (x_1, x_2, \dots) \mapsto Ax := (0, x_1, x_2, \dots) \in l_2$$

   is the bounded linear *left shift operator*

   $$l_2 \ni y := (y_1, y_2, \dots) \mapsto By := (y_2, y_3, y_4, \dots) \in l_2$$

   (see Examples 6.17), and, applying the prior theorem, one can show that

   $$\sigma(A) = \{\lambda \in \mathbb{C} \mid |\lambda| \le 1\}$$

   with

   $$\sigma_r(A) = \{\lambda \in \mathbb{C} \mid |\lambda| < 1\} \quad \text{and} \quad \sigma_c(A) = \{\lambda \in \mathbb{C} \mid |\lambda| = 1\}$$

and, for the bounded linear *left shift operator*

$$l_2 \ni x = (x_1, x_2, \dots) \mapsto Bx := (x_2, x_3, x_4, \dots) \in l_2,$$
$$\sigma(B) = \{\lambda \in \mathbb{C} \mid |\lambda| \le 1\}$$

with

$$\sigma_p(B) = \{\lambda \in \mathbb{C} \mid |\lambda| < 1\} \quad \text{and} \quad \sigma_c(B) = \{\lambda \in \mathbb{C} \mid |\lambda| = 1\}.$$

Remark 6.26. More generally, this description applies to the spectra of the *right* and *left shift operators* considered on the complex Banach spaces l_p ($1 \le p < \infty$) or $(c_0, \|\cdot\|_\infty)$. [16] (cf. Section 5.8, Problem 9).

4. For the complex Hilbert space $L_2(a, b)$ ($-\infty \le a < b \le \infty$), the adjoint to the closed densely defined linear operator A of multiplication by a *continuous function* $m : (a, b) \to \mathbb{C}$ with the maximal domain

$$D(A) := \left\{ x \in L_2(a, b) \mid \int_a^b |m(t)x(t)|^2 \, dt < \infty \right\}$$

is the operator A^* of multiplication by the conjugate function $\overline{m(\cdot)}$ with

$$D(A^*) = \left\{ y \in L_2(a, b) \mid \int_a^b |\overline{m(t)}y(t)|^2 \, dt < \infty \right\} = D(A)$$

(see Examples 6.17) and, consistently with the prior theorem,

$$\sigma(A) = \{m(t) \mid t \in (a, b)\}, \ \sigma(A^*) = \{\overline{m(t)} \mid t \in (a, b)\} = \overline{\sigma(A)}$$

(the bar is understood as conjugation in the complex plane).

Exercise 6.60. Verify 3(b).

6.10 Symmetry and Self-Adjointness

Symmetric and *self-adjoint operators* are of special interest for various applications and play an important role, in particular, in quantum mechanics (see, e. g., [66]).

6.10.1 Definitions, Examples, Properties

Definition 6.13 (Symmetric and Self-Adjoint Operators). A densely defined linear operator $(A, D(A))$ in a Hilbert space $(X, (\cdot, \cdot), \|\cdot\|)$ is called *symmetric* if

$$\forall x, y \in D(A) : \ (Ax, y) = (x, Ay),$$

i. e., if A is the restriction to $D(A)$ of its adjoint A^*:

$$\forall x \in D(A) \subseteq D(A^*) : \ Ax = A^*x.$$

If furthermore, $A = A^*$, i. e., $D(A^*) = D(A)$ and

$$\forall x, y \in D(A) : \ (Ax, y) = (x, Ay),$$

the symmetric operator $(A, D(A))$ is called *self-adjoint*.

Remarks 6.27.

- As follows from the definition and the *Closedness of Adjoint Operator Corollary* (Corollary 6.8), every self-adjoint operator is necessarily *closed*.
- As the following examples demonstrate, a symmetric operator need not be self-adjoint. To prove that a symmetric operator $(A, D(A))$ is self-adjoint, one needs to show that

$$D(A^*) \subseteq D(A).$$

  Symmetric operators that are not self-adjoint may be extendable to self-adjoint (see, e. g., [2]).
- A symmetric operator A with $D(A) = X$ is necessarily *self-adjoint* and *bounded* (see the *Hellinger–Toeplitz Theorem*[9] (Theorem 6.28, Section 6.12, Problem 22)).

Examples 6.19.

1. Let $(X, (\cdot, \cdot), \| \cdot \|)$ be a Hilbert space over $\mathbb{F}$.
    (a) the bounded linear operator A of multiplication by an arbitrary number $\lambda \in \mathbb{F}$
        is self-adjoint iff $\bar{\lambda} = \lambda$, i. e., $\lambda \in \mathbb{R}$.
        In particular, the *zero operator* 0 ($\lambda = 0$) and the *identity operator* I ($\lambda = 1$) are *self-adjoint* (see Examples 6.17).
    (b) Any *orthogonal projection operator* P is a bounded self-adjoint operator (see the *Self-Adjoint Characterization of Orthogonal Projections Proposition* (Proposition 6.28, Section 6.12, Problem 26)).
2. On the Hilbert space $l_2^{(n)}$ ($n \in \mathbb{N}$), the bounded linear operator A of multiplication by an $n \times n$ matrix $[a_{ij}]$ with entries from $\mathbb{F}$ is self-adjoint *iff*

$$[\overline{a_{ij}}]^T = [a_{ij}],$$

    i. e., the matrix $[a_{ij}]$ is *symmetric* (or *Hermitian*):

$$a_{ij} = \overline{a_{ji}}, \ i, j = 1, \dots, n,$$

    (see Examples 6.17).

---

9 Otto Toeplitz (1881–1940), Ernst David Hellinger (1883–1950).

3. In the Hilbert space l_2,
   (a) The closed densely defined linear operator of multiplication by a numeric sequence $(a_n)_{n \in \mathbb{N}}$,

   $$(x_n)_{n \in \mathbb{N}} \mapsto Ax := (a_n x_n)_{n \in \mathbb{N}}$$

   with the maximal domain

   $$D(A) := \{(x_n)_{n \in \mathbb{N}} \in l_2 \mid (a_n x_n)_{n \in \mathbb{N}} \in l_2\}$$

   is self-adjoint *iff*

   $$(\overline{a_n})_{n \in \mathbb{N}} = (a_n)_{n \in \mathbb{N}},$$

   i. e., the multiplier sequence $(a_n)_{n \in \mathbb{N}}$ is *real-termed*:

   $$a_n \in \mathbb{R}, \; n \in \mathbb{R},$$

   (see Examples 6.17);
   (b) The densely defined linear operator of multiplication by a real-termed sequence $(a_n)_{n \in \mathbb{N}}$,

   $$(x_n)_{n \in \mathbb{N}} \mapsto Ax := (a_n x_n)_{n \in \mathbb{N}}$$

   with the domain $D(A) := c_{00}$ is symmetric, but not self-adjoint;
   (c) The bounded linear *right shift operator*

   $$l_2 \ni x = (x_1, x_2, \dots) \mapsto Ax := (0, x_1, x_2, \dots) \in l_2$$

   is *not symmetric* since its adjoint is the *left shift operator*

   $$l_2 \ni y := (y_1, y_2, \dots) \mapsto A^* y := (y_2, y_3, y_4, \dots) \in l_2$$

   (see Examples 6.17).
4. In the Hilbert space $L_2(a, b)$ $(-\infty \le a < b \le \infty)$, the closed densely defined linear operator A of multiplication by a *continuous function* $m : (a, b) \to \mathbb{F}$ with the maximal domain

   $$D(A) := \left\{ x \in L_2(a, b) \; \middle| \; \int_a^b |m(t)x(t)|^2 \, dt < \infty \right\}$$

   is self adjoint *iff*

   $$\overline{m(t)} = m(t), \; t \in (a, b),$$

   i. e., the multiplier function $m(\cdot)$ is real valued:

   $$m(t) \in \mathbb{R}, \; t \in (a, b)$$

   (see Examples 6.17).

Proposition 6.13 (Properties of Self-Adjoint Operators). *Let $(A, D(A))$ be a self-adjoint operator in a Hilbert space $(X, (\cdot, \cdot), \| \cdot \|)$. Then*

1. *the orthogonal decomposition*

$$X = \ker A \oplus \overline{R(A)}$$

holds;

2. *if $B \in L(X)$ and $B = B^*$, for any $\lambda, \mu \in \mathbb{R}$, $\lambda \neq 0$, $(\lambda A + \mu B)^* = \lambda A + \mu B$.*

Exercise 6.61. Prove.

Remark 6.28. In particular, for an arbitrary self-adjoint operator $(A, D(A))$ in a Hilbert space $(X, (\cdot, \cdot), \| \cdot \|)$ and any $\mu \in \mathbb{R}$, the operator $A + \mu I$ is self-adjoint, i. e.,

$$(A + \mu I)^* = A + \mu I.$$

6.10.2 Spectrum and Eigenvectors of a Self-Adjoint Operator

Proposition 6.14 (Points of Regular Type). *If $(A, D(A))$ is a closed symmetric operator in a complex Hilbert space $(X, (\cdot, \cdot), \| \cdot \|)$, for each $\lambda \in \mathbb{C} \setminus \mathbb{R}$,*

1. *the inverse operator $(A - \lambda I)^{-1} : R(A - \lambda I) \to D(A)$ exists and is bounded;*
2. *the range $R(A - \lambda I)$ is a closed subspace of $(X, (\cdot, \cdot), \| \cdot \|)$ (i. e., $R(A - \lambda I) = \overline{R(A - \lambda I)}$).*

Proof.

1. Let $\lambda \in \mathbb{C} \setminus \mathbb{R}$ be arbitrary. Then $\lambda = a + ib$ with some $a, b \in \mathbb{R}$, $b \neq 0$ and, for any $x \in D(A - \lambda I) = D(A)$,

$$\|(A - \lambda I)x\|^2 = ((A - aI)x - ibx, (A - aI)x - ibx)$$
$$= \|(A - aI)x\|^2 + ib((A - aI)x, x) - ib(x, (A - aI)x) + b^2\|x\|^2$$

since the operator $A - aI$ is *self-adjoint* (see Remark 6.28);

$$= \|(A - aI)x\|^2 + ib(x, (A - aI)x) - ib(x, (A - aI)x) + b^2\|x\|^2$$
$$= \|(A - aI)x\|^2 + b^2\|x\|^2 \geq b^2\|x\|^2,$$

which, by the *Existence of Bounded Inverse Proposition* (Proposition 5.6, Section 5.8, Problem 3), proves the fact that the inverse operator $(A - \lambda I)^{-1} : R(A - \lambda I) \to D(A)$ exists, and is bounded.

2. Since the operator $A - \lambda I$ is *closed*, by the *Closedness of Inverse Operator Proposition* (Proposition 4.23, Section 5.8, Problem 20), so is the inverse $(A - \lambda I)^{-1}$. Being both bounded and closed, by the *Characterization of Closedness for Bounded Linear Operators* (Proposition 4.5), the operator $(A - \lambda I)^{-1}$ has closed domain $D((A - \lambda I)^{-1}) = R(A - \lambda I)$, which concludes the proof of part 2 and the entire statement.

□

Theorem 6.23 (Spectrum of a Self-Adjoint Operator). *For a self-adjoint linear operator* $(A, D(A))$ *in a complex Hilbert space* $(X, (\cdot, \cdot), \| \cdot \|)$,

$$\sigma(A) \subseteq \mathbb{R} \quad and \quad \sigma_r(A) = \emptyset.$$

Proof. For an arbitrary $\lambda \in \mathbb{C}$, by the *Properties of Adjoint Operators* (Theorem 6.19), the *Orthogonal Sum Decompositions Theorem* (Theorem 6.21), and in view of $A^* = A$,

$$X = \ker((A - \lambda I)^*) \oplus \overline{R(A - \overline{\lambda}I)} = \ker(A - \overline{\lambda}I) \oplus \overline{R(A - \overline{\lambda}I)}.$$

If $\lambda \in \mathbb{C} \backslash \mathbb{R}$, then also $\overline{\lambda} \in \mathbb{C} \backslash \mathbb{R}$, and hence, by the *Points of Regular Type Proposition* (Proposition 6.14),

$$\ker(A - \overline{\lambda}I) = \{0\}, \quad R(A - \lambda I) = \overline{R(A - \lambda I)} = X,$$

and the inverse operator $(A - \lambda I)^{-1}$ is *bounded*.

Therefore, $(A - \lambda I)^{-1} \in L(X)$, which proves that $\lambda \in \rho(A)$, and hence,

$$\sigma(A) \subseteq \mathbb{R}.$$

For a $\lambda \in \mathbb{R}$, if we assume that $\lambda \in \sigma_r(A)$, then, by the *Spectrum of Adjoint Operator Theorem* (Theorem 6.22)

$$\lambda = \overline{\lambda} \in \sigma_p(A^*) = \sigma_p(A),$$

which, since the sets $\sigma_p(A)$ and $\sigma_r(A)$ are *disjoint*, is a *contradiction*, proving that

$$\sigma_r(A) = \emptyset. \qquad \square$$

Remark 6.29. Since, for an arbitrary self-adjoint operator $(A, D(A))$ in a real Hilbert space, its *complexification* $(\tilde{A}, D(\tilde{A}))$ is a self-adjoint operator in a complex Hilbert space and, for an arbitrary real number $\lambda \in \mathbb{R}$,

$$\lambda \in \rho(\tilde{A}) \iff \exists (A - \lambda I)^{-1} \in L(X).$$

(see Problem 6.12, Problem 23), by the prior theorem, the spectrum, the resolvent set, and the resolvent function of such an operator can be naturally defined as follows:

$$\sigma(A) := \sigma(\tilde{A}), \quad \rho(A) := \mathbb{R} \backslash \sigma(A) = \mathbb{R} \cap \rho(\tilde{A}),$$

and

$$\rho(A) \ni \lambda \mapsto R(\lambda, A) := (A - \lambda I)^{-1} \in L(X).$$

Proposition 6.15 (Eigenvectors of a Self-Adjoint Operator). *The eigenvectors associated with distinct eigenvalues of a self-adjoint operator* $(A, D(A))$ *in a Hilbert space* $(X, (\cdot, \cdot), \| \cdot \|)$ *are orthogonal.*

Exercise 6.62. Prove.

6.10.3 Bounded Self-Adjoint Operators

By *Gelfand's Spectral Radius Theorem* (Theorem 5.8), the spectrum of an arbitrary bounded self-adjoint linear operator on a (real or complex) Hilbert space (for the real case, see Remark 6.29 and Section 6.12, Problem 23) is a *nonempty compact subset* of the real line. Even more can be said according to the following:

Theorem 6.24 (Spectral Bounds). *Let A be a bounded self-adjoint linear operator on a Hilbert space* $(X, (\cdot, \cdot), \| \cdot \|)$ *(i. e., $A \in L(X)$ and $A = A^*$). Then*

$$m := \inf_{\|x\|=1} (Ax, x) \quad and \quad M := \sup_{\|x\|=1} (Ax, x)$$

are well-defined real numbers, and
1. $\sigma(A) \subseteq [m, M]$;
2. $m, M \in \sigma(A)$;
3. $\|A\| = \max\{-m, M\} = \sup_{\|x\|=1} |(Ax, x)|$.

For proof, see [9, Section 6.3] or [2, Volume 1, Section 20].

Remark 6.30. Thus, for any bounded self-adjoint operator A on a (real or complex) Hilbert space $(X, (\cdot, \cdot), \| \cdot \|)$, by the prior theorem and *Gelfand's Spectral Radius Theorem* (Theorem 5.8) (see Remarks 5.12), the *spectral radius* $r(A)$ attains the greatest possible value $\|A\|$.

6.11 Compact Operators

6.11.1 Fredholm Alternative in a Hilbert Space Setting

The concept of *adjoint operator* in a Hilbert space underlies the following version of the *Fredholm Alternative* (Theorem 5.20):

Theorem 6.25 (Fredholm Alternative in a Hilbert Space Setting). *Let A be a compact linear operator on a Hilbert space* $(X, (\cdot, \cdot), \| \cdot \|)$ *(i. e., $A \in K(X)$). Then*
1. $\ker(I - A)$ *is a finite-dimensional subspace;*
2. $R(I - A)$ *is a closed subspace;*
3. $R(I - A) = \ker(I - A^*)^\perp$;
4. $\ker(I - A) = \{0\}$ *iff $R(I - A) = X$;*
5. $\dim \ker(I - A) = \dim R(I - A)^\perp = \operatorname{codim} R(I - A)$.

In particular, either the equation

$$(I - A)x = y \tag{6.14}$$

has a unique solution for each y ∈ X or the equation

$$(I - A)x = 0 \tag{6.15}$$

has a nontrivial solution x ≠ 0, in which case equation (6.14) has solutions iff y ∈ R(I − A) = ker(I − A)$^\perp$, i. e., iff*

$$(y, u) = 0$$

for every solution u ∈ X of the equation

$$(I - A^*)u = 0.$$

Remark 6.31. To be precise, *Fredholm alternative* is the concluding statement concerning Equations (6.14) and (6.15).

Proof.
1. Let us reason *by contradiction*, assuming that the subspace ker(I − A) is *infinite-dimensional*. Then there exists a countably infinite linearly independent subset $\{x_n\}_{n \in \mathbb{N}}$ of ker(I−A), which, in view of the *Gram–Schmidt process* (see Section 6.6), without loss of generality, can be regarded as *orthonormal*.
    Considering that

    $$(I - A)x_n = 0, \ n \in \mathbb{N},$$

    and hence, equivalently,

    $$Ax_n = x_n, \ n \in \mathbb{N},$$

    by the *Pythagorean Theorem* (Theorem 6.4), we infer that

    $$\forall \, m, n \in \mathbb{N}: \ \|Ax_m - Ax_n\| = \|x_m - x_n\| = \sqrt{\|x_m\|^2 + \|x_n\|^2} = \sqrt{2},$$

    which, by the *Sequential Characterization of Compact Operators* (Proposition 5.2), *contradicts* the *compactness* of the operator A.

    Exercise 6.63. Explain.

    The obtained contradiction proves that the subspace ker(I − A) is *finite-dimensional*.
2. To prove that the subspace R(I − A) is *closed*, let us first show that

    $$\exists \, c > 0 \, \forall \, x \in \ker(I - A)^\perp : \ \|(I - A)x\| \geq c\|x\| \tag{6.16}$$

    or, equivalently,

    $$\exists \, c > 0 \, \forall \, x \in \ker(I - A)^\perp : \ \inf_{x \in \ker(I-A)^\perp, \, \|x\|=1} \|(I - A)x\| > 0.$$

    Let us reason *by contradiction*, assuming that (6.16) does not hold, which means that there exists a sequence $(x_n)_{n \in \mathbb{N}}$ in ker(I − A)$^\perp$ with $\|x_n\| = 1$, $n \in \mathbb{N}$, such that

    $$\|(I - A)x_n\| < 1/n. \tag{6.17}$$

Exercise 6.64. Explain.

Since, the sequence $(x_n)_{n\in\mathbb{N}}$ is *bounded* in the Hilbert space $(X, (\cdot, \cdot), \|\cdot\|)$, there exists a subsequence $(x_{n(k)})_{k\in\mathbb{N}}$ *weakly convergent* to a vector x in $(X, (\cdot, \cdot), \|\cdot\|)$ (see Section 6.12, Problem 15 and, e. g., [16, 45]).

Since $A \in K(X)$, by the *Compact Operators and Weakly Convergent Sequences Proposition* (Proposition 6.29, Section 6.12, Problem 29),

$$Ax_{n(k)} \to Ax, \ k \to \infty, \ \text{in } (X, (\cdot, \cdot), \|\cdot\|). \tag{6.18}$$

By (6.17), (6.18), and *subadditivity* of norm,

$$\|x_{n(k)} - Ax\| = \|x_{n(k)} - Ax_{n(k)} + Ax_{n(k)} - Ax\| \le \|x_{n(k)} - Ax_{n(k)}\|$$
$$+ \|Ax_{n(k)} - Ax\| = \|(I - A)x_{n(k)}\| + \|Ax_{n(k)} - Ax\| \to 0, \ k \to \infty,$$

which implies that

$$x_{n(k)} \to Ax, \ k \to \infty, \ \text{in } (X, (\cdot, \cdot), \|\cdot\|).$$

Whence, by the *uniqueness of the weak limit* (see Section 6.12, Problem 15), we conclude that

$$x = Ax, \ \text{i.e, } x \in \ker(I - A),$$

and

$$x_{n(k)} \to x, \ k \to \infty, \ \text{in } (X, (\cdot, \cdot), \|\cdot\|).$$

Since $\ker(I - A)^\perp$ is a *closed subspace* of $(X, (\cdot, \cdot), \|\cdot\|)$ (see the *Orthogonal Complement Proposition* (Proposition 6.4)), by the *Sequential Characterization of Closed Sets* (Theorem 2.19),

$$x = \lim_{k\to\infty} x_{n(k)} \in \ker(I - A)^\perp.$$

Thus,

$$x \in \ker(I - A) \cap \ker(I - A)^\perp,$$

which, by the *Orthogonal Complement Proposition* (Proposition 6.4), implies that $x = 0$, *contradicting* the fact that, by *continuity* of norm (see Remarks 3.18),

$$\|x\| = \lim_{k\to\infty} \|x_{n(k)}\| = 1.$$

The obtained contradiction proves (6.16).

Now, consider an arbitrary sequence $(y_n)_{n\in\mathbb{N}}$ in $R(I - A)$ such that

$$y_n \to y, \ n \to \infty, \ \text{in } (X, (\cdot, \cdot), \|\cdot\|).$$

Then, there exists a sequence $(x_n)_{n \in \mathbb{N}}$ in X such that

$$y_n = (I - A)x_n, \ n \in \mathbb{N}.$$

Since by the *Kernel of a Bounded Linear Operator Proposition* (Proposition 4.2), $\ker(I - A)$ is a *closed subspace* of $(X, (\cdot, \cdot), \| \cdot \|)$, by the *Projection Theorem* (Theorem 6.7),

$$X = \ker(I - A) \oplus \ker(I - A)^{\perp},$$

and hence,

$$x_n = u_n + w_n, \ n \in \mathbb{N},$$

with some $u_n \in \ker(I - A)$ and $w_n \in \ker(I - A)^{\perp}$.
Therefore,

$$\forall n \in \mathbb{N} : \ y_n = (I - A)x_n = (I - A)(u_n + w_n) = (I - A)u_n + (I - A)w_n = (I - A)w_n.$$

Whence, by (6.16), we infer that

$$\forall m, n \in \mathbb{N} : \ \|y_m - y_n\| = \|(I - A)(w_m - w_n)\| \geq c\|w_m - w_n\|.$$

This, since the convergent sequence $(y_n)_{n \in \mathbb{N}}$ is *fundamental* (see the *Properties of Fundamental Sequences* (Theorem 2.22)), implies that the sequence $(w_n)_{n \in \mathbb{N}}$ is *fundamental* as well. By the *completeness* of the space $(X, (\cdot, \cdot), \| \cdot \|)$,

$$\exists x \in X : \ w_n \to x, \ n \to \infty, \ \text{in} \ (X, (\cdot, \cdot), \| \cdot \|),$$

and hence, by the *continuity* of the operator A (see the *Characterizations of Bounded Linear Operators* (Theorem 4.4)),

$$(I - A)x = \lim_{n \to \infty} (I - A)w_n = \lim_{n \to \infty} y_n = y.$$

Whence, we conclude that $y \in R(I-A)$ and, which, by the *Sequential Characterization of Closed Sets* (Theorem 2.19), completes the proof that the subspace $R(I - A)$ is *closed*.

3. By the *Orthogonal Sum Decompositions Theorem* (Theorem 6.21), the *Properties of Adjoint Operators* (Theorem 6.19), and part 2,

$$\ker\left(I - A^*\right)^{\perp} = \ker\left((I - A)^*\right)^{\perp} = \overline{R(I - A)} = R(I - A).$$

4. *"Only if"* part. Let us prove this part by *contradiction*, assuming that

$$\ker(I - A) = \{0\}, \ \text{but} \ R(I - A) \neq X.$$

By the *Kernel Characterization of Invertibility* (Proposition 4.11, Section 4.7, Problem 4), since

$$\ker(I - A) = \{0\},$$

the linear operator $I - A$ is *invertible*, i. e., *one-to-one*. This and the fact that

$$R(I - A) \neq X,$$

imply that, for the subspaces

$$X_n := (I - A)^n X, \ n \in \mathbb{Z}_+,$$

where $(I - A)^0 := I$, all of which, by part 2, are *closed* in $(X, (\cdot, \cdot), \| \cdot \|)$, the *proper inclusions*

$$X = X_0 \supset X_1 \supset X_2 \supset \cdots \tag{6.19}$$

hold.

Exercise 6.65. Explain why the subspaces X_n, $n \in \mathbb{Z}_+$, are *closed* and why *proper inclusions* (6.19) hold.

For each $n \in \mathbb{N}$, choosing a vector

$$x_n \in X_n \cap X_{n+1}^{\perp} \quad \text{with} \quad \|x_n\| = 1, \tag{6.20}$$

we obtain a *bounded* sequence $(x_n)_{n \in \mathbb{N}}$ such that, for all $n = 2, 3, \ldots$ and $m = 1, \ldots, n - 1$,

$$Ax_m - Ax_n = (I - A)x_n - (I - A)x_m + x_m - x_n = x_m + y_{m,n},$$

where, in view of inclusions (6.19),

$$y_{m,n} := (I - A)x_n - (I - A)x_m - x_n \in X_{m+1}.$$

Exercise 6.66. Explain.

Since, considering (6.20), $x_m \in X_{m+1}^{\perp}$, by the *Pythagorean Theorem* (Theorem 6.4), we infer that

$$\|Ax_m - Ax_n\| = \sqrt{\|x_m\|^2 + \|y_{m,n}\|^2} \geq \|x_m\| = 1,$$

which, by the *Sequential Characterization of Compact Operators* (Proposition 5.2), *contradicts* the *compactness* of the operator A, and completes the proof of the "*only if*" part.

Exercise 6.67. Explain.

"If" part. Suppose that

$$R(I - A) = X.$$

Then, by the *Orthogonal Sum Decompositions Theorem* (Theorem 6.21) and the *Properties of Adjoint Operators* (Theorem 6.19),

$$\ker(I - A^*) = \ker((I - A)^*) = R(I - A)^\perp = X^\perp = \{0\}. \tag{6.21}$$

Whence, since, by the *Properties of Adjoint Operators* (Theorem 6.19), the adjoint operator A is *compact*, i. e., $A^* \in K(X)$, by the *"only if"* part, we infer that

$$R(I - A^*) = X.$$

By the *Orthogonal Sum Decompositions Theorem* (Theorem 6.21) and the *Properties of Adjoint Operators* (Theorem 6.19),

$$\ker(I - A) = R((I - A)^*)^\perp = R(I - A^*)^\perp = X^\perp = \{0\}, \tag{6.22}$$

which completes the proof of the *"if"* part.

5. To prove this part, we first show that

$$\dim \ker(I - A) \geq \dim R(I - A)^\perp. \tag{6.23}$$

Let us reason *by contradiction*, assuming that

$$\dim \ker(I - A) < \dim R(I - A)^\perp.$$

Then, as follows from the *Isomorphism Theorem* (Theorem 3.5), there exists an *isomorphic embedding*

$$T : \ker(I - A) \to R(I - A)^\perp$$

of $\ker(I - A)$ in $R(I - A)^\perp$, a linear operator, which is *one-to-one* but *not onto*. The operator T can be extended to a linear operator $\hat{T} : X \to R(I - A)^\perp$ defined on the entire space as follows:

$$\hat{T}x = \hat{T}(u + w) := Tu,$$

where, by the *Projection Theorem* (Theorem 6.7),

$$x = u + w$$

with certain unique $u \in \ker(I - A)$ and $w \in \ker(I - A)^\perp$.

Exercise 6.68. Explain why the operator $\hat{T}$ is *linear*.

In particular,

$$\hat{T}w := T0 = 0, \; w \in \ker(I - A)^{\perp}. \tag{6.24}$$

Clearly,

$$R(\hat{T}) = R(T) \subset R(I - A)^{\perp}. \tag{6.25}$$

Since

$$R(I - A)^{\perp} = \ker((I - A)^*) = \ker(I - A^*)$$

(see (6.21)) and, by the *Properties of Adjoint Operators* (Theorem 6.19), the adjoint operator A is *compact*, i. e., $A^* \in K(X)$, we infer by part 1 and the *proper inclusion*

$$R(\hat{T}) \subset R(I - A)^{\perp} = \ker(I - A^*) \tag{6.26}$$

that the range $R(\hat{T})$ is *finite-dimensional*, i. e., the operator $\hat{T}$ is of *finite rank*, and hence, *compact*, i. e., $\hat{T} \in K(X)$ (see Examples 5.17). Furthermore, by the *Properties of Adjoint Operators* (Theorem 6.19), $A + \hat{T} \in K(X)$.
Also,

$$\ker(I - (A + \hat{T})) = \{0\}. \tag{6.27}$$

Indeed, since, for any

$$x = u + w \in X$$

with certain unique $u \in \ker(I - A)$ and $w \in \ker(I - A)^{\perp}$, considering (6.24) and (6.25),

$$(I - (A + \hat{T}))x = (I - (A + \hat{T}))(u + w) = (I - A)w - \hat{T}u \in R(I - A) \oplus R(T), \tag{6.28}$$

where $(I - A)w \in R(I - A)$, and $\hat{T}u \in R(T) \subset R(I - A)^{\perp}$.

Exercise 6.69. Explain.

Since $(I - A)w \perp \hat{T}u$,

$$(I - (A + \hat{T}))x = (I - A)w - \hat{T}u = 0 \iff (I - A)w = 0 \text{ and } \hat{T}u := Tu = 0.$$

Exercise 6.70. Explain.

The operator $I - A$ is *one-to-one* on $\ker(I - A)^{\perp}$, and the operator T is *one-to-one* on $\ker(I - A)$.

Exercise 6.71. Explain.

Hence, we infer that $u = w = 0$, which implies that $x = u + w = 0$, and completes the proof of (6.27).

By part 4, when applied to the compact operator $A + \hat{T}$, we conclude that (6.27) implies that

$$R(I - (A + \hat{T})) = X. \tag{6.29}$$

On the other hand, since, by *proper inclusion* (6.26), there exists a vector

$$y \in R(I - A)^{\perp} \setminus R(\hat{T}),$$

and hence, in view of (6.28),

$$y \notin R(I - (A + \hat{T})),$$

which *contradicts* (6.29), proving (6.23).

Exercise 6.72. Explain.

Since

$$R(I - A)^{\perp} = \ker(I - A^*) \quad \text{and} \quad R(I - A^*)^{\perp} = \ker(I - A)$$

(see (6.21) and (6.22)), applying (6.23) to the *compact* adjoint operator A^* (see the *Properties of Adjoint Operators* (Theorem 6.19)) in place of A, we arrive at

$$\dim R(I - A)^{\perp} = \dim \ker(I - A^*) \geq \dim R(I - A^*)^{\perp} = \dim \ker(I - A). \tag{6.30}$$

Inequalities (6.23) and (6.30) jointly imply that

$$\dim \ker(I - A) = \dim R(I - A)^{\perp} = \operatorname{codim} R(A).$$

The *Fredholm alternative* concerning Equations (6.14) and (6.15) follows immediately from parts 4 and 3.

Exercise 6.73. Explain. $\qquad\qquad\qquad\qquad\qquad\qquad\qquad\qquad$ ☐

6.11.2 Compact Self-Adjoint Operators, Spectral Theorem

Here, we generalize the *Spectral Theorem for Symmetric Matrices*, which states that with any $n \times n$ ($n \in \mathbb{N}$) symmetric matrix A with real entries associated is an *orthonormal eigenbasis*, i. e., an orthonormal basis consisting of the eigenvectors of A, for $\mathbb{R}^n$ (see, e. g., [34, 49, 54]).

Let us first prove the following fundamental statement:

Proposition 6.16 (Existence of Nonzero Eigenvalues). *For any nonzero compact self-adjoint linear operator A on a Hilbert space $(X, (\cdot, \cdot), \| \cdot \|)$ (i.e., $A \neq 0$, $A \in K(X)$, and $A = A^*$), there exists a nonzero eigenvalue*

$$\lambda = \|A\| \quad or \quad \lambda = -\|A\|,$$

and hence,

$$\sigma_p(A) \setminus \{0\} \neq \emptyset.$$

Proof. By the *Spectral Bounds Theorem* (Theorem 6.24),

$$\sup_{\|x\|=1} |(Ax, x)| = \|A\| \neq 0.$$

Considering the fact that, by the *self-adjointness* of A,

$$\forall x \in X : (Ax, x) \in \mathbb{R}$$

(see the *Quadratic Form of a Symmetric Operator Proposition* (Proposition 6.27, Section 6.12, Problem 21)), there exists a sequence $(x_n)_{n \in \mathbb{N}}$ in X with $\|x_n\| = 1$, $n \in \mathbb{N}$, such that

$$(Ax_n, x_n) \to \lambda, \ n \to \infty, \tag{6.31}$$

where $\lambda = \|A\|$, or $\lambda = -\|A\|$.

Exercise 6.74. Explain.

By the *Sequential Characterization of Compact Operators* (Proposition 5.2), there is a subsequence $(x_{n(k)})_{k \in \mathbb{N}}$ such that

$$Ax_{n(k)} \to x, \ k \to \infty, \ \text{in} \ (X, (\cdot, \cdot), \| \cdot \|) \tag{6.32}$$

for some $x \in X$.

Since, considering the *self-adjointness* of A, and in view of $\lambda \in \mathbb{R}$ and $\|x_{n(k)}\| = 1$, $k \in \mathbb{N}$,

$$\|Ax_{n(k)} - \lambda x_{n(k)}\|^2 = \|Ax_{n(k)}\|^2 - 2\,\mathrm{Re}(Ax_{n(k)}, \lambda x_{n(k)}) + \|\lambda x_{n(k)}\|^2$$
$$= \|Ax_{n(k)}\|^2 - 2\lambda(Ax_{n(k)}, x_{n(k)}) + |\lambda|^2, \ k \in \mathbb{N},$$

by (6.31), (6.32) and the *continuity* of norm, we have

$$\lim_{k \to \infty} \|Ax_{n(k)} - \lambda x_{n(k)}\|^2 = \|x\|^2 - 2\lambda^2 + \lambda^2 = \|x\|^2 - \lambda^2, \tag{6.33}$$

which implies that

$$\|x\| \geq |\lambda|.$$

Since

$$\|Ax_{n(k)}\| \le \|A\| = |\lambda|, \ k \in \mathbb{N},$$

by (6.32) and the *continuity* of norm, we infer that

$$\|x\| = \lim_{k \to \infty} \|Ax_{n(k)}\| \le |\lambda|.$$

Hence,

$$\|x\| = |\lambda| = \|A\| \ne 0$$

and, by (6.33),

$$\lim_{k \to \infty} \|Ax_{n(k)} - \|x\|x_{n(k)}\| = 0,$$

which, in view of (6.32), implies that

$$\lim_{k \to \infty} x_{n(k)} = \frac{x}{\|x\|} \text{ in } (X, (\cdot, \cdot), \|\cdot\|).$$

Exercise 6.75. Explain.

Therefore, in view of (6.32) and by the *continuity* of A (see the *Characterizations of Bounded Linear Operators* (Theorem 4.4)), for the vector $e := \frac{x}{\|x\|}$ with $\|e\| = 1$,

$$Ae = \lambda e,$$

which completes the proof. □

Remark 6.32. As the example of the *integration operator* on the complex Banach space $(C[a, b], \|\cdot\|_\infty)$ $(-\infty < a < b < \infty)$ demonstrates, generally, for a nonzero compact operator, nonzero eigenvalues need not exist (see Examples 5.19, 3).

Theorem 6.26 (Spectral Theorem for Compact Self-Adjoint Operators). *Let A be a compact self-adjoint operator on a Hilbert space $(X, (\cdot, \cdot), \|\cdot\|)$ (i. e., $A \in K(X)$, and $A = A^*$).*

1. *If the space X is finite-dimensional, then the spectrum of A consists of a finite number of real eigenvalues with finite geometric multiplicities.*
2. *If the space X is infinite-dimensional, the spectrum $\sigma(A)$ of A consists of 0 and a countable set of nonzero real eigenvalues, each eigenvalue, if any, being of finite geometric multiplicity.*
   The set $\sigma(A) \setminus \{0\}$ being countably infinite, for its arbitrary countable arrangement $\{\lambda_n\}_{n=1}^\infty$,

$$\lambda_n \to 0, \ n \to \infty.$$

The operator A being of finite rank, its spectrum $\sigma(A)$ consists of 0 and a finite number of nonzero eigenvalues with finite geometric multiplicities.

3. If the space X is finite- or infinite-dimensional, provided $A \neq 0$, there exists a countable set $\{e_i\}_{i \in I}$ $(I = \{1, \dots, N\}$ with some $N \in \mathbb{N}$, or $I = \mathbb{N})$ of orthonormal eigenvectors:

$$Ae_i = \lambda_i e_i, \ i \in I,$$

where $\lambda_i \in \mathbb{R} \setminus \{0\}$, forming an orthonormal eigenbasis for $(\overline{R(A)}, (\cdot, \cdot), \| \cdot \|)$, which can be extended to an orthonormal eigenbasis for $(X, (\cdot, \cdot), \| \cdot \|)$ that is countable, provided the space $(X, (\cdot, \cdot), \| \cdot \|)$ is separable.

4. If $A \neq 0$, then

$$Ax = \sum_{i=1}^{N} \lambda_i(x, e_i)e_i, \ x \in X,$$

where, provided the space is X finite-dimensional, the sum is finite, and, provided the space X is infinite-dimensional and $N = \infty$, i. e., when the orthonormal eigenbasis $\{e_i\}_{i \in \mathbb{N}}$ for $(\overline{R(A)}, (\cdot, \cdot), \| \cdot \|)$ is countably infinite,

$$A = \sum_{i=1}^{\infty} \lambda_i P_i,$$

with

$$P_i x := (x, e_i)e_i, \ i \in \mathbb{N}, x \in X,$$

being the orthogonal projection onto the one-dimensional subspace span$(\{e_i\})$, and the operator series converging uniformly.

Proof. Let a compact self-adjoint operator $A \in K(X)$ be arbitrary.

Parts 1 and 2 immediately follow from the *Spectrum Theorem* (Theorem 5.11), the *Riesz–Schauder Theorem* (Theorem 5.21), and the *Spectrum of a Self-Adjoint Operator Theorem* (Theorem 6.23) (for the real case, see Remark 6.29 and Section 6.12, Problem 23).

Exercise 6.76. Explain.

Let us prove parts 3 and 4.

3. Suppose that $A \neq 0$. Then, by the *Existence of Nonzero Eigenvalues Proposition* (Proposition 6.16), there exists an $e_1 \in X$ such that $\|e_1\| = 1$ and

$$Ae_1 = \lambda_1 e_1,$$

where

$$\lambda_1 \in \mathbb{R} \quad \text{with} \quad |\lambda_1| = \|A\| \neq 0.$$

Let $X_1 := X$, $A_1 := A$, and

$$X_2 := \mathrm{span}(\{e_1\})^{\perp}.$$

Considering the *closedness* of the finite-dimensional subspace $\mathrm{span}(\{e_1\})$ (see the *Closedness of Finite-Dimensional Subspaces Theorem* (Theorem 3.12)), by the *Projection Theorem* (Theorem 6.7),

$$X = \mathrm{span}(\{e_1\}) \oplus X_2.$$

Since

$$\forall x \in X_2 : \ (x, e_1) = 0,$$

in view of the *self-adjointness* of A_1, we have

$$(A_1 x, e_1) = (x, A_1 e_1) = (x, \lambda_1 e_1) = \lambda_1 (x, e_1) = 0.$$

Whence, we infer that the closed orthogonally complementary subspaces $\mathrm{span}(\{e_1\})$ and $X_2 := \mathrm{span}(\{e_1\})^{\perp}$ *reduce* A_1, i. e., are both *invariant* relative to A_1:

$$A_1 X_1 \subseteq X_1 \quad \text{and} \quad A_1 X_2 \subseteq X_2.$$

Let A_2 be the restriction of the operator A_1 to X_2. Then A_2 is a *compact self-adjoint operator* on the Hilbert space $(X_2, (\cdot, \cdot), \| \cdot \|)$.

Exercise 6.77. Explain.

If $A_2 \neq 0$, then, similarly, there exists an $e_2 \in X_2$ such that $\|e_2\| = 1$, and

$$A_2 e_2 = \lambda_2 e_2,$$

where

$$\lambda_2 \in \mathbb{R} \quad \text{with} \quad 0 \neq |\lambda_2| = \|A_2\| \leq \|A_1\| = |\lambda_1|.$$

Exercise 6.78. Explain.

In the same manner, we define

$$X_3 := \mathrm{span}(\{e_1, e_2\})^{\perp},$$

which is $\mathrm{span}(\{e_2\})^{\perp}$ in $(X_2, (\cdot, \cdot), \| \cdot \|)$, and A_3 to be the restriction of A_2, and hence of A_1, to X_3, which is a *compact self-adjoint operator* on the Hilbert space $(X_3, (\cdot, \cdot), \| \cdot \|)$.

If $A_3 \neq 0$, we can also choose an eigenvector $e_3 \in X_3$ with $\|e_3\| = 1$ and the corresponding eigenvalue λ_3 with

$$\lambda_3 \in \mathbb{R} \quad \text{with} \quad 0 \neq |\lambda_3| = \|A_3\| \leq \|A_2\| = |\lambda_2| \leq |\lambda_1|.$$

Continuing in this fashion, we may stop after a finite number of steps if, for some $N = 2, 3, \ldots$, the restriction A_N of A_1, to the closed subspace

$$X_N = \mathrm{span}(\{e_1, e_2, \ldots, e_{N-1}\})^{\perp}$$

is the *zero operator*. In particular, this occurs when the space X is *finite-dimensional*.

In this case, we obtain a finite orthonormal set of eigenvectors $\{e_1, e_2, \ldots, e_{N-1}\}$ corresponding to the nonzero eigenvalues $\{\lambda_1, \lambda_2, \ldots, \lambda_{N-1}\}$ with

$$|\lambda_{N-1}| \leq \cdots \leq |\lambda_2| \leq |\lambda_1|$$

and, by the *Spectral Bounds Theorem* (Theorem 6.24),

$$|\lambda_k| = \|A_k\| = \sup_{x \in X_k, \|x\|=1} |(Ax, x)|, \; k = 1, \ldots, N - 1.$$

Since, by the *Properties of Self-Adjoint Operators* (Theorem 6.13),

$$X = \ker A \oplus \overline{R(A)}, \tag{6.34}$$

considering that

$$\ker A = \mathrm{span}(\{e_1, e_2 \ldots, e_{N-1}\})^{\perp},$$

and that the finite-dimensional subspace $\mathrm{span}(\{e_1, e_2 \ldots, e_{N-1}\})$ is *closed* (see the *Closedness of Finite-Dimensional Subspaces Theorem* (Theorem 3.12)), by the *Second Orthogonal Complement of a Closed Subspace Proposition* (Proposition 6.6), we infer that

$$\overline{R(A)} = \mathrm{span}(\{e_1, e_2 \ldots, e_{N-1}\}),$$

which further implies that

$$R(A) = \mathrm{span}(\{e_1, e_2 \ldots, e_{N-1}\}).$$

Exercise 6.79. Explain.

Thus, in the considered case, the operator A is of *finite rank* and $\{e_1, e_2 \ldots, e_{N-1}\}$ is an orthonormal basis for $(R(A), (\cdot, \cdot), \| \cdot \|)$. In view of orthogonal decomposition (6.34), we obtain an *orthonormal eigenbasis* for $(X, (\cdot, \cdot), \| \cdot \|)$ via appending to the orthonormal eigenbasis $\{e_1, e_2 \ldots, e_{N-1}\}$ for $(R(A), (\cdot, \cdot), \| \cdot \|)$ an orthonormal basis for $(\ker A, (\cdot, \cdot), \| \cdot \|)$, if any (i. e., $\ker A \neq \{0\}$). By the *Orthogonal Dimension of a Separable Hilbert Space Theorem* (Theorem 6.15), such an orthonormal eigenbasis is countable, provided the space $(X, (\cdot, \cdot), \| \cdot \|)$ is *separable*.

If the space X is *infinite dimensional*, i. e., $\dim X > \aleph_0$ (see the *Basis of a Banach Space Theorem* (Theorem 3.14) and Remark 3.29), the process may continue indefinitely, in which case, as a result, we obtain a countably infinite set $\{e_i\}_{i \in \mathbb{N}}$ of orthonormal eigenvectors:

$$Ae_i = \lambda_i e_i, \ i \in \mathbb{N},$$

where

$$0 \neq |\lambda_{i+1}| \leq |\lambda_i|,$$

and, by part 2,

$$|\lambda_i| = \|A_i\| \to 0, \ i \to \infty.$$

Let $x \in X$ be arbitrary, then, by the *Finite Orthonormal Set Corollary* (Corollary 6.4), for any $n \in N$,

$$y_n := x - \sum_{i=1}^{n}(x, e_i)e_i \in \mathrm{span}(\{e_1, e_2 \ldots, e_n\})^{\perp} =: X_{n+1}$$

with

$$\|y_n\| \leq \|x\|, \ n \in \mathbb{N}.$$

Hence,

$$\left\| Ax - \sum_{i=1}^{n}(x, e_i)Ae_i \right\| = \|Ay_n\| = \|A_{n+1}y_n\| \leq \|A_{n+1}\|\|y_n\| \leq \|A_{n+1}\|\|x\|, \ n \in \mathbb{N}.$$

Whence, since, in view of the *self-adjointness* of A,

$$(x, e_i)Ae_i = (x, e_i)\lambda_i e_i = (x, \lambda_i e_i)e_i = (x, Ae_i)e_i = (Ax, e_i)e_i, \ i = 1, \ldots, n, \ n \in \mathbb{N},$$

we infer that

$$\left\| Ax - \sum_{i=1}^{n}(Ax, e_i)e_i \right\| \leq \|A_{n+1}\|\|x\| = |\lambda_{n+1}|\|x\| \to 0, \ n \to \infty,$$

i. e.,

$$Ax = \sum_{i=1}^{\infty}(Ax, e_i)e_i \text{ in } (X, (\cdot, \cdot), \|\cdot\|),$$

which implies that the orthonormal set $\{e_n\}_{n \in \mathbb{N}}$ is a *fundamental set* in the Hilbert space $(\overline{R(A)}, (\cdot, \cdot), \|\cdot\|)$, i. e.,

$$\overline{\mathrm{span}(\{e_n\}_{n=1}^{\infty})} = \overline{R(A)},$$

and hence, by the *Orthonormal Basis Characterizations* (Theorem 6.12), $\{e_n\}_{n \in \mathbb{N}}$ is an orthonormal basis for $(\overline{R(A)}, (\cdot, \cdot), \| \cdot \|)$.

Considering orthogonal decomposition (6.34), we obtain an *orthonormal eigenbasis* for $(X, (\cdot, \cdot), \| \cdot \|)$ via appending to the orthonormal eigenbasis $\{e_n\}_{n \in \mathbb{N}}$ for $(R(A), (\cdot, \cdot), \| \cdot \|)$ an orthonormal basis for $(\ker A, (\cdot, \cdot), \| \cdot \|)$, if any (i. e., $\ker A \neq \{0\}$). By the *Orthogonal Dimension of a Separable Hilbert Space Theorem* (Theorem 6.15), such an orthonormal eigenbasis is countable, provided the space $(X, (\cdot, \cdot), \| \cdot \|)$ is *separable*.

4. The fact that, for $A \neq 0$,

$$Ax = \sum_{i=1}^{N} \lambda_i (x, e_i) e_i, \ x \in X,$$

where $N \in \mathbb{N}$ or $N = \infty$, provided the orthonormal eigenbasis $\{e_i\}_{i \in I}$ for $(\overline{R(A)}, (\cdot, \cdot), \| \cdot \|)$ is *finite* $(I = \{1, \ldots, N\})$ or *countably infinite* $(I = \mathbb{N})$, respectively, immediately follows from part 3.

If the space is X *finite-dimensional*, the above sum is finite.

Exercise 6.80. Explain.

If the space is *infinite-dimensional* and $N = \infty$, i. e., when the orthonormal eigenbasis $\{e_i\}_{i \in \mathbb{N}}$ for $(\overline{R(A)}, (\cdot, \cdot), \| \cdot \|)$ is *countably infinite*, we have

$$A = \sum_{i=1}^{\infty} \lambda_i P_i x, \ x \in X, \tag{6.35}$$

where

$$P_i x := (x, e_i) e_i, \ i \in \mathbb{N}, x \in X,$$

is the orthogonal projection onto the one-dimensional subspace $\text{span}(\{e_i\})$.

The fact that the operator series in (6.35) *uniformly converges* to A follows by the *Approximation by Finite-Rank Operators Corollary* (Corollary 5.6) (see Remarks 5.31) when applied to the restriction of the operator A to the *separable* Hilbert space $(\overline{R(A)}, (\cdot, \cdot), \| \cdot \|)$ with the orthonormal eigenbasis $\{e_i\}_{i \in \mathbb{N}}$.

Exercise 6.81. Fill in the details. $\qquad \square$

Remarks 6.33.

– The *Spectral Theorem for Compact Self-Adjoint Operators* is also referred to in the literature as the *Hilbert–Schmidt Theorem* or the *Eigenvector Expansion Theorem* (see, e. g., [9]).

– Let A be a compact self-adjoint operator on a Hilbert space $(X, (\cdot, \cdot), \| \cdot \|)$.
By the *Spectral Theorem for Compact Self-Adjoint Operators*, there exists an orthonormal eigenbasis $\{e_i\}_{i \in I}$ $(I = \{1, \ldots, N\}$ with some $N \in \mathbb{N}$ or $I = \mathbb{N})$ for $(\overline{R(A)}, (\cdot, \cdot), \| \cdot \|)$.

Due to the orthogonal decomposition

$$X = \ker A \oplus \overline{R(A)}$$

(see the *Properties of Self-Adjoint Operators* (Theorem 6.13)),

$$\forall x \in X : \ x = \sum_{i=0}^{N} P_i x,$$

where P_0 is the orthogonal projection onto $\ker A$ and

$$Px = \sum_{i=1}^{N} P_i x, \ x \in X,$$

with

$$P_i x := (x, e_i) e_i, \ i \in I, x \in X,$$

being the orthogonal projection onto the one-dimensional subspace span($\{e_i\}$), is the orthogonal projection onto $\overline{R(A)}$.

Exercise 6.82. Explain.

Also, for any $\lambda \in \rho(A)$ (for the real case, see Remark 6.29 and Section 6.12, Problem 23),

$$R(\lambda, A)x = \sum_{i=0}^{N} \frac{1}{\lambda_i - \lambda} P_i x, \ x \in X,$$

where $\lambda_0 := 0$.

Exercise 6.83. Explain.

We immediately obtain the following:

Corollary 6.9 (Diagonal Matrix Representation of a Self-Adjoint Operator). *An arbitrary self-adjoint operator A on a finite-dimensional Hilbert space $(X, (\cdot, \cdot), \| \cdot \|) (X, \| \cdot \|)$ has a matrix representation, which is a diagonal matrix with real entries.*

Exercise 6.84. Prove (see Section 5.6.2.1).

Examples 6.20.
1. On the (real or complex) two-dimensional Hilbert space $l_2^{(2)}$, for the compact self-adjoint operator A of multiplication by the 2×2 real symmetric matrix

$$\begin{bmatrix} 1 & 1 \\ 1 & 1 \end{bmatrix}$$

(see Examples 5.17 and 6.19),

$$\sigma(A) = \sigma_p(A) = \{0, 2\}$$

(see Examples 5.2 and 5.11), each eigenvalue being of geometric multiplicity 1, the set

$$B := \left\{ e_1 := \frac{1}{\sqrt{2}} \begin{bmatrix} 1 \\ -1 \end{bmatrix}, e_2 := \frac{1}{\sqrt{2}} \begin{bmatrix} 1 \\ 1 \end{bmatrix} \right\}$$

being an orthonormal eigenbasis for $l_2^{(2)}$ (see Examples 5.11), and

$$\forall x := (x_1, x_2) \in l_2^{(2)} : x = (x, e_1)e_1 + (x, e_2)e_2,$$
$$Ax = 2(x, e_2)e_2,$$
$$R(\lambda, A)x = -\frac{1}{\lambda}(x, e_1)e_1 + \frac{1}{2 - \lambda}(x, e_2)e_2, \; \lambda \neq 0, 2,$$

where $(\cdot, \cdot)$ designates the inner product in $l_2^{(2)}$.

Relative to the orthonormal eigenbasis B, the matrix representation of A is the diagonal matrix

$$[A]_B = \begin{bmatrix} 0 & 0 \\ 0 & 2 \end{bmatrix}$$

(see Examples 5.11).

2. On the (real or complex) infinite-dimensional separable Hilbert space l_2,
   (a) for the compact self-adjoint operator A of multiplication by the vanishing sequence $(1/n)_{n \in \mathbb{N}}$,

   $$\sigma(A) = \{1/n\}_{n \in \mathbb{N}} \cup \{0\} \text{ with } \sigma_p(A) = \{1/n\}_{n \in \mathbb{N}} \text{ and } \sigma_c(A) = \{0\}$$

   (see Examples 5.19), each eigenvalue being of geometric multiplicity 1, the standard orthonormal basis $\{e_n\}_{n \in \mathbb{N}}$ being a countable orthonormal eigenbasis for l_2, and

   $$\forall x := (x_n)_{n \in \mathbb{N}} \in l_2 : x = \sum_{n=1}^{\infty} (x, e_n)e_n = \sum_{n=1}^{\infty} x_n e_n,$$
   $$Ax = \sum_{n=1}^{\infty} \frac{1}{n}(x, e_n)e_n = \sum_{n=1}^{\infty} \frac{x_n}{n} e_n,$$
   $$R(\lambda, A)x = \sum_{n=1}^{\infty} \frac{1}{n - \lambda}(x, e_n)e_n = \sum_{n=1}^{\infty} \frac{x_n}{n - \lambda} e_n, \; \lambda \in \rho(A).$$

   (b) For the compact self-adjoint operator A of multiplication by the vanishing sequence $(0, 1/2, 1/3, \ldots)$,

   $$\sigma(A) = \sigma_p(A) = \{0, 1/2, 1/3, \ldots\}$$

(see Examples 5.19), each eigenvalue being of geometric multiplicity 1, the standard orthonormal basis $\{e_n\}_{n\in\mathbb{N}}$ being a countable orthonormal eigenbasis for l_2, and

$$\forall x := (x_n)_{n\in\mathbb{N}} \in l_2 : x = \sum_{n=1}^{\infty}(x, e_n)e_n = \sum_{n=1}^{\infty} x_n e_n,$$

$$Ax = \sum_{n=2}^{\infty}\frac{1}{n}(x, e_n)e_n = \sum_{n=2}^{\infty}\frac{x_n}{n}e_n,$$

$$R(\lambda, A)x = \frac{1}{0-\lambda}(x, e_1)e_1 + \sum_{n=2}^{\infty}\frac{1}{n-\lambda}(x, e_n)e_n$$

$$= -\frac{x_1}{\lambda}e_1 + \sum_{n=2}^{\infty}\frac{x_n}{n-\lambda}e_n, \ \lambda \in \rho(A).$$

(c) For the compact self-adjoint operator A of multiplication by the vanishing sequence $(0, 1/2, 0, 1/4, \ldots)$,

$$\sigma(A) = \sigma_p(A) = \{0, 1/2, 1/4, \ldots\}$$

(see Examples 5.17 and 5.2), each nonzero eigenvalue being of geometric multiplicity 1, the eigenvalue 0 being of geometric multiplicity dim ker $A = \aleph_0$, the standard orthonormal basis $\{e_n\}_{n\in\mathbb{N}}$ being a countable orthonormal eigenbasis for l_2, and

$$\forall x := (x_n)_{n\in\mathbb{N}} \in l_2 : x = \sum_{n=1}^{\infty}(x, e_n)e_n = \sum_{n=1}^{\infty} x_n e_n,$$

$$Ax = \sum_{n=1}^{\infty}\frac{1}{2n}(x, e_{2n})e_{2n} = \sum_{n=1}^{\infty}\frac{x_{2n}}{2n}e_{2n},$$

$$R(\lambda, A)x = \frac{1}{0-\lambda}\sum_{n=1}^{\infty}(x, e_{2n-1})e_{2n-1} + \sum_{n=1}^{\infty}\frac{1}{2n-\lambda}(x, e_{2n})e_{2n}$$

$$= -\frac{1}{\lambda}\sum_{n=1}^{\infty}x_{2n-1}e_{2n-1} + \sum_{n=1}^{\infty}\frac{x_{2n}}{2n-\lambda}e_{2n}, \ \lambda \in \rho(A).$$

3. On a (real or complex) nonseparable Hilbert space $(X, (\cdot, \cdot), \|\cdot\|)$, for the projection operator P onto a one-dimensional subspace,

$$\sigma(P) = \sigma_p(P) = \{0, 1\},$$

(see Examples 5.17 and 5.2), the eigenvalue 1 being of geometric multiplicity 1, the eigenvalue 0 being of geometric multiplicity dim ker $P > \aleph_0$, and

$$\forall x \in X : x = (I - P)x + Px,$$

$$R(\lambda, P)x = -\frac{1}{\lambda}(I - P) + \frac{1}{1-\lambda}P, \ \lambda \neq 0, 1,$$

where $I - P$ is the orthogonal projection onto ker P (cf. Examples 5.4).

Exercise 6.85. Verify.

6.12 Problems

In the subsequent problems, $\mathbb{F}$ stands for the scalar field of real or complex numbers (i. e., $\mathbb{F} = \mathbb{R}$ or $\mathbb{F} = \mathbb{C}$).

1. Prove

   Proposition 6.17 (Inner Product Separation Property). *For vectors x and y in an inner product space* $(X, (\cdot, \cdot))$,

   $$x = y \iff \forall z \in X : (x, z) = (y, z).$$

2. Let $(X, (\cdot, \cdot), \| \cdot \|)$ be an inner product space. Prove that

   $$\forall y \in X : \|y\| = \sup_{\|x\|=1} |(x, y)| = \sup_{\|x\|=1} |(y, x)|$$

   $$= \sup_{\|x\|\leq1} |(x, y)| = \sup_{\|x\|\leq1} |(y, x)|.$$

   Hint. Apply the *Cauchy–Schwarz Inequality* (Theorem 6.2).

3. Prove

   Proposition 6.18 (Equality in Triangle Inequality). *Let* $(X, (\cdot, \cdot), \| \cdot \|)$ *be an inner product space. For vectors* $x, y \in X$,

   $$\|x + y\| = \|x\| + \|y\|$$

   iff either $x = 0$ *or* $y = 0$ *or* $x, y \neq 0$ *and have the same direction, i. e.,*

   $$\exists \lambda > 0 : y = \lambda x.$$

4. Prove

   Proposition 6.19 (Characterization of Convergence in Inner Product Spaces). *For a sequence* $(x_n)_{n\in\mathbb{N}}$ *in an inner product space* $(X, (\cdot, \cdot), \| \cdot \|)$,

   $$x_n \to x \in X, \ n \to \infty, \ in \ (X, (\cdot, \cdot), \| \cdot \|)$$

   iff
   (1) $\forall y \in X : (x_n, y) \to (x, y), \ n \to \infty,$ *and*
   (2) $\|x_n\| \to \|x\|, \ n \to \infty.$

5. (Cartesian Product of Inner Product Spaces)
   Let $(X_1, (\cdot, \cdot)_1, \| \cdot \|_1)$ and $(X_2, (\cdot, \cdot)_2, \| \cdot \|_2)$ be inner product spaces over $\mathbb{F}$.

(a) Show that the Cartesian product $X = X_1 \times X_2$ is an inner product space relative to the inner product

$$X_1 \times X_2 \ni x := \langle x_1, x_2 \rangle, y := \langle y_1, y_2 \rangle \mapsto (x, y) := (x_1, y_1)_1 + (x_2, y_2)_2 \in \mathbb{F}$$

generating the *product norm*

$$X_1 \times X_2 \ni x := \langle x_1, x_2 \rangle \mapsto \|x\|_{X_1 \times X_2} = \sqrt{\|x_1\|_1^2 + \|x_2\|_2^2}$$

(see Section 3.5, Problem 7).

(b) Show that the product space $(X_1 \times X_2, (\cdot, \cdot), \| \cdot \|_{X_1 \times X_2})$ is Hilbert space *iff* each space $(X_i, (\cdot, \cdot)_i, \| \cdot \|_i)$, $i = 1, 2$, is a Hilbert space.

6. (Complexification of Real Inner Product Spaces)

Let $(X, (\cdot, \cdot)_X, \| \cdot \|_X)$ be a real inner product space. Its *complexification* $X^{\mathbb{C}}$ (see Section 3.5, Problem 1) is a complex inner product space relative to the inner product

$$X^{\mathbb{C}} \ni \langle x_1, y_1 \rangle, \langle x_2, y_2 \rangle \mapsto (\langle x_1, y_1 \rangle, \langle x_2, y_2 \rangle) := (x_1, x_2)_X + (y_1, y_2)_X$$
$$- i[(x_1, y_2)_X - (y_1, x_2)_X].$$

The inner product $(\cdot, \cdot)$ extends the inner product $(\cdot, \cdot)_X$ in the sense that

$$\forall x, y \in X : (\langle x, 0 \rangle, \langle y, 0 \rangle) = (x, y)_X.$$

In particular,

$$\forall x \in X : \|\langle x, 0 \rangle\| = \sqrt{(\langle x, 0 \rangle, \langle x, 0 \rangle)} = \|x\|_X,$$

where

$$\forall \langle x, y \rangle \in X^{\mathbb{C}} : \|\langle x, y \rangle\| := \sqrt{(\langle x, y \rangle, \langle x, y \rangle)} = \sqrt{\|x\|_X^2 + \|y\|_X^2}$$

is the inner product norm on $X^{\mathbb{C}}$ generated by $(\cdot, \cdot)$, i. e., the *isomorphic embedding*

$$X \ni x \mapsto Tx := (x, 0) \in X^{\mathbb{C}}$$

is norm-preserving, and hence, is an *isometric isomorphism* from $(X, (\cdot, \cdot)_X, \| \cdot \|_X)$ to $(X_{\mathbb{R}}^{\mathbb{C}}, (\cdot, \cdot), \| \cdot \|)$ (see Remark 3.21), where $X_{\mathbb{R}}^{\mathbb{C}}$ is the associated with $X^{\mathbb{C}}$ real space (see Remarks 3.1).

(a) Show that the mapping $(\cdot, \cdot)$ is an *inner product* on $X^{\mathbb{C}}$.

(b) Show that $(X_{\mathbb{R}}^{\mathbb{C}}, (\cdot, \cdot), \| \cdot \|)$ is a Hilbert space *iff* $(X, (\cdot, \cdot)_X, \| \cdot \|_X)$ is a Hilbert space.

7. Prove

Proposition 6.20 (Convex Hull's Structure). *For a nonempty set S in a vector space X over* $\mathbb{F}$, *conv(S) is the set of all convex combinations of its elements:*

$$\text{conv}(S) = \left\{ \sum_{k=1}^{n} \lambda_k x_k \,\middle|\, x_1, \ldots, x_n \in S, \lambda_1, \ldots, \lambda_n \in [0, 1] \text{ with } \sum_{k=1}^{n} \lambda_k = 1, n \in \mathbb{N} \right\}.$$

8. Prove the following proposition and corollary:

   Proposition 6.21 (Convexity of the Set of the Nearest Points). *Let $(X, \| \cdot \|)$ be a normed vector space. Then for an arbitrary $x \in X$, the set N_x of all nearest to x points in a convex set $C \subseteq X$, if nonempty, is convex.*

   Corollary 6.10 (Cardinality of the Set of Nearest Points). *Let $(X, \| \cdot \|)$ be a normed vector space. Then for an arbitrary $x \in X$, the set N_x of all nearest to x points in a convex subset $C \subseteq X$ is either empty, a singleton, or uncountably infinite.*

9. Show that, in an inner product space $(X, (\cdot, \cdot), \| \cdot \|)$, for nonempty sets M and N,
   (a) $M \subseteq N \Rightarrow M^\perp \supseteq N^\perp$ and
   (b) $M^\perp = \overline{M}^\perp$.

10. Prove the following:

    Proposition 6.22 (Coincidence of Orthogonal Complements). *For a nonempty set M in an inner product space $(X, (\cdot, \cdot), \| \cdot \|)$,*

    $$M^\perp = \overline{\mathrm{span}(M)}^\perp.$$

11. In the real space $l_2^{(3)}$, for the subspace

    $$Y := \{(x, y, z) \in \mathbb{R}^3 \mid x + y + z = 0\},$$

    (a) determine $Y^\perp$, and
    (b) for $x := (1, 2, 3)$, find the unique decomposition

    $$x = y + z, \ y \in Y, z \in Y^\perp.$$

12. In the (real or complex) space l_2, for the subspace,

    $$Y := \{(x_n)_{n \in \mathbb{N}} \in l_2 \mid x_{2n} = 0, \ n \in \mathbb{N}\},$$

    (a) determine $Y^\perp$, and
    (b) for $x := (1/n)_{n \in \mathbb{N}}$, find the unique decomposition

    $$x = y + z, \ y \in Y, z \in Y^\perp.$$

13. Prove

    Proposition 6.23 (Second Orthogonal Complement of a Set). *For a nonempty set M in a Hilbert space $(X, (\cdot, \cdot), \| \cdot \|)$,*

    $$(M^\perp)^\perp = \overline{\mathrm{span}(M)}.$$

    Hint. Use the *Coincidence of Orthogonal Complements Proposition* (Proposition 6.22, Problem 10) and *Second Orthogonal Complement of a Closed Subspace Proposition* (Proposition 6.6).

14. * Prove

    Proposition 6.24 (Norm Characterization of Orthogonal Projections). *A nonzero projection operator P (P ≠ 0) on a Hilbert space $(X, (\cdot,\cdot), \|\cdot\|)$ is an orthogonal projection iff*

    $$\|P\| = 1.$$

15. Weak Convergence in a Hilbert Space
    A sequence $(x_n)_{n\in\mathbb{N}}$ is said to *weakly converge* (be *weakly convergent*) to a vector x in a Hilbert space $(X, (\cdot,\cdot), \|\cdot\|)$ if

    $$\forall y \in X : (x_n, y) \to (x, y), \ n \to \infty.$$

    The vector x is called the *weak limit* of the sequence $(x_n)_{n\in\mathbb{N}}$.
    (a) Prove that any convergent sequence in a Hilbert space is weakly convergent to the same limit (see Problem 4).
    (b) Give an example showing that a weakly convergent sequence in a Hilbert space need not be convergent.
    (c) Prove that the weak limit of a sequence in a Hilbert space, if existent, is *unique*.
    (d) Prove the following:

    Proposition 6.25 (Boundedness of Weakly Convergent Sequences). *Any weakly convergent sequence $(x_n)_{n\in\mathbb{N}}$ in a Hilbert space $(X, (\cdot,\cdot), \|\cdot\|)$ is bounded, i. e.,*

    $$\exists C > 0 \ \forall n \in \mathbb{N} : \|x_n\| \leq C.$$

    Hint. Consider the sequence of bounded linear functionals

    $$f_n(x) := (x, x_n), \ x \in X,$$

    on $(X, (\cdot,\cdot), \|\cdot\|)$ and apply the Uniform Boundedness Principle (Theorem 4.6) and the *Riesz Representation Theorem* (Theorem 6.8).

    (e) Give an example showing that a bounded sequence in a Hilbert space need not be weakly convergent.
16. Prove

    Theorem 6.27 (Generalized Pythagorean Theorem). *For a finite orthogonal set $\{x_1, \ldots, x_n\}$ $(n \in \mathbb{N})$ in an inner product space $(X, (\cdot,\cdot), \|\cdot\|)$,*

    $$\left\|\sum_{i=1}^{n} x_i\right\|^2 = \sum_{i=1}^{n} \|x_i\|^2.$$

17. In the space l_2,

(a) apply the *Gram–Schmidt process* to *orthonormalize* the set

$$\{x_1 := (1,1,0,0,\dots), x_2 := (1,0,1,0,\dots), x_3 := (1,1,1,0,\dots)\};$$

(b) for the orthonormal set $\{e_1, e_2, e_3\}$ obtained in (a), find the *orthogonal projection* of $x := (1,0,0,\dots)$ on $Y := \mathrm{span}(\{e_1, e_2\})$ and $\rho(x, Y)$.

18. Prove that an arbitrary *orthonormal sequence* $\{e_i\}_{i\in\mathbb{N}}$ in an infinite-dimensional Hilbert space $(X, (\cdot,\cdot), \|\cdot\|)$ *weakly converges* to 0, i. e.,

$$\forall y \in X: \ (e_n, y) \to (0, y) = 0, \ n \to \infty,$$

(see Problem 15).

19. * Prove that, if $\{e_i\}_{i\in\mathbb{N}}$ is an *orthonormal basis* for a separable Hilbert space $(X, (\cdot,\cdot), \|\cdot\|)$ and $\{e'_j\}_{j\in\mathbb{N}}$ is an *orthonormal sequence* such that

$$\sum_{i=1}^{\infty} \|e_i - e'_i\|^2 < \infty,$$

then $\{e'_j\}_{j\in\mathbb{N}}$ is also an *orthonormal basis* for $(X, (\cdot,\cdot), \|\cdot\|)$.

Hint. First, show that $|(e_j - e'_j, e_i)| = |(e_i - e'_i, e'_j)|$, $i, j \in \mathbb{N}$, then apply *Parseval's identity*.

20. Prove

Proposition 6.26 (Norm of a Bounded Linear Operator). *Let A be a bounded linear operator on a inner product space $(X, (\cdot,\cdot), \|\cdot\|)$ (i. e., $A \in L(X)$). Then*

$$\|A\| := \sup_{\|x\|=1} \|Ax\| = \sup_{\|x\|=1} \sup_{\|y\|=1} |(Ax, y)| = \sup_{\|x\|=1, \|y\|=1} |(Ax, y)|.$$

Hint. Use the result of Problem 2.

21. Prove

Proposition 6.27 (Quadratic Form of a Symmetric Operator). *Let $(A, D(A))$ be a symmetric linear operator in a complex Hilbert space $(X, (\cdot,\cdot), \|\cdot\|)$. Then*

$$\forall x \in D(A): \ (Ax, x) \in \mathbb{R},$$

the mapping

$$D(A) \ni x \mapsto (Ax, x)$$

being called the quadratic form *of A.*
In particular, this is true if the operator A is self-adjoint.

22. Prove

Theorem 6.28 (Hellinger–Toeplitz Theorem). *Let $(X, (\cdot, \cdot), \| \cdot \|)$ be a Hilbert space and $A : X \to X$ be a symmetric linear operator, i. e.,*

$$\forall x, y \in X : \quad (Ax, y) = (x, Ay),$$

then $A \in L(X)$.

Hint. Show that the operator A is *self-adjoint* and apply the *Closed Graph Theorem* (Theorem 4.10) (see Remarks 6.27).

23. (Complexification of Linear Operators in Real Hilbert Spaces)
    Let $(\tilde{A}, D(\tilde{A}))$ be the *complexification* of an arbitrary linear operator $(A, D(A))$ in a real Hilbert space $(X, (\cdot, \cdot)_X, \| \cdot \|_X)$ (see Section 4.7, Problem 1).
    (a) Show that

    $$\tilde{A} \in L(X^{\mathbb{C}}) \iff A \in L(X).$$

    (b) Show that

    $$\tilde{A} \in K(X^{\mathbb{C}}) \iff A \in K(X).$$

    (c) Show that, if the operator $(A, D(A))$ is densely defined in the space $(X, (\cdot, \cdot)_X, \| \cdot \|_X)$, then the operator $(\tilde{A}, D(\tilde{A}))$ is densely defined in the complexification $(X^{\mathbb{C}}, (\cdot, \cdot), \| \cdot \|)$ of $(X, (\cdot, \cdot)_X, \| \cdot \|_X)$, which is a complex Hilbert space (see Problem 6), and

    $$(\tilde{A})^* = \tilde{A^*},$$

    where $\tilde{A^*}$ is the complexification of the adjoint operator A^*.
    In particular, if the operator $(A, D(A))$ is *self-adjoint*, then its complexification $(\tilde{A}, D(\tilde{A}))$ is a *self-adjoint operator*.
    (d) For an arbitrary real number $\lambda \in \mathbb{R}$,

    $$\lambda \in \rho(\tilde{A}) \iff \exists (A - \lambda I)^{-1} \in L(X).$$

    (e) For an arbitrary real number $\lambda \in \mathbb{R}$,

    $$\lambda \in \sigma_p(\tilde{A}) \iff \lambda \text{ is an eigenvalue of } A.$$

24. Let A be a bounded self-adjoint linear operator on a Hilbert space $(X, (\cdot, \cdot), \| \cdot \|)$ (i. e., $A \in L(X)$, and $A = A^*$). Show that the operators A^*A and AA^* are self-adjoint.
25. Let A and B be bounded self-adjoint linear operators on a Hilbert space $(X, (\cdot, \cdot), \| \cdot \|)$ (i. e., $A, B \in L(X)$, $A = A^*$, and $B = B^*$). Prove that the product AB is self-adjoint *iff* A and B *commute*, i. e.,

    $$AB = BA.$$

26. Prove the following:

Proposition 6.28 (Self-Adjoint Characterization of Orthogonal Projections).
A projection operator P on a Hilbert space $(X, (\cdot, \cdot), \|\cdot\|)$ is an orthogonal projection iff P is self-adjoint, i. e.,

$$\forall x, y \in X: \ (Px, y) = (x, Py).$$

27. Prove that, if a sequence $(A_n)_{n\in\mathbb{N}}$ of bounded self-adjoint linear operators on a Hilbert space $(X, (\cdot, \cdot), \|\cdot\|)$ *weakly converges* to an operator $A \in L(X)$, i. e.,

$$\forall x, y \in X: \ (A_n x, y) \to (Ax, y), \ n \to \infty$$

(see Problem 15), then the limit operator A is *self-adjoint*.

28. Prove that, if a sequence $(P_n)_{n\in\mathbb{N}}$ of orthogonal projections on a Hilbert space $(X, (\cdot, \cdot), \|\cdot\|)$ *weakly converges* to an orthogonal projection P, i. e.,

$$\forall x, y \in X: \ (P_n x, y) \to (Px, y), \ n \to \infty$$

(see Problem 27), then it *strongly converges* to P, i. e.,

$$\forall x \in X: \ P_n x \to Px, \ n \to \infty, \ \text{in} \ (X, (\cdot, \cdot), \|\cdot\|).$$

Hint. Use the result of the prior problem to show that

$$\forall x \in X: \ \|P_n x\| \to \|Px\|, \ n \to \infty,$$

and then apply the result of Problem 4.

29. Prove

Proposition 6.29 (Compact Operators and Weakly Convergent Sequences). *Let $(x_n)_{n\in\mathbb{N}}$ be a sequence weakly convergent to a vector x in a Hilbert space $(X, (\cdot, \cdot), \|\cdot\|)$ (see Problem 15) and A be a compact linear operator on X (i. e., $A \in K(X)$). Then*

$$Ax_n \to Ax, \ n \to \infty, \ \text{in} \ (X, (\cdot, \cdot), \|\cdot\|),$$

i. e., compact operators transform weakly convergent sequences to convergent.

Hint. Using the *Boundedness of Weakly Convergent Sequences Proposition* (Proposition 6.25, Problem 15) and the *Sequential Characterization of Compact Operators* (Proposition 5.2), show that every subsequence $(x_{n(k)})_{k\in\mathbb{N}}$ of $(x_n)_{n\in\mathbb{N}}$ contains a subsequence $(x_{n(k(j))})_{j\in\mathbb{N}}$ such that

$$Ax_{n(k(j))} \to Ax, \ j \to \infty, \ \text{in} \ (X, (\cdot, \cdot), \|\cdot\|),$$

and then apply the *Characterization of Convergence* (Theorem 2.10).

30. Let A be a compact linear operator on an infinite-dimensional Hilbert space $(X, (\cdot, \cdot), \| \cdot \|)$ (i. e., $A \in K(X)$). Prove that, for an arbitrary *orthonormal sequence* $\{e_i\}_{i \in \mathbb{N}}$,

$$Ae_n \to 0, \; n \to \infty, \; \text{in } (X, (\cdot, \cdot), \| \cdot \|).$$

Hint. Use the results of Problems 18 and 29.

A The Axiom of Choice and Equivalents

A.1 The Axiom of Choice

> To choose one sock from each of infinitely many pairs of socks requires the Axiom of Choice, but for shoes the Axiom is not needed.
> Bertrand Russell

Here we give a concise discourse on the celebrated *Axiom of Choice*, its equivalents, and *ordered sets*.

A.1.1 The Axiom of Choice

Expository Reference to a Set by Cantor. *By a set X, we understand "a collection into a whole of definite, well-distinguished objects, called the elements of X, of our perception or of our thought."*

Axiom of Choice (1904). *For each nonempty collection $\mathscr{F}$ of nonempty sets, there exists a function $f : \mathscr{F} \to \bigcup_{X \in \mathscr{F}} X$ such that[1]*

$$\mathscr{F} \ni X \mapsto f(X) \in X,$$

or equivalently, for each nonempty collection $\{X_i\}_{i \in I}$ of nonempty sets, there exists a function $f : I \to \bigcup_{i \in I} X_i$ such that

$$I \ni i \mapsto f(i) \in X_i.$$

The function f is called a choice function on $\mathscr{F}$, respectively, on I.

See, e. g., [29, 33, 36, 50].

A.1.2 Controversy

The *Axiom of Choice* allows one to prove the following counterintuitive statements:

Theorem A.1 (Vitali Theorem (1905)). *There exists a set in $\mathbb{R}$, which is not Lebesgue measurable.[2]*

See, e. g., [15, 46].

---

1 Due to Ernst Zermelo (1871–1953).
2 Giuseppe Vitali (1875–1932).

https://doi.org/10.1515/9783110600988-007

Theorem A.2 (Banach–Tarski Paradox (1924)). *Given a solid ball in 3-dimensional space, there exists a decomposition of the ball into a finite number of disjoint pieces, which can be reassembled, using only rotations and translations, into two identical copies of the original ball. The pieces involved are nonmeasurable, i.e., one cannot meaningfully assign volumes to them.*[3]

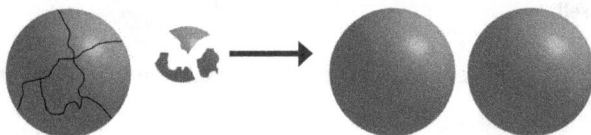

A.1.3 Timeline

1904: Ernst Zermelo formulates the *Axiom of Choice* in order to prove the *Well-Ordering Principle* (Theorem A.7).

1939: Kurt Gödel[4] proves that, if the other standard set-theoretic *Zermelo–Fraenkel*[5] *Axioms* (see, e. g., [29, 33, 36, 50]) are consistent, they do not disprove the *Axiom of Choice*.

1963: Paul Cohen[6] completes the picture by showing that, if the other standard *Zermelo–Fraenkel Axioms* are consistent, they do not yield a proof of the *Axiom of Choice*, i. e., the *Axiom of Choice* is *independent*.

A.2 Ordered Sets

Here, we discuss various types of *order* on a set.

Definition A.1 (Partially Ordered Set). A *partially ordered set* is a nonempty set X with a *binary relation* $\leq$ of *partial order*, which satisfies the following *partial order axioms*:

1. For any $x \in X$, $x \leq x$.   *Reflexivity*
2. For any $x, y \in X$, if $x \leq y$ and $y \leq x$, then $x = y$.   *Antisymmetry*
3. For any $x, y, z \in X$, if $x \leq y$ and $y \leq z$, then $x \leq z$.   *Transitivity*

If $x \leq y$, we say that x is a *predecessor* of y and that y is a *successor* of x.

Notation. $(X, \leq)$.

---

3 Alfred Tarski (1901–1983).
4 Kurt Gödel (1906–1978).
5 Abraham Fraenkel (1891–1965).
6 Paul Cohen (1934–2007).

Examples A.1.
1. An arbitrary nonempty set X is partially ordered by the *equality* (*coincidence*) relation $=$.
2. The set $\mathbb{R}$ of real numbers is partially ordered by the usual order $\leq$.
3. The *power set* $\mathscr{P}(X)$ of a nonempty set X (see Section 1.1.1) is partially ordered by the set-theoretic inclusion $\subseteq$.

Remark A.1. Elements x, y of a partially ordered set $(X, \leq)$ are called *comparable* if $x \leq y$ or $y \leq x$. In a partially ordered set, incomparable elements may occur.

Exercise A.1.
(a) Verify the prior examples and remark.
(b) Give two more examples.
(c) Give an example of a partially ordered set $(X, \leq)$ in which no two distinct elements are comparable.

Remarks A.2.
$-$ If $\leq$ is partial order on X, then the relation $\geq$, defined as follows:

$$x \geq y \iff y \leq x,$$

is also a partial order on X.
$-$ If $x \leq y$ and $x \neq y$, then we write $x < y$ or $y > x$.

Definition A.2 (Upper and Lower Bounds). Let Y be a nonempty subset of a *partially ordered set* $(X, \leq)$.
$-$ An element $x \in X$ is called an *upper bound* of Y if

$$\forall y \in Y : y \leq x.$$

$-$ An element $x \in X$ is called a *lower bound* of Y if

$$\forall y \in Y : x \leq y.$$

Remark A.3. Upper/lower bounds of a set Y need not exist.

Exercise A.2. Give corresponding examples.

Definition A.3 (Maximal and Minimal Elements). Let Y be a nonempty subset of a *partially ordered set* $(X, \leq)$.
$-$ An element $x \in Y$ is called a *maximal element* of Y if

$$\nexists y \in Y : x < y,$$

i. e., x has no successors in Y.

– An element $x \in Y$ is called a *minimal element* of Y if

$$\not\exists y \in Y : y < x,$$

i. e., x has no predecessors in Y.

Remarks A.4.
– If an element $x \in Y$ is not comparable with all other elements of Y, it is automatically both maximal and minimal element of Y.
– A *maximal element* of Y need not be greater than all other elements in Y.
– A *minimal element* of Y need not be less than all other elements in Y.
– Maximal and minimal elements of Y need not exist nor be unique.

Exercise A.3. Give corresponding examples.

Definition A.4 (Greatest and Least Elements). Let Y be a nonempty subset of a *partially ordered set* $(X, \leq)$.
– An element $x \in Y$ is called the *greatest element* (also the *last element*) of Y if it is an *upper bound* of Y, i. e.,

$$\forall y \in Y : y \leq x.$$

– An element $x \in Y$ is called the *least element* (also the *first element*) of Y if it is a *lower bound* of Y, i. e.,

$$\forall y \in Y : x \leq y.$$

Remark A.5. The greatest/least element of a set need not exist.

Exercise A.4. Give corresponding examples.

Exercise A.5. Let $\mathscr{P}(X)$ be the power set of a set X consisting of more than one element and partially ordered by the set-theoretic inclusion $\subseteq$ and

$$\mathscr{Y} := \{A \in \mathscr{P}(X) \mid A \neq \emptyset, X\}.$$

(a) What are the *lower* and *upper bounds* of $\mathscr{Y}$?
(b) What are the *minimal* and *maximal elements* of $\mathscr{Y}$?
(c) What are the *least* and *greatest elements* of $\mathscr{Y}$?

Proposition A.1 (Properties of the Greatest and Least Elements). *Let Y be a nonempty subset of a partially ordered set $(X, \leq)$.*
(1) *If the greatest/least element of Y exists, it is unique.*
(2) *If the greatest/least element of Y exists, it is also the unique maximal/minimal element of Y.*

Exercise A.6.

(a) Prove.

(b) Give an example showing that a maximal/minimal element of Y need not be its greatest/least element.

Definition A.5 (Least Upper and Greatest Lower Bounds). Let Y be a nonempty subset of a *partially ordered set* $(X, \leq)$.

– If the set U of all upper bounds of Y is nonempty, and has the least element u, u is called the *least upper bound* (the *supremum*) of Y, and we write $u = \sup Y$.

– If the set L of all lower bounds of Y is nonempty, and has the greatest element l, l is called the *greatest lower bound* (the *infimum*) of Y, and we write $l = \inf Y$.

Remark A.6. For a subset Y in a partially ordered set $(X, \leq)$, $\sup Y$ and $\inf Y$ need not exist, and when they do, they need not belong to Y.

Exercise A.7. Give corresponding examples.

Definition A.6 (Totally Ordered Set). A *totally ordered set* (also a *linearly ordered set* or a *chain*) is a partially ordered set $(X, \leq)$ in which any two elements are comparable, i. e.,

$$\forall x, y \in X : x \leq y \text{ or } y \leq x.$$

Examples A.2.

1. Any partial order on a *singleton* is merely *equality*, and hence, a total order.

2. The set $\mathbb{R}$ of real numbers is totally ordered by the usual order $\leq$.

Exercise A.8.

(a) Verify the prior examples.

(b) When is the *power set* $\mathscr{P}(X)$ of a nonempty set X partially ordered by the set-theoretic inclusion $\subseteq$ a chain?

(c) Show that, for a nonempty subset Y of a *totally ordered set* $(X, \leq)$, a *maximal/minimal element* of Y, when it exists, is the *greatest/least element* of Y.

Definition A.7 (Well-Ordered Set). A *well-ordered set* is a totally ordered set $(X, \leq)$ in which every nonempty subset has the first element.

Examples A.3.

1. Any partial order on a *singleton* is merely *equality*, and hence, a well order.

2. $(\mathbb{N}, \leq)$, as well as its arbitrary nonempty subset, is well ordered.

3. $(\mathbb{Z}, \leq)$ is totally ordered, but *not* well ordered.

Remarks A.7.

-   As follows from *Zermelo's Well-Ordering Principle* (Theorem A.7), every nonempty set can be well ordered.
-   Each infinite well-ordered set $(X, \leq_X)$ is similar to $(\mathbb{N}, \leq)$, in the sense that each nonempty subset $Y \subseteq X$ has the *first element* and each element $x \in X$, except for the last one, if any, has the *unique immediate successor* (the *unique next element*) $s(x) \in X$, i. e., there exists a *unique* element $s(x) \in X$ such that

$$x <_X s(x) \text{ and } \nexists y \in X : x <_X y <_X s(x). \tag{A.1}$$

This fact affords the possibility of inductive proofs (*transfinite induction*), and constructions over such sets similar to those over $\mathbb{N}$.

Exercise A.9.
(a) Verify the prior examples and (A.1).
(b) Show that any *nonempty finite set* can be well ordered.
(c) Prove that, if a nonempty subset Y of a well-ordered set $(X, \leq)$ has an upper bound, then it has the *least upper bound*.
(d) Prove that the usual order $\leq$ of the real line $\mathbb{R}$ restricted to any *uncountable* subset $Y \subseteq \mathbb{R}$ is a total order, but not a well order on Y.

A.3 Equivalents

The Axiom of Choice is obviously true, the well-ordering principle obviously false, and who can tell about Zorn's lemma?
Jerry Bona

Here, we are to prove the equivalence of the *Axiom of Choice* to following three fundamental set-theoretic principles.

Theorem A.3 (Hausdorff Maximal Principle). *In a partially ordered set, there exists a maximal chain.*

Theorem A.4 (Hausdorff Maximal Principle (Precise Version)). *In a partially ordered set, every chain is contained in a maximal chain.*

Theorem A.5 (Zorn's Lemma). *In a partially ordered set whose every chain has an upper bound, there is a maximal element.*[7]

Theorem A.6 (Zorn's Lemma (Precise Version)). *For each element x in a partially ordered set $(X, \leq)$ whose every chain has an upper bound, there is a maximal element u in $(X, \leq)$ such that $x \leq u$.*

---

7 Max Zorn (1906–1993).

Theorem A.7 (Zermelo's Well-Ordering Principle). *Every nonempty set can be well or-dered.*

Proof. We prove the following closed chain of implications:

$$AC \Rightarrow HMP \Rightarrow ZL \Rightarrow ZWOP \Rightarrow AC,$$

where the abbreviations *AC*, *HMP*, *ZL*, and *ZWOP* stand for the *Axiom of Choice*, the *Hausdorff Maximal Principle*, *Zorn's Lemma*, and *Zermelo's Well-Ordering Principle*, re-spectively.

AC ⟹ HMP
Assuming the *Axiom of Choice*, let C be an arbitrary chain in a partially ordered set $(X, \leq)$ and let $\mathscr{C}$ be the collection of all chains in $(X, \leq)$ containing C partially ordered by the set-theoretic inclusion $\subseteq$.

Observe that $\mathscr{C} \neq \emptyset$ since, clearly, $C \in \mathscr{C}$.

Our goal is to prove that $(\mathscr{C}, \subseteq)$ has a *maximal element U*.

Let f be a *choice function* assigning to every nonempty subset A of X one of its elements $f(A)$.

For each $A \in \mathscr{C}$, let $\hat{A}$ be the set of all those elements in X whose adjunction to A produces a chain belonging $\mathscr{C}$:

$$\hat{A} := \{x \in X \mid A \cup \{x\} \in \mathscr{C}\}.$$

Clearly, $A \subseteq \hat{A}$, the equality holding *iff* A is a maximal element in $(\mathscr{C}, \subseteq)$.

Consider a function $g : \mathscr{C} \mapsto \mathscr{C}$ defined as follows:

$$\mathscr{C} \ni A \mapsto g(A) := \begin{cases} A \cup \{f(\hat{A} \setminus A)\} & \text{if } A \subset \hat{A}, \\ A & \text{if } A = \hat{A}. \end{cases}$$

Observe that, for each $A \in \mathscr{C}$, the set $g(A)$ differs from A by *at most one element*.

Thus, to prove that $U \in \mathscr{C}$ is a maximal element in $(\mathscr{C}, \subseteq)$, one needs to show that $U = \hat{U}$, i. e., that $g(U) = U$.

Let us introduce the following temporary definition:

Definition A.8 (Tower). We call a subcollection $\mathscr{I}$ of $\mathscr{C}$ a *tower* if it satisfies the follow-ing conditions:
(1) $C \in \mathscr{I}$.
(2) If $A \in \mathscr{I}$, then $g(A) \in \mathscr{I}$.
(3) If $\mathscr{D}$ is a chain in $(\mathscr{I}, \subseteq)$, then $\bigcup_{A \in \mathscr{D}} A \in \mathscr{I}$.

Observe that *towers* exist since, as is easily seen, $\mathscr{C}$ is a *tower* itself.

Furthermore, the intersection of all towers $\mathscr{I}_0$ is also a tower and is, in fact, the *smallest tower*, the *nonemptiness* of $\mathscr{I}_0$ being ensured by condition (1) of the above definition.

Let us show that $\mathcal{I}_0$ is a *chain* in $(\mathscr{C}, \subseteq)$.

We call a set $B \in \mathcal{I}_0$ *comparable* if, for each $A \in \mathcal{I}_0$ either $A \subseteq B$ or $B \subseteq A$. Thus, proving that $\mathcal{I}_0$ is a chain amounts to showing that all its elements are comparable.

Observe that there exists at least one comparable set in $\mathcal{I}_0$, which is C, since it is contained in any other set in $\mathcal{I}_0$.

Let B be an arbitrary comparable set in $\mathcal{I}_0$. Suppose that $A \in \mathcal{I}_0$ and A is a *proper subset* of B. Then $g(A) \subseteq B$. Indeed, since B is comparable, either $g(A) \subseteq B$, or $B \subset g(A)$. In the latter case, A is a proper subset of a proper subset of $g(A)$, which contradicts the fact that $g(A) \setminus A$ is *at most a singleton*.

Furthermore, consider the collection $\mathscr{U}(B)$ of all those $A \in \mathcal{I}_0$ for which either $A \subseteq g(B)$, or $g(B) \subseteq A$.

It is not difficult to make sure that $\mathscr{U}(B)$ is a *tower* (the verification of the least trivial condition (2) uses the argument of the preceding paragraph).

Since $\mathscr{U}(B)$ is a tower and $\mathscr{U}(B) \subseteq \mathcal{I}_0$, $\mathscr{U}(B) = \mathcal{I}_0$ necessarily.

All these considerations imply that, for each comparable set $B \in \mathcal{I}_0$, the set $g(B)$ is also comparable. The latter, jointly with the facts that a) the set C is comparable and that b) by condition (3), the union of the sets of a chain of comparable sets is also comparable, imply that all comparable sets of $\mathcal{I}_0$ constitute a tower, and hence, they exhaust the entire $\mathcal{I}_0$.

Since $\mathcal{I}_0$ is a chain in $\mathcal{I}$, by condition (3), the set

$$U := \bigcup_{A \in \mathcal{I}_0} A \in \mathcal{I} \subseteq \mathscr{C}$$

and, obviously, U contains every set of $\mathcal{I}_0$. In particular, by condition (2) applied to the tower $\mathcal{I}_0$, $g(U) \subseteq U$, which implies that $g(U) = U$, proving that U is a maximal chain containing C.

Hence, the *Axiom of Choice* implies the *Hausdorff Maximal Principle*, in fact, its precise version.

$HMP \Rightarrow ZL$

Assume the *Hausdorff Maximal Principle* and let x be an arbitrary element in a partially ordered set $(X, \leq)$ whose every chain has an upper bound.

By the *HMP*, there exists a *maximal chain* U in $(X, \leq)$ containing the trivial chain $\{x\}$ (see Examples A.2), i.e., $x \in U$. By the premise of *Zorn's Lemma*, U has an *upper bound* u in $(X, \leq)$. In particular, this implies that $x \leq u$.

From the maximality of the chain U, it follows that u is a *maximal element* in $(X, \leq)$. Otherwise, there exists such an element $v \in X$ that $u < v$, which, by *transitivity* of partial order, implies that the chain U can be extended to a larger chain $U \cup \{v\}$, which *contradicts* the maximality of U.

Thus, the *Hausdorff Maximal Principle* implies *Zorn's Lemma*, in fact, its precise version.

ZL ⟹ *ZWOP*

Assume *Zorn's Lemma* and let X be an arbitrary nonempty set.

Let

$$\mathscr{W} := \{A \subseteq X \mid \exists \leq_A \text{ a well order on } A\}.$$

Observe that $\mathscr{W} \neq \emptyset$ since it contains all *singletons* (more generally, all *nonempty finite subsets*) of X (see Examples A.3 and Exercise A.9).

Exercise A.10. Explain.

The following defines a *partial order* on $\mathscr{W}$:

$$(A, \leq_A) \preceq (B, \leq_B)$$

iff

$$A \subseteq B \text{ and } \forall x, y \in A : x \leq_A y \iff x \leq_B y \text{ and } \forall x \in A \, \forall y \in B \setminus A : x <_B y. \qquad \text{(A.2)}$$

Exercise A.11. Verify that $\preceq$ is a *partial order* on $\mathscr{W}$.

For an arbitrary *chain* $\mathscr{C}$ in $(\mathscr{W}, \preceq)$, let

$$D := \bigcup_{C \in \mathscr{C}} C \subseteq X.$$

We define a *total order* $\leq_D$ on D as follows:

For any $x, y \in D$, exists $C_x, C_y \in \mathscr{C}$ such that $x \in C_x, y \in C_y$. Considering that $(\mathscr{C}, \preceq)$ is a chain, $C_x \subseteq C_y$ or $C_y \subseteq C_x$, and hence, $x, y \in C$, where $C := \max\{C_x, C_y\}$, and we define

$$x \leq_D y \iff x \leq_C y.$$

Exercise A.12. Verify

(a) that $\leq_D$ is *well defined*, i. e., does not depend on the choice of $(C, \leq_C) \in \mathscr{C}$ that contains both x and y and

(b) that $(D, \leq_D)$ is a *chain*.

Thus, $(D, \leq_D)$ is a *chain*.

Let E be an arbitrary nonempty subset of D. For any $x \in E$, there is a $(C_x, \leq_{C_x}) \in \mathscr{C}$ such that $x \in C_x$. Then $E \cap C_x$ is a nonempty subset of C_x and, since $(C_x, \leq_{C_x})$ is *well ordered*. Without loss of generality, we can regard x as the least element of $E \cap C_x$. Suppose $y \in E$ and $y <_D x$. Then, considering that $(\mathscr{C}, \preceq)$ is a *chain*, there is a $(C, \leq_C) \in \mathscr{C}$ containing both x and y with two possibilities:

1. $(C, \leq_C) \preceq (C_x, \leq_{C_x})$, in which case $y \in E \cap C_x$ implying that $y <_{C_x} x$, and hence, contradicting the choice of x.

2. $(C_x, \leq_{C_x}) \preceq (C, \leq_C)$, in which case, if $y \in C_x$, then $y \in E \cap C_x$ implying that $y <_{C_x} x$ and hence, contradicting the choice of x. If $y \in C \setminus C_x$, then, by (A.2), $x <_C y$, contradicting the fact that $y <_D x$.

The obtained contradictions show that x is the *first element* of E in $(D, \leq_D)$, proving that $(D, \leq_D)$ is *well ordered*.

The set $(D, \leq_D)$ is an *upper bound* of $\mathscr{C}$ in $(\mathscr{W}, \preceq)$.

Exercise A.13. Verify.

The latter, by *Zorn's Lemma*, implies that $(\mathscr{W}, \preceq)$ has a *maximal element* $(M, \leq_M)$. Then $M := X$ since, otherwise, choosing any $u \in X \setminus M$, we extend the well order on M to a well order $\leq_N$ on $N := M \cup \{u\}$ as follows:

$$\forall x, y \in M : x \leq_N y \Leftrightarrow x \leq_M y \quad \text{and} \quad \forall x \in M : x <_N u,$$

which implies that $(M, \leq_M) \prec (N, \leq_N)$, contradicting the maximality of $(M, \leq_M)$ in $(\mathscr{W}, \preceq)$. Thus, X can be well ordered, i.e., *Zorn's Lemma* implies *Zermelo's Well-Ordering Principle*.

ZWOP $\Rightarrow$ *AC*

Exercise A.14. Prove.

Hint. For a nonempty collection of nonempty sets $\{X_i\}_{i \in I}$ consider a *well order* $\leq$ on $X := \bigcup_{i \in I} X_i$. □

As an extra demonstration of how a typical proof by *Zorn's Lemma* works, let us prove the implication

ZL $\Rightarrow$ *AC*

Assume *Zorn's Lemma* and let $\{X_i\}_{i \in I}$ be a nonempty collection of nonempty sets. Consider the collection $\mathscr{F}$ of all possible functions $f : I \supseteq D(f) \to \bigcup_{i \in I} X_i$, such that for any i in the *domain* $D(f)$ of f, $f(i) \in X_i$.

Such functions, obviously, exist on *singletons* (more generally, on *nonempty finite subsets*) of I.

Let us introduce a *partial order* $\preceq$ on $\mathscr{F}$ as follows:

$$f \preceq g \Leftrightarrow D(f) \subseteq D(g) \text{ and } f(x) = g(x), \ x \in D(f),$$

i.e., g is an *extension* of f.

Exercise A.15. Verify that $\preceq$ is a partial order on $\mathscr{F}$.

The premise of *Zorn's Lemma* holds in $(\mathscr{F}, \preceq)$.

Exercise A.16. Verify.

Hint. For any chain $\mathscr{C}$ in $(\mathscr{F}, \preceq)$, show that the function

$$D(u) := \bigcup_{g \in \mathscr{C}} D(g) \ni i \mapsto u(i) := g(i),$$

where $g \in \mathscr{C}$ is such that $i \in D(g)$ is arbitrary, is an *upper bound* of $\mathscr{C}$ in $(\mathscr{F}, \preceq)$.

Then, by *Zorn's Lemma*, there exists a *maximal choice function f* in $(\mathscr{F}, \preceq)$. This implies that the domain $D(f)$ of f is the entire indexing set I, and completes the proof.

Exercise A.17. Explain.

Bibliography

[1] R. Akerkar, *Nonlinear Functional Analysis*, Narosa Publishing House, New Delhi, London, 1999.

[2] N. I. Akhiezer and I. M. Glazman, *Theory of Linear Operators in Hilbert Space*, 2nd ed., Dover Publications, Inc., New York, 1993.

[3] W. Arveson, *A Short Course in Spectral Theory*, Graduate Texts in Mathematics, vol. 209, Springer-Verlag, New York, 2002.

[4] G. Bachman and L. Narici, *Functional Analysis*, 2nd ed., Dover Publications, Inc., Mineola, New York, 2000.

[5] A. V. Balakrishnan, *Applied Functional Analysis*, 2nd ed., Springer-Verlag, New York, 1981.

[6] Yu. M. Berezansky, G. F Us, and Z. G. Sheftel, *Functional Analysis. Lecture Course*, Vishcha Shkola, Kiev, 1990 (Russian).

[7] M. S. Birman and M. Z. Slomjak, *Spectral Theory of Self-Adjoint Operators in Hilbert Space*, D. Reidel Publishing Company, Dordrecht, Holland, 1987.

[8] H. Bresis, *Functional Analysis, Sobolev Spaces and Partial Differential Equations*, Universitext, Springer, New York, 2011.

[9] A. Bressan, *Lecture Notes on Functional Analysis. With Applications to Linear Partial Differential Equations*, Graduate Studies in Mathematics, vol. 143, American Mathematical Society, Providence, RI, 2013.

[10] G. Chacón, H. Rafeiro, and J. C. Vallejo, *Functional Analysis. A Terse Introduction*, Walter de Gruyter GmbH, Berlin/Boston, 2017.

[11] J. B. Conway, *A Course in Operator Theory*, Graduate Studies in Mathematics, vol. 21, American Mathematical Society, Providence, RI, 2000.

[12] Yu. L. Daletskiĭ and M. G. Krein, *The Stability of Solutions of Differential Equations in a Banach Space*, Nauka, Moscow, 1970 (Russian).

[13] L. Debnath and P. Mikusiński, *Introduction to Hilbert Spaces with Applications*, 2nd ed., Academic Press, San Diego, California, 1999.

[14] A. Ya. Dorogovtsev, *Mathematical Analysis. A Reference Handbook*, Vishcha Shkola, Kiev, 1985 (Russian).

[15] A. Ya. Dorogovtsev, *Elements of the General Theory of Measure and Integral*, Vishcha Shkola, Kiev, 1989 (Russian).

[16] N. Dunford and J. T. Schwartz with the assistance of W. G. Bade and R. G. Bartle, *Linear Operators. Part I: General Theory*, Interscience Publishers, New York, 1958.

[17] N. Dunford and J. T. Schwartz with the assistance of W. G. Bade and R. G. Bartle, *Linear Operators. Part II: Spectral Theory. Self Adjoint Operators in Hilbert Space*, Interscience Publishers, New York, 1963.

[18] C. H. Edwards and D. E. Penney, *Differential Equations and Linear Algebra*, 3rd ed., Pearson, Upper Saddle River, New Jersey, 2010.

[19] P. Enflo, *A counterexample to the approximation problem in Banach spaces*, Acta Math. **130** (1973), 309–17.

[20] K.-J. Engel and R. Nagel, *One-Parameter Semigroups for Linear Evolution Equations*, Graduate Texts in Mathematics, vol. 194, Springer-Verlag, New York, 2000.

[21] K.-J. Engel and R. Nagel, *A Short Course on Operator Semigroups*, Universitext, Springer, New York, 2006.

[22] I. M. Gelfand, *On normed rings*, Dokl. Akad. Nauk SSSR **23** (1939), 430–2 (Russian).

[23] I. M. Gelfand, *Normierte ringe*, Mat. Sb. **9** (1941), 3–24.

[24] I. M. Glazman and Yu. I. Lyubich, *Finite-Dimensional Linear Analysis*, Nauka, Moscow, 1969 (Russian).

https://doi.org/10.1515/9783110600988-008

[25] C. Goffman and G. Pedrick, *First Course in Functional Analysis*, 2nd ed., Chelsea Publishing Co., New York, 1983.

[26] I. Gohberg and S. Goldberg, *Basic Operator Theory*, Birkhäuser, Boston, 1981.

[27] S. Goldberg, *Unbounded Linear Operators: Theory and Applications*, Dover Publications, Inc., New York, 1985.

[28] A. Granas and J. Dugundji, *Fixed Point Theory*, Springer-Verlag, New York, 2003.

[29] P. R. Halmos, *Naive Set Theory*, Undergraduate Texts in Mathematics, Springer-Verlag, New York, Heidelberg, Berlin, 1974.

[30] P. R. Halmos, *Measure Theory*, Graduate Texts in Mathematics, vol. 18, Springer-Verlag, New York, Heidelberg, Berlin, 1974.

[31] P. R. Halmos, *A Hilbert Space Problem Book*, 2nd ed., Graduate Texts in Mathematics, vol. 19, Springer-Verlag, New York, Heidelberg, Berlin, 1982.

[32] F. Haslinger, *Complex Analysis. A Functional Analytic Approach*, De Gruyter Graduate, Walter de Gruyter GmbH, Berlin/Boston, 2018.

[33] F. Hausdorff, *Set Theory*, 2nd ed., Chelsea Publishing Co., New York, 1962.

[34] R. A. Horn and C. R. Johnson, *Matrix Analysis*, Cambridge University Press, New York, 1986.

[35] V. Hutson, J. S. Pym, and M. J. Cloud, *Applications of Functional Analysis and Operator Theory*, 2nd ed., Elsevier, Amsterdam, 2005.

[36] T. J. Jech, *The Axiom of Choice*, Dover Publications, Inc., Mineola, New York, 2008.

[37] T. F. Jordan, *Linear Operators for Quantum Mechanics*, Dover Publications, Inc., Mineola, New York, 2006.

[38] I. Kaplansky, *Set Theory and Metric Spaces*, Allyn and Bacon, Inc., Boston, 1972.

[39] T. Kato, *Perturbation Theory for Linear Operators*, reprint of the 1980 edition, Classics in Mathematics, Springer-Verlag, Berlin/Heidelberg, 1995.

[40] A. N. Kolmogorov and S. V. Fomin, *Elements of the Theory of Functions and Functional Analysis*, 6th ed., Nauka, Moscow, 1989 (Russian).

[41] A. G. Kurosh, *Lectures on General Algebra*, 2nd ed., Nauka, Moscow, 1973 (Russian).

[42] A. M. Lyapunov, *Stability of Motion*, Ph. D. Thesis, Kharkov, 1892, English translation, Academic Press, New York, London, 1966.

[43] L. A. Lyusternik and V. I. Sobolev, *A Short Course in Functional Analysis*, Vysshaya Shkola, Moscow, 1982 (Russian).

[44] B. D. MacCluer, *Elementary Functional Analysis*, Springer, New York, 2009.

[45] M. V. Markin, *Elementary Functional Analysis*, De Gruyter Graduate, Walter de Gruyter GmbH, Berlin/Boston, 2018.

[46] M. V. Markin, *Real Analysis. Measure and Integration*, De Gruyter Graduate, Walter de Gruyter GmbH, Berlin/Boston, 2019.

[47] M. V. Markin, *Integration for Calculus, Analysis, and Differential Equations: Techniques, Examples, and Exercises*, World Scientific Publishing Co. Pte. Ltd., New Jersey, London, Singapore, 2019.

[48] M. V. Markin, *On the mean ergodicity of weak solutions of an abstract evolution equation*, Methods Funct. Anal. Topology **24** (2018), no. 1, 53–70.

[49] A. N. Michel and C. J. Herget, *Applied Algebra and Functional Analysis*, Dover Publications, Inc., New York, 2011.

[50] G. H. Moore, *Zermelo's Axiom of Choice: Its Origins, Development, and Influence*, Dover Publications, Inc., Mineola, New York, 2013.

[51] M. H. Mortad, *An Operator Theory Problem Book*, World Scientific Publishing Co. Pte. Ltd., New Jersey, London, Singapore, 2019.

[52] J. R. Munkres, *Topology*, 2nd ed., Prentice Hall, Upper Saddle River, New Jersey, 2000.

[53] J. Muscat, *Functional Analysis. An Introduction to Metric Spaces, Hilbert spaces, and Banach Algebras*, Springer International Publishing, Switzerland, 2014.

[54] M. O'Nan, *Linear Algebra*, 2nd ed., Harcourt Brace Jovanovich, Inc., New York, 1976.

[55] R. Palais, *A simple proof of the Banach contraction principle*, J. Fixed Point Theory Appl. **2** (2007), no. 2, 221–3.

[56] C. W. Patty, *Foundations of Topology*, Waveland Press, Inc., Prospect Heights, Illinois, 1997.

[57] A. I. Plesner, *Spectral Theory of Linear Operators*, Nauka, Moscow, 1965 (Russian).

[58] M. Reed and B. Simon, *Methods of Modern Mathematical Physics. I. Functional Analysis*, 2nd ed., Academic Press, Inc., New York, London, 1980.

[59] W. Rudin, *Principles of Mathematical Analysis*, 3rd ed., International Series in Pure and Applied Mathematics, McGraw-Hill Book Co., New York, Auckland, Düsseldorf, 1976.

[60] M. Schechter, *Principles of Functional Analysis*, 2nd ed., Graduate Studies in Mathematics, vol. 36, American Mathematical Society, Providence, RI, 2002.

[61] R. A. Silverman, *Complex Analysis with Applications*, Dover Publications, Inc., New York, 2010.

[62] B. Simon, *Real Analysis. A Comprehensive Course in Analysis, Part 1*, American Mathematical Society, Providence, RI, 2015.

[63] B. Simon, *Operator Theory. A Comprehensive Course in Analysis, Part 4*, American Mathematical Society, Providence, RI, 2015.

[64] W. A. Sutherland, *Introduction to Metric and Topological Spaces*, 2nd ed., Oxford University Press Inc., New York, 2009.

[65] A. E. Taylor and D. C. Lay, *Introduction to Functional Analysis*, reprint of the 2nd ed., Robert E. Krieger Publishing Co., Inc., Melbourne, FL, 1986.

[66] G. Teschl, *Mathematical Methods in Quantum Mechanics. With Applications to Schrödinger Operators*, 2nd ed., Graduate Studies in Mathematics, vol. 157, American Mathematical Society, Providence, RI, 2014.

[67] V. A. Trenogin, *Functional Analysis*, Nauka, Moscow, 1980 (Russian).

[68] B. Z. Vulikh, *Introduction to Functional Analysis*, 2nd ed., Nauka, Moscow, 1967 (Russian).

[69] S. Willard, *General Topology*, Dover Publications, Inc., Mineola, New York, 2004.

[70] N. Young, *An Introduction to Hilbert Space*, Cambridge University Press, Cambridge, UK, 1997.

Index

https://doi.org/10.1515/9783110600988-009

www.ingramcontent.com/pod-product-compliance
Lightning Source LLC
Chambersburg PA
CBHW080647220326
41598CB00033B/5131